Operator Theory: Advances and Applications
Volume 237

Founded in 1979 by Israel Gohberg

Marinus A. Kaashoek
Leiba Rodman
Hugo J. Woerdeman
Editors

Advances in Structured Operator Theory and Related Areas

The Leonid Lerer Anniversary Volume

 Birkhäuser

Editors
Marinus A. Kaashoek
Department of Mathematics, FEW
VU University
Amsterdam
Netherlands

Leiba Rodman
Department of Mathematics
College of William and Mary
Williamsburg
USA

Hugo J. Woerdeman
Department of Mathematics
Drexel University
Philadelphia
USA

ISSN 0255-0156   ISSN 2296-4878 (electronic)
ISBN 978-3-0348-0798-2   ISBN 978-3-0348-0639-8 (eBook)
DOI 10.1007/978-3-0348-0639-8
Springer Basel Heidelberg New York Dordrecht London

Mathematics Subject Classification (2010): 47A68, 47B35, 39B42, 93B28, 15A15; 11C20, 47A13, 47A48

Printed on acid-free paper

Springer Basel is part of Springer Science+Business Media (www.birkhauser-science.com)

Contents

Preface

This volume is dedicated to Leonid Arie Lerer on the occasion of his seventieth birthday (April 19, 2013). Leonia, as he is known to his friends, is an expert in the theory of structured matrices and operators and related matrix-valued functions. He has been a great inspiration to many.

Leonid Lerer started his mathematical career in Kishinev, Ukraine, with Alek Markus and Israel Gohberg as research supervisors. He defended his Ph.D. thesis in 1969 in Kharkov, Ukraine. In December 1973 he immigrated to Israel. Since 1981 he has been a professor at the Technion in Haifa where at present he has the status of emeritus. He has educated six Ph.D. students and five masters students. His more than 80 papers cover a wide spectrum of topics, ranging from functional analysis and operator theory, linear and multilinear algebra, ordinary differential equations, to systems and control theory.

This anniversary volume begins with a picture of Leonid Lerer, his Curriculum Vitae and List of Publications, and personal notes written by former students, mathematical friends and colleagues.

The main part of the book consists of a selection of peer-reviewed research articles presenting recent results in areas that are close to Lerer's mathematical interests. This includes articles on Toeplitz, Wiener–Hopf, and Toeplitz plus Hankel operators, Bezout equations, inertia type results, matrix polynomials, in one and severable variables, and related areas in matrix and operator theory.

We present this book to Leonid Lerer on behalf of all the authors as a token of our respect and gratitude. We wish him many more years of good health and happiness.

March 2013

Rien Kaashoek
Leiba Rodman
Hugo Woerdeman

Leonid Arie Lerer in an undated photograph

Operator Theory:
Advances and Applications, Vol. 237, ix–xii
© 2013 Springer Basel

Leonid Lerer's Curriculum Vitae

Date and place of birth: April 19, 1943; USSR
Date of immigration: December 1973
Marital status: Married, 2 children

Academic degrees

1965   M.Sc. Mathematics, Magna Cum Laude
       Kishinev State University, Kishinev

1969   Ph.D. Mathematics
       Physico-Technical Institute of Low Temperatures of the
       Academy of Sciences of the Ukrainian SSR, Kharkov

Academic appointments

1969–73   Lecturer, Dept. of Physics and Mathematics
             Kishinev State University

1974–75   Senior Researcher,
             Technion Research & Development Foundation
             Technion, Haifa

1975–80   Senior Lecturer, Dept. of Mathematics, Technion, Haifa

1981–88   Assoc. Professor, Dept. of Mathematics, Technion, Haifa

1988 (Feb.)– Full Professor, Dept. of Mathematics, Technion, Haifa

Visiting academic appointments

Visiting Professor, Dept. of Mathematics and Computer Science
at Vrije Universiteit, Amsterdam:

> Feb. 1984–Feb. 1985, Mar.–Sept. 1990, Apr.–Nov. 2002,
> Feb.–Apr. 2003, Jul.–Aug. 2003

Invited Guest, Dept. of Mathematics and Computer Science
at Vrije Universiteit, Amsterdam:

> Jul.–Aug. 1978, 1980, 1982, 1983, 1985–93, 1995, 1996–97,
> 2000, 2004, Sept. 2005 (3–6 weeks each time)

Invited Guest, Institute for Mathematics and its Applications,
University of Minnesota, Minneapolis:

June 1992 (2 weeks)

Invited Guest, The Thomas Stieltjes Institute of Mathematics,
Vrije Universiteit, Amsterdam:

Jul.–Aug. 1994 (5 weeks)

Research field

Operator Theory, Systems Theory, Linear Algebra, Integral Equations

Public professional activities

- Editorial board of the international journal "Integral Equations and Operator Theory" (since its foundation in 1977).
- Editorial board of the book series "Operator Theory: Advances and Applications", Birkhäuser Verlag, Basel–Boston–Berlin.
- Special editor of "Linear Algebra and its Applications", 1988 and 2005.
- Editor of "Gohberg Memorial Volumes", Birkhäuser, Basel, 2009–.
- Editor of "Convolution equations and singular integral operators. Selected papers", Birkhäuser, Basel, 2009–2010.
- Organizing committee for:
    1. The third, forth, fifth, sixth, seventh, ninth, tenth, eleventh, thirteenth, fourteens and fifteenth Haifa Matrix Theory Conferences, 1985–2007.
    2. International Conference "Operator Theory: Advances and Applications", Calgary, Canada, 1988.
    3. Conference of the International Linear Algebra Society (ILAS), 2001.
    4. Workshops on Operator Theory and Applications, Amsterdam, 2002 and 2003.
- Organizer and Chairman of Invited Special Sessions at
    1. International Symposium on the Mathematical Theory of Networks and Systems, Amsterdam, The Netherlands, 1989.
    2. Second SIAM Conference on Linear Algebra in Signals, Systems and Control, San Francisco, USA, 1991.
    3. International Symposium on Mathematical Theory of Networks and Systems, St. Louis, Missouri, USA, 1991.
    4. The 14th Matrix Theory Conference, Haifa, 2007.
    5. The 15th Matrix Theory Conference, Haifa, 2009.
    6. International Workshop on Operator Theory and Applications, Williamsburg, USA, 2008.
- Program Committees for the International Symposia of the Mathematical Theory of Networks and Systems:
    1. MTNS – 89, Amsterdam, The Netherlands;
    2. MTNS – 93, Regensburg, Germany

Grants and awards

- Israel-U.S. Binational Science Foundation (BSF) Grant, 1988–1992.
- Fellowship of the Netherlands Organization for Scientific Research (NWO), March–June, 1990.
- Grants from the Technion V.P.R. Fund – annually since 1979.
- Grants from the Fund for Promotion of Research at the Technion – annually since 1981.
- Israel-U.S. Binational Science Foundation (BSF) Grant, 1995–1999.
- Fellowship of the Netherlands Organization for Scientific Research (NWO), April–September 2002.
- Grant for Promotion of Funded Research ($ 2,500) (for the proposal submitted to BSF, graded "very good" but not funded by this agency), 2005.
- ISF grant 121/09 for the period 2009–2012, $ 40,000 yearly (still in progress).

Graduate students

B.A. Kon	M.Sc. 1976 Thesis title: "Asymptotic distribution of the spectra of multidimensional Wiener–Hopf operators and multidimensional singular integral operators".
M. Tismenetsky	M.Sc. 1977 Thesis title: "Spectral analysis of polynomial pencils via the roots of the corresponding matrix equations".
B.A. Kon	Ph.D. 1981 Thesis title: "Operators of Wiener–Hopf types and resultants of analytic functions".
M.Tismenetsky	Ph.D. 1981 Thesis title: "Bezoutians, Toeplitz and Hankel matrices in the spectral theory of matrix polynomials".
H.J. Woerdeman	M.Sc. 1985 (Vrije Univ., Amsterdam) Thesis title: "Resultant operators for analytic matrix functions in a finitely connected domain".
J. Haimovici	Ph.D. 1991 Thesis title: "Operator equations and Bezout operators for analytic operator functions".
I. Karelin	M.Sc. 1993 Thesis title: "The algebraic Riccati equation and the spectral theory of matrix polynomials".
G. Gomez	Ph.D. 1996 Thesis title: "Bezout operators for analytic operator functions and inversion of structured operators".

I. Karelin	Ph.D. 2000
	Thesis title: "Factorization of rational matrix functions, generalized Bezoutians and matrix quadratic equations".
I. Margulis	M.Sc. 2008
	Thesis title: "Inertia theorems based on operator equations of Lyapunov type and their applications".
I. Margulis	Ph.D. in progress.

Memberships

- Israel Mathematical Society
- Society for Industrial and Applied Mathematics
- The New York Academy of Sciences
- European Mathematical Society

Invited talks at conferences

- Invited participation at 80 conferences.
- Plenary speaker at:
  1. *SIAM Conference on Linear Algebra in Signals*, Systems and Control, Boston, Mass., USA, 1986.
  2. *Workshop on Matrix and Operator Theory*, Rotterdam, The Netherlands, 1989.
  3. *Workshop on Linear Algebra for Control Theory*, Institute for Mathematics and its Applications, University of Minnesota, USA, June 1992.
  4. *Colloquium of the Royal Dutch Academy of Arts and Sciences on Challenges of a Generalized System Theory*, June 1992.
  5. *The Second Conference of ILAS*, Lisbon, Portugal, August 1992.
  6. *The Third Conference of ILAS*, Rotterdam, The Netherlands, August 1994.
  7. *International Workshop on Operator Theory and Applications (IWOTA-95)*, Regensburg, Germany, August 1995.
  8. *Workshop on Operator Theory and Analysis on the occasion of the 60th birthday of M.A. Kaashoek*, Vrije Universiteit, Amsterdam, November 1997.
  9. *AMS-IMS-SIAM Summer Research Conference on Structured Matrices in Operator Theory*, Numerical Analysis, Control, Signal and Image Processing, July 1999.
  10. *Workshop on Operator Theory, System Theory and Scattering Theory*, Beer-Sheba, June 2005.

Leonid (Arie) Lerer
Department of Mathematics
Technion-Israel Institute of Technology
Haifa 32000, Israel
e-mail: llerer@tx.technion.ac.il

Operator Theory:
Advances and Applications, Vol. 237, xiii–xix
© 2013 Springer Basel

Leonid Lerer's List of Publications

Theses

1. M.Sc. "Localization of zeroes of polynomials and some properties of normal matrices", Kishinev State University, 1965.
2. Ph.D. "Some problems in the theory of linear operators and in the theory of bases in locally convex spaces", Physico-Technical Institute of Low Temperatures of the Academy of Sciences of the Ukrainian SSR, Kharkov, 1969.

Original papers in professional journals

1. L. Lerer, On the diagonal elements of normal matrices, *Mat. Issled.* **2** (1967), 153–163.
2. L. Lerer, About the spectral theory of bounded operators in a locally convex space, *Mat. Issled.* **2** (1967), 206–214.
3. L. Lerer, On completeness of the system of root vectors of a Fredholm operator in a locally convex space, *Mat. Issled.* **3** (1968), 31–60.
4. L. Lerer, Basic sequences in a Montel space, *Mat. Zametki* **6** (1969), 329–334. Announcement of results: *Mat. Issled.* **3** (1968), 235–236.
5. L. Lerer, Certain criteria for stability of bases in locally convex spaces, I, *Mat. Issled.* **4** (1969), 35–55.
6. L. Lerer, Certain criteria for stability of bases in locally convex spaces, II, *Mat. Issled.* **4** (1969), 42–57.
7. L. Lerer, The stability of bases in locally convex spaces, *Dokl. Akad. Nauk U.S.S.R.* **184** (1969), 30–33; *Soviet Math. Dokl.* **10** (1969), 24–28.
8. L. Lerer, On a class of perturbation for operators that admit reduction, *Mat. Issled.* **6** (1971), 168–173.
9. S. Buzurniuk, L. Lerer and E. Shevchik, An operator modification of Seidel Method, *Tezisi Dokl. Nauchn. Konf.* Kishinev, (1972), 22–23.
10. L. Lerer, The asymptotic distribution of the spectra of finite truncations of Wiener–Hopf operators, *Dokl. Akad. Nauk U.S.S.R.* **207** (1972), 1035–1038; *Soviet Math. Dokl.* **13** (1972), 1651–1655.
11. L. Lerer, The asymptotic distribution of the spectra. I. General theorems and the distribution of the spectrum of truncated Wiener–Hopf operators, *Mat. Issled.* **8** (1972), 141–164.

12. L. Lerer, On the asymptotic distribution of the spectra. II. The distribution of the spectrum of truncated dual operators, *Mat. Issled.* **8** (1973), 84–95.

13. I. Gohberg and L. Lerer, Resultants of matrix polynomials, *Bull. Amer. Math. Soc.* **82** (1976), 565–567.

14. I. Gohberg and L. Lerer, Singular integral operators as a generalization of the resultant matrix, *Applicable Anal.* **7** (1977/78), 191–205.

15. I. Gohberg, L. Lerer and L. Rodman, Factorization indices for matrix polynomials, *Bull. Amer. Math. Soc.* **84** (1978), 275–277.

16. L. Lerer, On approximating the spectrum of convolution type operators. I. Wiener–Hopf matricial integral operator, *Israel J. Math.* **30** (1978), 339–362.

17. I. Gohberg and L. Lerer, On resultant operators of a pair of analytic functions, *Proc. Amer. Math. Soc.* **72** (1978), 65–73.

18. I. Gohberg, L. Lerer and L. Rodman, On canonical factorization of operator polynomials, spectral divisors and Toeplitz matrices, *Integral Equations Operator Theory* **1** (1978), 176–214.

19. I. Gohberg and L. Lerer, Factorization indices and Kronecker indices, *Integral Equations Operator Theory* **2** (1979), 199–243. Erratum: *Integral Equations Operator Theory* **2** (1979), 600–601.

20. I. Gohberg, L. Lerer and L. Rodman, Stable factorization of operator polynomials, I. Spectral divisors simply behaved at infinity, *J. Math. Anal. Appl.* **74** (1980), 401–431.

21. I. Gohberg, L. Lerer and L. Rodman, Stable factorization of operator polynomials, II. Main results and applications to Toeplitz operators, *J. Math. Anal. Appl.* **75** (1980), 1–40.

22. I. Gohberg, M.A. Kaashoek, L. Lerer and L. Rodman, Common multiples and common divisors of matrix polynomials, I. Spectral method, *Indiana Univ. Math. J.* **30** (1981), 321–356.

23. L. Lerer and M. Tismenetsky, The Bezoutian and the eigenvalue separation problem, *Integral Equations Operator Theory* **5** (1982), 386–445.

24. I. Gohberg, M.A. Kaashoek, L. Lerer and L. Rodman, Common multiples and common divisors of matrix polynomials, II. Vandermonde and resultant, *Linear and Multilinear Algebra* **12** (1982/83), 159–203.

25. I. Gohberg and L. Lerer, On non-square sections of Wiener–Hopf operators, *Integral Equations Operator Theory* **5** (1982), 518–532. Errata: *Integral Equations Operator Theory* **6** (1983), 904.

26. I. Gohberg, L. Lerer and L. Rodman, Wiener–Hopf factorization of piecewise matrix polynomials, *Linear Algebra Appl.* **52/53** (1983), 315–350.

27. L. Lerer, L. Rodman and M. Tismenetsky, Bezoutian and Schur–Cohn problem for operator polynomials, *J. Math. Anal. Appl.* **103** (1984), 83–102.

28. P. Lancaster, L. Lerer and M. Tismenetsky, Factored forms for solutions of $AX - XB = C$ and $X - AXB = C$ in companion matrices, *Linear Algebra Appl.* **62** (1984), 19–49.

29. I. Gohberg, M.A. Kaashoek, L. Lerer and L. Rodman, Minimal divisors of rational matrix functions with prescribed zero and pole structure, in: *Topics in operator theory systems and networks* (*Rehovot*, 1983), Oper. Theory Adv. Appl. **12**, Birkhäuser Verlag, Basel, 1984, pp. 241–275.

30. L. Lerer and M. Tismenetsky, On the location of spectrum of matrix polynomials, *Contemp. Math.* **47** (1985), 287–297.

31. B. Kon and L. Lerer, Resultant operators for analytic functions in a simple connected domain, *Integral Equations Operator Theory* **9** (1986), 106–120.

32. I. Gohberg, M.A. Kaashoek and L. Lerer, Minimality and irreducibility of the time invariant linear boundary value systems, *Internat. J. Control* **44** (1986), 363–379.

33. I. Gohberg, M.A. Kaashoek, L. Lerer and L. Rodman, On Toeplitz and Wiener–Hopf operators with contourwise rational matrix and operator symbols, in: *Constructive methods of Wiener–Hopf factorization*, Oper. Theory Adv. Appl. **21**, Birkhäuser Verlag, Basel, 1986, pp. 75–126.

34. L. Lerer and M. Tismenetsky, Generalized Bezoutian and the inversion problem for block matrices, I. General scheme, *Integral Equations Operator Theory* **9** No. 6 (1986), 790–819.

35. L. Lerer and H.J. Woerdeman, Resultant operators and the Bezout equation for analytic matrix functions, I. *J. Math. Anal. Appl.* **125** (1987), 531–552.

36. L. Lerer and H.J. Woerdeman, Resultant operators and the Bezout equation for analytic matrix functions, II. *J. Math. Anal. Appl.* **125** (1987), 553–567.

37. I. Gohberg, M.A. Kaashoek and L. Lerer, On minimality in the partial realization problem, *Systems Control Lett.* **9** (1987), 97–104.

38. I. Gohberg and L. Lerer, Matrix generalizations of M.G. Kreĭn theorems on orthogonal polynomials, in: *Orthogonal matrix-valued polynomials and applications* (*Tel Aviv*, 1987/88), Oper. Theory Adv. Appl. **34**, Birkhäuser Verlag, Basel, 1987/88, pp. 137–202.

39. L. Lerer and M. Tismenetsky, Generalized Bezoutian and matrix equations, *Linear Algebra Appl.* **99** (1988), 123–160.

40. I. Gohberg, M.A. Kaashoek and L. Lerer, Nodes and realizations of rational matrix functions: minimality theory and applications, in: *Topics in operator theory and interpolation*, Oper. Theory Adv. Appl. **29**, Birkhäuser Verlag, Basel, 1988, pp. 181–232.

41. L. Lerer, The matrix quadratic equation and factorization of matrix polynomials, in: *The Gohberg anniversary collection, Vol. I* (*Calgary, AB,* 1988), Oper. Theory Adv. Appl. **40**, Birkhäuser Verlag, Basel, 1989, pp. 279–324.

42. L. Lerer, L. Rodman and M. Tismenetsky, Inertia theorems for matrix polynomials, *Linear and Multilinear Algebra* **30** (1991), 157–182.

43. I. Gohberg, M.A. Kaashoek and L. Lerer, A directional partial realization problem, *Systems Control Lett.* **17** (1991), 305–314.

44. I. Gohberg, M.A. Kaashoek and L. Lerer, Minimality and realization of discrete time-varying systems, in: *Time-variant systems and interpolation*, Oper. Theory Adv. Appl. **56**, Birkhäuser Verlag, Basel, 1992, pp. 261–296.

45. L. Lerer and L. Rodman, Spectrum separation and inertia for operator polynomials, *J. Math. Anal. Appl.* **169** (1992), 260–282.

46. I. Gohberg, M.A. Kaashoek and L. Lerer, Minimal rank completion problems and partial realization, in: Recent Advances in Math. Theory of Systems, Control, Networks and Signal Processing, I, Mita Press, Tokyo, 1992, 65–70.

47. L. Lerer and L. Rodman, Sylvester and Lyapunov equations and some interpolation problems for rational matrix functions, *Linear Algebra Appl.* **185** (1993), 83–117.

48. I. Koltracht, B.A. Kon and L. Lerer, Inversion of structured operators, *Integral Equations Operator Theory* **20** (1994), 410–448.

49. G. Gomez and L. Lerer, Generalized Bezoutian for analytic operator functions and inversion of structured operators, in: *Systems and Networks: Mathematical Theory and Applications* (U. Helmke, R. Mennicken, J. Saures, eds.), Akademie Verlag, 1994, pp. 691–696.

50. L. Lerer and L. Rodman, Bezoutians and factorizations of rational matrix functions and matrix equations, in: *Systems and Networks: Mathematical Theory and Applications* (U. Helmke, R. Mennicken, J. Saures, eds.), Akademie Verlag, 1994, pp. 761–766.

51. I. Haimovici and L. Lerer, Bézout operators for analytic operator functions. I. A general concept of Bézout operator, *Integral Equations Operator Theory* **21** (1995), 33–70.

52. L. Lerer and L. Rodman, Inertia of operator polynomials and stability of differential equations, *J. Math. Anal. Appl.* **192** (1995), 579–606.

53. L. Lerer and L. Rodman, Common zero structure of rational matrix functions, *J. Funct. Anal.* **136** (1996), 1–38.

54. I. Gohberg, M.A. Kaashoek and L. Lerer, Factorization of banded lower triangular infinite matrices, *Linear Algebra Appl.* **247** (1996), 347–357.

55. L. Lerer and L. Rodman, Symmetric factorization and localization of zeroes of rational matrix functions, *Linear and Multilinear Algebra* **40** (1996), 259–281.

56. L. Lerer and L. Rodman, Bezoutian of rational matrix functions, *J. Funct. Anal.* **141** (1996), 1–36.

57. L. Lerer and A.C.M. Ran, J-pseudo-spectral and J-inner-pseudo-outer factorization for matrix polynomials, *Integral Equations Operator Theory* **29** (1997), 23–51.

58. L. Lerer and L. Rodman, Inertia theorems for Hilbert space operators based on Lyapunov and Stein equations, *Math. Nachr.* **198** (1999), 131–148.

59. L. Lerer and L. Rodman, Bezoutian of rational matrix functions, matrix equations and factorizations of rational matrix functions, *Linear Algebra Appl.* **302-303** (1999), 105–133.

60. I. Karelin and L. Lerer, Generalized Bezoutian, matrix quadratic equations and factorization of rational matrix functions, in: *Recent advances in operator theory (Groningen, 1998)*, Oper. Theory Adv. Appl. **122**, Birkhäuser Verlag, Basel, 2001, pp. 303–321.

61. I. Karelin, L. Lerer and A.C.M. Ran, J-symmetric factorization and the algebraic Riccati equation, in: *Recent advances in operator theory (Groningen, 1998)*, Oper. Theory Adv. Appl. **124**, Birkhäuser Verlag, Basel, 2001, pp. 319–360.

62. I. Karelin and L. Lerer, Matrix quadratic equations and column/row factorization of matrix polynomials, *Int. J. Appl. Math. Comput. Sci.* **11** (2001), 1285–1310.

63. L. Lerer and L. Rodman, Inertia bounds for matrix polynomials and applications, in: *Linear operators and matrices*, Oper. Theory Adv. Appl. **130**, Birkhäuser Verlag, Basel, 2002, pp. 255–276.

64. L. Lerer and A. Ran, A new inertia theorem for Stein equations, inertia of invertible block Toeplitz matrices and matrix orthogonal polynomials, *Integral Equations Operator Theory* **47** (2003), 339–360.

65. L. Lerer, M.A. Petersen and A.C.M. Ran, Existence of minimal nonsquare J-symmetric factorizations for self-adjoint rational matrix functions, *Linear Algebra Appl.* **379** (2004), 159–178.

66. I. Gohberg, I. Haimovici, M.A. Kaashoek and L. Lerer, The Bezout integral operator: main property and underlying abstract scheme, in: *The state space method generalizations and applications*, Oper. Theory Adv. Appl. **161**, Birkhäuser Verlag, Basel, 2005, pp. 225–270.

67. I. Gohberg, M.A. Kaashoek and L. Lerer, Quasi-commutativity of entire matrix functions and the continuous analogue of the resultant, in: *Modern operator theory and applications*, Oper. Theory Adv. Appl. **170**, Birkhäuser Verlag, Basel, 2007, pp. 101–106.

68. I. Gohberg, M.A. Kaashoek and L. Lerer, The continuous analogue of the resultant and related convolution operators, in: *The Extended Field of Operator Theory*, Oper. Theory Adv. Appl. **171**, Birkhäuser, Basel, 2007, 107–127.

69. I. Gohberg, M.A. Kaashoek and L. Lerer, On a class of entire matrix function equations, *Linear Algebra Appl.* **425** (2007), 434–442.

70. I. Gohberg, M.A. Kaashoek and L. Lerer, The inverse problem for Kreĭn orthogonal matrix functions, (Russian) *Funktsional. Anal. i Prilozhen.* **41** (2007), 44–57; translation in *Funct. Anal. Appl.* **41** (2007), 115–125.

71. L. Lerer, I. Margulis and A.C.M. Ran, Inertia theorems based on operator Lypunov equations, *Oper. Matrices* **2** (2008), 153–166.

72. I. Gohberg, M.A. Kaashoek and L. Lerer, The resultant for matrix polynomials and quasi commutativity, *Indiana Univ. Math. J.* **57** (2008), 2793–2813.

73. D. Alpay, I. Gohberg, M.A. Kaashoek, L. Lerer and A. Sakhnovich, Kreĭn systems, in: *Modern analysis and applications. The Mark Kreĭn Centenary Conference. Vol. 2: Differential operators and mechanics*, Oper. Theory Adv. Appl. **191**, Birkhäuser Verlag, Basel, 2009, pp. 19–36.

74. M.A. Kaashoek and L. Lerer, Quasi-commutativity of regular matrix polynomials: resultant and Bezoutian, in: *Topics in operator theory. Volume 1. Operators, matrices and analytic functions*, Oper. Theory Adv. Appl. **202**, Birkhäuser Verlag, Basel, 2010, pp. 297–314.

75. M.A. Kaashoek, L. Lerer and I. Margulis, Kreĭn orthogonal entire matrix functors and related Lypunov equations: A state space approach, *Integral Equations Operator Theory* **65** (2009), 223–242.

76. D. Alpay, I. Gohberg, M.A. Kaashoek, L. Lerer and A.L. Sakhnovich, Kreĭn systems and canonical systems on a finite interval: accelerants with a jump discontinuities at the origin and continuous potentials, *Integral Equations Operator Theory* **68** (2010), 115–150.

77. L. Lerer and A.C.M. Ran, The discrete algebraic Riccati equation and Hermitian block-Toeplitz matrices, in: *A panorama of modern operator theory and related topics*, Oper. Theory Adv. Appl. **218**, Birkhäuser/ Springer Basel AG, Basel, 2012, pp. 495–512.

78. M.A. Kaashoek and L. Lerer, The band method and inverse problems for orthogonal functions of Szegö–Kreĭn type, *Indag. Math. (N. S.)* **23** (2012), 900–920.

79. M.A. Kaashoek and L. Lerer, On a class of matrix polynomial equations, *Linear Algebra Appl.* **439** (2013), 613–620.

Edited books

1. *Convolution equations and singular integral equations.* Edited by L. Lerer, V. Olshevsky and I. Spitkovsky, Oper. Theory Adv. Appl. **206**, Birkhäuser Verlag, Basel, 2010.

2. *A panorama of modern operator theory and related topics. The Israel Gohberg memorial volume.* Edited by Harry Dym, Marinus A. Kaashoek, Peter Lancaster, Heinz Langer and Leonid Lerer. Oper. Theory Adv. Appl. **218**, Birkhäuser/Springer Basel AG, Basel, 2012.

Other publications
(Reports listed were not fully incorporated in papers published elsewhere)

1. I. Gohberg, L. Lerer and L. Rodman, On factorization, indices and completely decomposable matrix polynomials, Technical report 80–47, Tel-Aviv University, 1980, 72 pages.

2. Feldman, I.A. Wiener–Hopf operator equations and their application to the transport equation. Translated from the Russian by C.G. Lekkerker, L. Lerer and R. Troelstra. *Integral Equations Operator Theory* **3** (1980), 43–61.

3. I. Gohberg, M.A. Kaashoek, L. Lerer and L. Rodman, Common multiples and common divisors of matrix polynomials, II. Vandermonde and resultant matrices, Technical report 80–53, Tel-Aviv University, 1981, 122 pages.

4. L. Lerer and H.J. Woerdeman, Resultant operators and the Bezout equation for analytic matrix functions, Rapport No. 299, Vrije Universiteit, Amsterdam, 1985, 54 pages.

5. H. Bart, M.A. Kaashoek and L. Lerer, Review on "Matrix Polynomials" by I. Gohberg, P. Lancaster and L. Rodman, *Linear Algebra Appl.* **64** (1985), 167–172.

6. L. Lerer and M. Tismenetsky, Toeplitz classification of matrices and inversion formulas, II. Block-Toeplitz and perturbed block-Toeplitz matrices, Technical Report, IBM SC, Haifa, 1986, 38 pages.

7. L. Lerer and A.C.M. Ran, On a new inertia theorem and orthogonal polynomials, *Proceedings of the Sixteenth International Symposium on Mathematical Theory of Networks and Systems (MTNS)*, Leuven, Belgium, 2004.

8. A. Berman, L. Lerer and R. Loewy, Preface to the 2005 Haifa Matrix Theory Conference at the Technion, *Linear Algebra Appl.* **416** (2006), 15–16.

9. M.A. Kaashoek and L. Lerer, Gohberg's mathematical work in the period 1998–2008, in: *Israel Gohberg and his Friends*, eds. H. Bart, T. Hempfling, M.A. Kaashoek, Birkhäuser Verlag, Basel, 2008, pp. 111–115.

10. L. Lerer, V. Olshevsky and I. Spitkovsky, Introduction, in: *Convolution equations and singular integral operators*, Oper. Theory Adv. Appl. **206**, Birkhäuser Verlag, Basel, 2010, pp. ix–xxii.

Operator Theory:
Advances and Applications, Vol. 237, 1–7

Leonia Lerer's Mathematical Work and Amsterdam Visits

M.A. Kaashoek

It is a great pleasure to congratulate professor Lerer on the occasion of his 70th birthday and to wish him many happy returns.

I will refer to professor Lerer as Leonia, the name used by his Dutch friends. Arie would have been an alternative. The latter is a very common Dutch name, but few know about its Hebrew meaning. Leonid is out of the question; too many political recollections.

Leonia Lerer, Bil Helton, Harm Bart, Israel Gohberg, Joe Pincus
VU University campus, 1979

Leonia and I met for the first time in October 1976 at the Mathematisches Forschungsinstitut Oberwolfach, a wonderful conference resort in the southwestern part of Germany (at that time Bundesrepublik Deutschland), in the black forest near Freiburg. It was a very special meeting, organized by Gohberg, Gramsch and Neubauer, with a select group of 39 participants, including a strong US delegation consisting of Kevin Clancey, Lewis Coburn, Chandler Davis, Ron Douglas, Bill

Helton, and Joe Pincus. From Israel, besides Gohberg, three other participants: Harry Dym, Paul Furhrman, and Leonia Lerer. Directly after this meeting Ron Douglas and Leonia came to Amsterdam. It was Leonia's first visit. Three years later, again directly after an Oberwolfach meeting, we had a mini-conference in Amsterdam with the five persons on the picture at the previous page as the main lecturers. If I remember it correctly, this was Leonia's third visit to Amsterdam.

Many other visits were to follow. There were short visits of about 3 to 5 weeks, in total more than 20, often supported by the Dutch National Science Foundation. Apart from that, having a sabbatical at the Technion, Leonia held visiting professorships at the VU University, for four periods:

- February 1984–February 1985,
- March–September 1990,
- April–November 2002,
- February–April and July–August 2003.

In his first period as visiting professor, Leonia supervised the master thesis of Hugo Woerdeman, now professor at Drexel University and co-editor of this volume. The joint work of Leonia and Hugo resulted in two papers on "Resultant operators and the Bezout equation for analytic matrix functions," both appeared in 1987 in the Journal of Math. Anal. Appl. Later in an acknowledgment in his PhD thesis, Hugo wrote: *I am indebted to professor Lerer, who introduced me in a very stimulating way to mathematical research.*

When one lives in the Netherlands for so many years as Leonia did, one learns the local customs and uses the local means of transportation:

On the bicycle with Israel Gohberg.

Leonia and I have 24 joint papers, of these 24 papers 19 were written jointly with Israel Gohberg. Our many meetings, in the Netherlands as well as in Israel, and our joint work belong to the gratifying experiences of my life. Andre Ran is Leonia's second co-author at the VU Amsterdam; together Leonia and Andre wrote 7 joint papers on topics involving J-spectral factorization and inertia theorems.

What did Leonia talk about at Oberwolfach in 1976? Here I present the abstract of his talk as it appears in the *Tagungsbericht*:

```
Lerer, L.E.: On resultants of analytic matrix functions
The classical result on the resultant matrix for two poly-
nomials is generalized to the case of two arbitrary matrix-
functions which are holomorphic in some anulus-like or
strip-like domain and contionuous on its closure.
For two such matrix functions A(z) and B(z) we construct
resultant operators defined on suitable Banach spaces so
that the dimension of their kernel is equal to the number of
common zeros of A(z) and B(z). It turns out that some kind
of singular integral operators play the rôle of such resultant
operators. The present approach gives a new look on the
singular integral operators as a natural generalization
of the classical resultant matrix.
```

The abstract has three elements. First element: from polynomials to analytic functions; second element: from resultant matrices to resultant operators, and third element: singular integral operators as generalized resultant matrices. It was a beautiful lecture.

Leonia's work has a wonderful mix of matrix and operator theory on the one hand and matrix function theory on the other hand. It reminds me of a statement Paul Halmos once made in an interview. He said: *I still have this religion that if you know the answer to every matrix question, somehow you answer every operator question.* I do not believe in this statement, and I think Leonia does not either; there is much more two way traffic between the two fields. But Leonia's talk at Oberwolfach certainly provided some support for the Halmos religion.

Resultants and Bezout operators form a main theme in Leonia's work. About one third of his papers after 1976 have the word resultant or Bezout in the title. His talk in Oberwolfach was based on his first papers after immigration to Israel. They appeared in the Bulletin and Proceedings of the AMS, both in 1976, and both co-authored by Gohberg.

His work in this area is partially motivated by mathematical system and control theory with location of zeros and problems of stability as main themes. The famous Anderson-Jury paper in IEEE Transactions Automatic Control from 1976 served as a source of inspiration for his later work on generalized Bezout operators. As a further illustration of Leonia's work I will discuss in more detail his 1994 paper, with Israel Koltracht and Ben Kon as co-authors. I consider this article as one of Leonia's top papers. It has a short title:

Integr Equat Oper Th   0378/620X/94/040410-39$1.50+0.20/0
Vol. 20 (1994)   (c) Birkhäuser Verlag, Basel

INVERSION OF STRUCTURED OPERATORS

I. Koltracht*, B.A. Kon and L. Lerer**

The KKL-paper deals with bounded linear integro-differential operators A on $L^2[0, \omega]$ of which the action is given by

$$(Af)(x) = \frac{d}{dx} \int_0^\omega \left(\frac{\partial}{\partial t} \Phi(x,t) \right) f(t) \, dt,$$

$$\Phi(x,t) = \frac{1}{2} \sum_{j=1}^n \int_{x+t}^{2\omega - |x-t|} c_j \left(\frac{s+x-t}{2} \right) \overline{b_j \left(\frac{s-x+t}{2} \right)} \, ds.$$

Here $b_j, c_j \in L^2[0, \omega]$, $j = 1, 2, \ldots, n$. If A is as above, we say that A belongs to the *KKL class*, and we shall refer to $b_j, c_j \in L^2[0, \omega]$, $j = 1, 2, \ldots, n$, as the *defining data*.

The paper has two beautiful theorems. To state the first theorem we need the Volterra integral operator on $L^2[0, \omega]$ which is defined by

$$(Vf)(x) = -i \int_0^x f(t) \, dt, \quad 0 \le x \le \omega.$$

Furthermore, with the defining data $b_j, c_j \in L^2[0, \omega]$, $j = 1, 2, \ldots, n$, and the Volterra integral operator we associate the Lyapunov equation

$$XV - V^*X = \sum_{j=1}^n c_j \langle \cdot, b_j \rangle. \tag{1}$$

Note that the right-hand side of the Lyapunov equation (1) is an operator of finite rank. If this operator is zero, then X is also equal to zero. Thus, using terminology common for structured matrices, an operator X satisfying (1) is an operator with a *relatively small displacement*. The authors proved that these operators X with a relative small displacement are precisely the integro-differential operators A introduced above. This is the first beautiful theorem.

Theorem 1. [KKL-1994] *An operator A on $L^2[0, \omega]$ belongs to the KKL class if and only if for some $b_j, c_j \in L^2[0, \omega]$, $j = 1, 2, \ldots, n$ the operator $X = A$ satisfies the identity* (1).

Now assume that A satisfies the Lyapunov equation (1), and let A be invertible. Multiplying (1) from the left and from the right by the inverse of A yields another Lyapunov operator identity, which is analogous to (1):

$$VA^{-1} - A^{-1}V^* = \sum_{j=1}^{n} \gamma_j \langle \cdot, \beta_j \rangle, \text{ where } A\beta_j = b_j \text{ and } A\gamma_j = c_j.$$

A variant of the first theorem now yields the second which can be viewed as an operator analogue of the finite-dimensional inversion theorems for Toeplitz matrices due to Gohberg–Semencul and Gohberg–Heinig.

Theorem 2. [KKL-1994] *Assume A belongs to the KKL class with defining data $b_j, c_j \in L^2[0, \omega]$, $j = 1, 2, \ldots, n$, and let A be a Fredholm operator. If the equations*

$$A\beta_j = b_j \quad and \quad A\gamma_j = c_j \quad (j = 1, \ldots, n)$$

are solvable, then A has a bounded inverse and A^{-1} is given by

$$(A^{-1}f) = \frac{d}{dx} \int_0^{\omega} \left(\frac{\partial}{\partial t} k(x, t) \right) f(t) \, dt,$$

where

$$k(x, t) = \sum_{j=1}^{n} \int_0^{\min(x,t)} \gamma_j(x - s)\overline{\beta_j(t - s)} \, ds.$$

The KKL paper present lots of examples, another characteristic of Leonia's work. Here we mention the following two, both are taken from the KKL paper.

Example 1. Let $k \in L^1[-\omega, \omega]$, and let A be the operator from the KKL class defined by the following data:

$$b_1(x) = 1, \quad b_2(x) = 1 + \int_0^{\omega-x} \overline{k(s)} \, ds,$$

$$c_1(x) = \int_{x-\omega}^0 k(s) \, ds, \quad c_2(x) = 1, \quad 0 \leq x \leq \omega.$$

Then A is the convolution operator given by

$$(Af)(t) = f(t) + \int_0^{\omega} k(t - s)f(s) \, ds.$$

Example 2. Let $\phi, \psi \in L^2[0, \omega]$, and let A be the operator from the KKL class defined by the following data:

$$b_1(x) = \overline{\phi(x)}, \quad b_2(x) = \overline{\psi(x)},$$
$$c_1(x) = 1 - \psi(\omega - x) \quad c_2(x) = 1 - \psi(\omega - x), \quad 0 \leq x \leq \omega.$$

Then A is the *Bezout operator* defined by the entire functions

$$F(z) = 1 + iz \int_o^{\omega} e^{izt} \phi(t) \, dt \quad and \quad G(z) = 1 + iz \int_o^{\omega} e^{izt} \phi(t) \, dt.$$

Other themes in the work of Leonia are:
- Operators in locally convex spaces
- Asymptotic distribution of spectra and related limit theorems

- Rational matrix functions
- Spectral theory of matrix and operator polynomials
- Inverse problems for Szegő–Krein orthogonal polynomials and their continuous analogs
- Partial realization problems
- Minimality of partial realization problems in discrete time, including multivariable systems

The majority of my joint papers with Leonia belong to the areas described by the last four bullets.

Leonia and Israel Gohberg

Gohberg had a profound influence on the work of Leonia. He was Leonia's mathematical (grand-)father, and according to MatSciNet Gohberg is Leonia's top co-author. Gohberg's passing away in October 2009 was a great personal loss for Leonia, as for many of us. For Leonia it meant a set back for a long period.

Leonia and Alek Markus
Amsterdam (November 29, 2002)

Alek Markus is Leonia's doctor-father. Of course, the mathematical son followed his mathematical fathers in many ways, but not always. On one particular non-mathematical point, it was the other way around: the mathematical son was leading and his mathematical fathers were following. Of the three, Leonia was the first to immigrate from Kishinev to Israel. Gohberg followed later and Markus many years later.

I conclude with best wishes: for Leonia personally, for Bertha his wife, and for his two daughters Hannah and Safira, and for the new family member Gal, the husband of Safira. It is my sincere hope that both of us will have the time and energy to continue our joint work.

M.A. Kaashoek
Department of Mathematics
Faculty of Sciences
VU University
De Boelelaan 1081a
NL-1081 HV Amsterdam, The Netherlands
e-mail: `ma.kaashoek@vu.nl`

Operator Theory:
Advances and Applications, Vol. 237, 9–10

Leonia Lerer and My First Visit to Israel

H. Bart

In 1981, my wife Greetje and I went to Israel for the first time. I was to attend the Toeplitz Memorial Conference in Tel-Aviv organized by Israel Gohberg. But before going there we went to Haifa. There, at the Technion, I gave my very first lecture in Israel. The topic was 'New methods for solving integral equations'. Leonia Lerer was our host. We had already met him before during one of his visits to Amsterdam.

For both Greetje and me, coming to Israel meant something special. Being children of traditionally protestant parents we were raised with the stories from the Bible. So we were excited to be at the places we had heard about so much. We made several trips, some on our own, but at least one with Leonia. He took us to Rosh HaNikra, a white chalk cliff face located on the coast of the Mediterranean Sea, which opens up into spectacular interconnected grottos. Leonia's daughter Hanna went with us, doing the driving. Bertha, Leonia's wife could not join us because she was pregnant. A little later Saffira was born.

We had good contacts as families. Several times Leonia was our guest in The Netherlands, sometimes on his own, sometimes accompanied by other family members. In preparing this note, my wife and I went through our old pictures in order to find traces of such get togethers. The photographs we found brought back good memories. The picture above, showing Leonia in our garden talking with Roland Duduchava, was taken in 2003 during a celebration of Israel's Gohberg 75th birthday.

Leonia: congratulations with your seventieth birthday and best wishes to you and your family for the future. Also thanks for everything you did in connection with the many Haifa matrix theory conferences.

H. Bart
Econometric Institute
Erasmus University Rotterdam
P.O. Box 1738
NL-3000 DR Rotterdam, The Netherlands
e-mail: `bart@ese.eur.nl`

Operator Theory:
Advances and Applications, Vol. 237, 11–12
© 2013 Springer Basel

Through the Eyes of a Student

Irina Karelin

I began my graduate studies in 1989 with Professor Lerer. I did my aliah in March 1989. The main reason to come to Haifa was to study in Technion – Israel Institute of Technology. I was a graduate of Kharkov University (Ukraine) in Mathematics and wanted to continue my studies. My degree thesis was in Operator theory area and I of course expected to find somebody who is working in this area and will agree to be my supervisor for master studies. I came to the Department of Mathematics in Technion to Professor Abraham Berman. He asked me about my studies in Kharkov and referred me to Professor Lerer.

We talked about my previous studies and my final student work. Leonid told me that the subject of my work is rather close to one of his areas of interest, that he has some problems to be solved and proposed me to do my master degree under his supervision. I was happy to accept his offer and we started to work together. To approach to the subject of your future research Leonid said you have to read a lot. He proposed me a long list of articles and some books for reading. Among these works were the books *Matrix Polynomials* by Gohberg, Lancaster, Rodman, *The Theory of Matrices* by Lancaster and Tismenetsky, articles by Lerer and his joint works with Rodman, Tismenetsky and some other authors. In the beginning all seemed to me unknown and new but I always could turn to Leonid with any question. I understood it later that all that Leonid advised to do was well thought out, methodically correct. All chapters and articles he chose much contributed to my entering into the subject.

At the same time during my first semester in Technion I attended Leonid's lectures for graduate students on realizations and factorizations of rational matrix functions. I liked him as a lecturer. It was not only my impression. I heard this opinion from other students. His teaching manner was calm and pleasing. His explanations always were very clear and accurate. Atmosphere he created was always good and friendly.

The subject of my research was chosen. It was related to factorizations of column and row reduced matrix polynomials. At the beginning our meetings were not so frequent, later on we met much more often. But every meeting was very useful and productive for me. Sometimes my research progress was good and smooth, sometimes I felt frustrated from unsuccessful attempts to find a solution of some problem. Even in such crisis moments of my work I knew that great mathematician and very good man, Professor Lerer always encourages and supports me, turns me towards right direction, gives useful advice.

Leonid himself worked very intensively and also had many unsolved problems for his students. When I decided to study for PhD degree and Leonid agreed to continue with me there was no question what to do. He proposed me a number of opportunities for my research. The theme of my work was chosen. It was connected to factorization of rational matrix functions and Riccati equations. Leonid's style of supervision was as it had to be. He didn't generally tell me, and I suppose to other students as well, what to do. He encouraged our independent research, didn't restrict our freedom, but he always might warn us of some things and tried to steer us towards others. So when I prepared my conference report his assistance was very important and fruitful. He helped me to choose more profitable results for the presentation, to build the report in the right form, to provide possible questions.

Many times I run tutorials in the courses where Leonid taught. Students appreciated Leonid as a lecturer very much. I heard from students that they liked his explanations, liked his logical and consistent presentation of material, illustrated with many examples and solved problems. Working with Leonid as his teaching assistant I acquired wide experience for my future professional life.

Papers of Leonid always had sharp form and profound content. Statements and proofs of results were exact, explanations at the same time were sufficient and never redundant. Leonid also asked his students exact statements and logical, full explanations. When I prepared my PhD thesis Leonid read the written text and returned it back to me many times until the version of the thesis was good.

Today I work in industry as an algorithm developer. While working with Professor Lerer I acquired many skills very useful in my professional activities. Leonid taught me to do thorough literature review, helped me to develop my mathematical thinking, my writing skills, taught me to be an independent researcher.

Dear Professor Lerer! I want to express my deep gratitude and admiration! Happy birthday!

Irina Karelin
CmosaiX, Ltd.
Kibutz Yagur 30065, Israel
e-mail: irina@cmosaix.com

Operator Theory:
Advances and Applications, Vol. 237, 13–13

Reminiscences on Visits to Haifa

André C.M. Ran

My first contacts with Leonid Lerer date from the middle of the eighties, when I was still a PhD student at the VU university. Leonia visited Amsterdam regularly, and we maintained warm contacts without actually working together. This changed when I visited Israel for a somewhat longer time in 1996. We started a collaboration which produced a first paper in 1997, and I became a frequent visitor to Haifa after this. All in all, we wrote six papers together, several with students of Leonia.

To visit the Technion and work with Leonia was always a great pleasure. The mathematics was wonderful, and the warm personality of Leonia made visiting there a joy. Leonia was aware of the sensitivities of my family regarding travel to Israel, and so he made sure I was met at the airport by a trusted driver. I was also not allowed to leave campus without a local to guide me.

In early 1998 I visited Haifa for the Tenth Haifa Matrix Meeting. Before leaving I talked to my father, who was terminally ill. A few days into my visit in Haifa my sister called me to inform me that my father had passed away quietly on January 7. I will always be grateful for the support and warmth that I received from Leonia on that day and the following one. Obviously, I had to arrange a very quick return to the Netherlands, and Leonia helped and comforted me as much as possible.

Fortunately, I could return to Haifa for several more Haifa Matrix Meetings and an ILAS Meeting over the course of the years, most of them combined with some extra time to work together with Leonia.

I would like to close with a heartfelt thanks to Leonia for his inspiring lead in our joint work and for his warm friendship.

André C.M. Ran
Afdeling wiskunde, FEW
Vrije Universiteit
De Boelelaan 1081 a
NL-1081 HV Amsterdam
The Netherlands
e-mail: a.c.m.ran@vu.nl

Operator Theory:
Advances and Applications, Vol. 237, 15–15

My First Research Experience

Hugo J. Woerdeman

My first exposure to mathematical research was under the supervision of Professor Leonid ("Leonia") Lerer during the academic year 1984–1985. From the very beginning we had a very amicable relationship, and what will always stand out in my memory is his consistently calm and friendly demeanor. Leonia was always encouraging, and he never lost his patience, even during the slow periods. He also managed to enjoy life to the fullest – among other things, he savored the fresh Dutch raw herring during lunch. It was a great environment for me to enter this completely new endeavor. I learned about resultants of analytic functions, singular integral operators, and many other things I had never heard of, and was expected to generalize the scalar case to the matrix case. So I learned about realizations, Jordan triples, matrix factorizations, and a lot of other good stuff. So, how did we manage to make a breakthrough? At some point Leonia encouraged me to work through a specific example, and I remember generating pages and pages of computations with matrices that very quickly became quite large. And then at some point things clicked: I could see the big picture and focus in all those pages of scribbles on the steps where something real had happened. And the rest is history as they say, as after a year of hard work we were able to put two papers together resulting in my first two publications.

So on this occasion of your 70th birthday, Leonia, I would like to thank you very much for the excellent guidance you gave me during my initial research experience, and for the many years of friendship that followed. I will never forget the way my first mathematical breakthrough came about, and it has been a guide for me ever since. Thank you!

Hugo J. Woerdeman
Department of Mathematics
Drexel University
Philadelphia, PA 19104, USA
e-mail: hugo@math.drexel.edu

Operator Theory:
Advances and Applications, Vol. 237, 17–39
© 2013 Springer Basel

Interpolation in Sub-Bergman Spaces

Joseph A. Ball and Vladimir Bolotnikov

Dedicated to Leonia Lerer, a long-time friend and colleague

Abstract. A general interpolation problem with operator argument is studied for functions f from sub-Bergman spaces associated with an analytic function S mapping the open unit disk $\mathbb{D}$ into the closed unit disk.

Mathematics Subject Classification (2010). 30E05, 47A57, 46E22.

Keywords. De Branges–Rovnyak space, Schur-class function.

1. Introduction

Given a Hilbert space $\mathcal{Y}$ and a positive integer n, we denote by $\mathcal{A}_n(\mathcal{Y})$ the standard weighted Bergman space of $\mathcal{Y}$-valued functions f analytic on the open unit disk $\mathbb{D}$ and with finite norm $\|f\|_{\mathcal{A}_n(\mathcal{Y})}$:

$$\mathcal{A}_n(\mathcal{Y}) = \left\{ f(z) = \sum_{j \geq 0} f_j z^j : \|f\|^2_{\mathcal{A}_n(\mathcal{Y})} := \sum_{j \geq 0} \mu_{n,j} \|f_j\|^2_{\mathcal{Y}} < \infty \right\} \qquad (1.1)$$

where the weights $\mu_{n,j}$'s are defined by

$$\mu_{n,j} := \frac{1}{\binom{j+n-1}{j}} = \frac{j!(n-1)!}{(j+n-1)!}. \qquad (1.2)$$

It is clear from (1.1) that the spaces $\mathcal{A}_1(\mathcal{Y})$ and $\mathcal{A}_2(\mathcal{Y})$ are respectively the standard vector-valued Hardy space $H^2(\mathcal{Y})$ and the unweighted Bergman space $A_2^2(\mathcal{Y})$ of the unit disk. It then follows that $\mathcal{A}_n(\mathcal{Y})$ is the reproducing kernel Hilbert space with reproducing kernel

$$k_n(z, \zeta) = \frac{1}{(1 - z\bar{\zeta})^n} I_{\mathcal{Y}}$$

and that for $n > 1$, the $\mathcal{A}_n(\mathcal{Y})$-norm equals

$$\|f\|^2_{\mathcal{A}_n(\mathcal{Y})} = \int_{\mathbb{D}} (n-1) \|f(z)\|^2_{\mathcal{Y}} (1 - |z|^2)^{n-2} dA(z) < \infty$$

where dA is the planar Lebesgue measure normalized so that $A(\mathbb{D}) = 1$.

For Hilbert spaces $\mathcal{U}$ and $\mathcal{Y}$, we denote by $\mathcal{L}(\mathcal{U}, \mathcal{Y})$ the space of bounded linear operators mapping $\mathcal{U}$ into $\mathcal{Y}$ (abbreviated to $\mathcal{L}(\mathcal{U})$ in case $\mathcal{U} = \mathcal{Y}$) and we define the operator-valued *Schur class* $\mathcal{S}(\mathcal{U}, \mathcal{Y})$ to be the class of analytic functions S on $\mathbb{D}$ whose values $S(z)$ are contraction operators in $\mathcal{L}(\mathcal{U}, \mathcal{Y})$. Each Schur-class function $S \in \mathcal{S}(\mathcal{U}, \mathcal{Y})$ induces a contractive multiplication operator $M_S : f \mapsto Sf$ from $\mathcal{A}_n(\mathcal{U})$ into $\mathcal{A}_n(\mathcal{Y})$ for every $n \geq 1$; in fact the Schur-class is exactly the class of contractive multipliers on $\mathcal{A}_n(\mathcal{Y})$ for each $n = 1, 2, \ldots$ (see, e.g., [14, Proposition 4.1]). The latter property is equivalent to the positivity of the kernels

$$K_{S,n}(z, \zeta) = \frac{I_{\mathcal{Y}} - S(z)S(\zeta)^*}{(1 - z\bar{\zeta})^n} \quad (n \geq 1) \tag{1.3}$$

on $\mathbb{D} \times \mathbb{D}$ for all $n \geq 1$. Thus the positivity of $K_{S,n}$ for some $n \geq 1$ already guarantees the membership $S \in \mathcal{S}(\mathcal{U}, \mathcal{Y})$ so that $K_{S,n}$ is positive for all $n \geq 1$.

Thus, with any function $S \in \mathcal{S}(\mathcal{U}, \mathcal{Y})$, one can associate a family of positive kernels (1.3) and thereby a family of reproducing kernel Hilbert space $\mathcal{H}(K_{S,n})$. The spaces $\mathcal{H}(K_{S,1})$ were introduced by de Branges and Rovnyak [11, 12] as a convenient and natural setting for canonical operator models. Further developments on de Branges–Rovnyak spaces can be found in [18]. The study of $\mathcal{H}(K_{S,2})$ was initiated in [20, 21] and there are very few publications concerning the case where $n > 2$ (see, e.g., [19]).

The general complementation theory applied to the contractive operator $M_S : \mathcal{A}_n(\mathcal{U}) \to \mathcal{A}_n(\mathcal{Y})$ (see, e.g., [18]) provides the characterization of $\mathcal{H}(K_{S,n})$ as the operator range $\mathcal{H}(K_{S,n}) = \operatorname{Ran}(I - M_S M_S^*)^{\frac{1}{2}} \subset \mathcal{A}_n(\mathcal{Y})$ with the lifted norm

$$\|(I - M_S M_S^*)^{\frac{1}{2}} f\|_{\mathcal{H}(K_{S,n})}^2 = \|(I - \pi)f\|_{\mathcal{A}_n(\mathcal{Y})} \tag{1.4}$$

where π is the orthogonal projection onto $\operatorname{Ker}(I - M_S M_S^*)^{\frac{1}{2}}$. It follows that $\|f\|_{\mathcal{H}(K_{S,n})} \geq \|f\|_{\mathcal{A}_n(\mathcal{Y})}$ for every $f \in \mathcal{H}(K_{S,n})$ and thus the spaces $\mathcal{H}(K_{S,n})$ are contractively included in $\mathcal{A}_n(\mathcal{Y})$. For this reason, the spaces $\mathcal{H}(K_{S,n})$ are termed *sub-Hardy* (in [18] for $n = 1$) and *sub-Bergman* (in [20], [21] for $n = 2$). Upon setting $f = (I - M_S M_S^*)^{\frac{1}{2}} h$ in (1.4) we get

$$\|(I - M_S M_S^*)h\|_{\mathcal{H}(K_{S,n})} = \langle(I - M_S M_S^*)h, h\rangle_{\mathcal{A}_n(\mathcal{Y})}. \tag{1.5}$$

The purpose of this paper is to study an interpolation problem of Nevanlinna–Pick type in the space $\mathcal{H}(K_{S,n})$. To formulate the problem we need several definitions.

A pair (E, T) consisting of operators $T \in \mathcal{L}(\mathcal{X})$ and $E \in \mathcal{L}(\mathcal{X}, \mathcal{Y})$ is called an *output pair*. An output pair (E, T) is called *n-output-stable* if the associated *n*-observability operator

$$\mathcal{O}_{n,E,T} : x \mapsto E(I - zT)^{-n}x = \sum_{j=0}^{\infty} (\mu_{n,j}^{-1} ET^j x) z^j \tag{1.6}$$

maps $\mathcal{X}$ into $\mathcal{A}_n(\mathcal{Y})$ and is bounded. For an n-output stable pair (E, T), we define the tangential functional calculus $f \mapsto (E^* f)^{\wedge L}(T^*)$ on $\mathcal{A}_n(\mathcal{Y})$ by

$$(E^* f)^{\wedge L}(T^*) = \sum_{j=0}^{\infty} T^{*j} E^* f_j \quad \text{if} \quad f(z) = \sum_{j=0}^{\infty} f_j z^j \in \mathcal{A}_n(\mathcal{Y}). \qquad (1.7)$$

The computation

$$\left\langle \sum_{j=0}^{\infty} T^{*j} E^* f_j, \, x \right\rangle_{\mathcal{X}} = \sum_{j=0}^{\infty} \langle f_j, \, ET^j x \rangle_{\mathcal{Y}}$$

$$= \sum_{j=0}^{\infty} \mu_{n,j} \cdot \langle f_j, \, \mu_{n,j}^{-1} ET^j x \rangle_{\mathcal{Y}} = \langle f, \, \mathcal{O}_{n,E,T} x \rangle_{\mathcal{A}_n(\mathcal{Y})}$$

shows that the n-output stability of (E, T) is exactly what is needed to verify that the infinite series in the definition (1.7) of $(E^* f)^{\wedge L}(T^*)$ converges in the weak topology on $\mathcal{X}$. The same computation shows that tangential evaluation with operator argument amounts to the adjoint of $\mathcal{O}_{n,E,T}$ (in the metric of $\mathcal{A}_n(\mathcal{Y})$):

$$(E^* f)^{\wedge L}(T^*) = \mathcal{O}_{n,E,T}^* f \quad \text{for} \quad f \in \mathcal{A}_n(\mathcal{Y}). \qquad (1.8)$$

Since $\mathcal{H}(K_{S,n})$ is included in $\mathcal{A}_n(\mathcal{Y})$, evaluation (1.7) applies to functions in $\mathcal{H}(K_{S,n})$. In this paper we study the following interpolation problem.

Problem 1.1. *Given a Schur-class function $S \in \mathcal{S}(\mathcal{U}, \mathcal{Y})$, given an n-output stable pair $(E, T) \in \mathcal{L}(\mathcal{X}, \mathcal{Y}) \times \mathcal{L}(\mathcal{X})$ and given $\mathbf{x} \in \mathcal{X}$,*

(i) *Find all $f \in \mathcal{H}(K_{S,n})$ such that*

$$(E^* f)^{\wedge L}(T^*) := \mathcal{O}_{n,E,T}^* f = \mathbf{x}. \qquad (1.9)$$

(ii) **(norm-constrained version):** *Find all $f \in \mathcal{H}(K_{S,n})$ satisfying (1.9) with $\|f\|_{\mathcal{H}(K_{S,n})} \leq 1$.*

The Hardy-space special case of this problem (where $n = 1$ and $\mathcal{H}(K_S) := \mathcal{H}(K_{S,1})$) is the classical de Branges–Rovnyak space and has been studied by the authors and S. ter Horst in [5, 6]. The set of all functions in $H^2(\mathcal{Y})$ satisfying a condition of the form $\mathcal{O}_{n,E,T}^* f = 0$ (i.e., condition (1.9) with $S = 0$ and $\mathbf{x} = 0$) is one way to describe a generic shift-invariant subspace of $H^2(\mathcal{Y})$ and the description of the set of all solutions amounts to a calculation of the Beurling–Lax representer $\Theta(z)$ for the space $\mathcal{M}$ in terms of the interpolation data $\{E, T, S = 0, \mathbf{x} = \mathbf{0}\}$. Another special case is to allow a general n but still insist that $S = 0$ and $\mathbf{x} = \mathbf{0}$; this special case recovers the Beurling–Lax representation theorem for shift-invariant subspaces $\mathcal{M}$ in the weighted Bergman space obtained by the authors in [4].

2. Interpolation in reproducing kernel Hilbert spaces: A brief survey

The following operator interpolation problem with norm constraint is well known in the literature: *Given Hilbert space operators $A \in \mathcal{L}(\mathcal{Y}, \mathcal{X})$ and $B \in \mathcal{L}(\mathcal{U}, \mathcal{X})$, describe all $X \in \mathcal{L}(\mathcal{U}, \mathcal{Y})$ that satisfy the conditions*

$$AX = B \quad and \quad \|X\| \le 1. \tag{2.1}$$

The solvability criterion is known as the Douglas factorization lemma [13]: *There is an $X \in \mathcal{L}(\mathcal{U}, \mathcal{Y})$ satisfying (2.1) if and only if $AA^* \ge BB^*$.* If this is the case, then (see, e.g., [6]) $X \in \mathcal{L}(\mathcal{U}, \mathcal{Y})$ satisfies conditions (2.1) if and only if the operator

$$\begin{bmatrix} I_{\mathcal{U}} & B^* & X^* \\ B & AA^* & A \\ X & A^* & I_{\mathcal{Y}} \end{bmatrix} : \begin{bmatrix} \mathcal{U} \\ \mathcal{X} \\ \mathcal{Y} \end{bmatrix} \to \begin{bmatrix} \mathcal{U} \\ \mathcal{X} \\ \mathcal{Y} \end{bmatrix} \quad \text{is positive semidefinite.} \tag{2.2}$$

On the other hand, if $AA^* \ge BB^*$, then there exist (unique) contractions $X_1 \in \mathcal{L}(\mathcal{U}, \overline{\mathrm{Ran}}A)$ and $X_2 \in \mathcal{L}(\mathcal{Y}, \overline{\mathrm{Ran}}A)$ such that

$$(AA^*)^{\frac{1}{2}} X_1 = B, \quad (AA^*)^{\frac{1}{2}} X_2 = A, \quad \mathrm{Ker} X_1 = \mathrm{Ker} B, \quad \mathrm{Ker} X_2 = \mathrm{Ker} A. \tag{2.3}$$

Applying Schur complement arguments to the positive semidefinite operator in (2.2) leads us to the following more explicit description of the set of all solutions to the problem (2.1) (see, e.g., [6] for the proof).

Lemma 2.1. *Let $AA^* \ge BB^*$. Then an operator X satisfies condition (2.2) (and therefore, also conditions (2.1)) if and only if it is of the form*

$$X = X_2^* X_1 + (I - X_2^* X_2)^{\frac{1}{2}} Q (I - X_1^* X_1)^{\frac{1}{2}} \tag{2.4}$$

where X_1 and X_2 are defined in (2.3) and where the parameter Q is an arbitrary contraction from $\overline{\mathrm{Ran}}(I - X_1^* X_1)$ into $\overline{\mathrm{Ran}}(I - X_2^* X_2)$.

Remark 2.2. It follows from (2.4) that there is a unique X subject to conditions (2.1) if and only if X_1 is isometric on $\mathcal{U}$ or X_2 is isometric on $\mathcal{Y}$. Furthermore, for each Q in (2.4) and each $u \in \mathcal{U}$, we have

$$\|Xu\|^2 = \|X_2^* X_1 u\|^2 + \|(I - X_2^* X_2)^{\frac{1}{2}} Q (I - X_1^* X_1)^{\frac{1}{2}} u\|^2,$$

so that $X_2^* X_1$ is the minimal norm solution to the problem (2.1) (see [6, Section 2]).

The left tangential Nevanlinna–Pick interpolation problem for the reproducing kernel Hilbert space $\mathcal{H}(K)$ can be formulated as follows. We are given vectors $y_1, \ldots, y_r \in \mathcal{Y}$ and points $\omega_1, \ldots, \omega_r \in \mathbb{D}$ along with numbers $x_1, \ldots, x_r \in \mathbb{C}$ and seek $f \in \mathcal{H}(K)$ (possibly also with $\|f\|_{\mathcal{H}(K)} \le 1$) satisfying the left tangential Nevanlinna–Pick interpolation conditions

$$\langle f(\omega_i), y_i \rangle = x_i \text{ for } i = 1, \ldots, r. \tag{2.5}$$

This problem can be reformulated more abstractly as follows. We introduce the $\mathcal{L}(\mathbb{C}^r, \mathcal{Y})$-valued function

$$z \mapsto F(z) := \begin{bmatrix} K(z, \omega_1) y_1 & \cdots & K(z, \omega_r) y_r \end{bmatrix}. \tag{2.6}$$

Then F induces a multiplication operator $M_F \colon \mathbb{C}^r \to \mathcal{H}(K)$

$$M_F \colon \begin{bmatrix} c_1 \\ \vdots \\ c_r \end{bmatrix} \to F(z) \begin{bmatrix} c_1 \\ \vdots \\ c_r \end{bmatrix} = c_1 K(z, \omega_1) y_1 + \cdots + c_r K(z, \omega_r) y_r.$$

Then a standard reproducing-kernel computation gives us that $M_F^* \colon \mathcal{H}(K) \to \mathbb{C}^r$ is given by

$$M_F^* \colon f \mapsto \begin{bmatrix} \langle f(\omega_1), y_1 \rangle_\mathcal{Y} \\ \vdots \\ \langle f(\omega_r), y_r \rangle_\mathcal{Y} \end{bmatrix}$$

and the Nevanlinna–Pick problem with interpolation conditions (2.5) can be reformulated as follows: *for given F as in (2.6) and $\mathbf{x} \in \mathbb{C}^r$, find $f \in \mathcal{H}(K)$ (possibly also with $\|f\|_{\mathcal{H}(K)} \leq 1$) so that $M_F^* f = \mathbf{x}$.*

We now formulate our abstract left tangential Nevanlinna–Pick interpolation problem as follows. Let $K(z, \zeta)$ be an $\mathcal{L}(\mathcal{Y})$-valued positive kernel on a Cartesian product set $\Omega \times \Omega$ and let $\mathcal{H}(K)$ be the associated reproducing kernel Hilbert space, that is, the unique inner product space of $\mathcal{Y}$-valued functions on Ω that contains the functions $z \mapsto K_\omega(z) := K(z, \omega) y$ for all fixed $\omega \in \Omega$ and $y \in \mathcal{Y}$ which in turn have the reproducing property for $\mathcal{H}(K)$:

$$\langle f, K_\omega y \rangle_{\mathcal{H}(K)} = \langle f(\omega), y \rangle_\mathcal{Y} \quad \text{for all} \quad f \in \mathcal{H}(K).$$

For $\mathcal{X}$ an auxiliary Hilbert space, we let $\mathcal{M}(\mathcal{X}, \mathcal{H}(K))$ (the space of *multipliers* from $\mathcal{X}$ into $\mathcal{H}(K)$) denote the space of $\mathcal{L}(\mathcal{X}, \mathcal{Y})$-valued functions such that the function $z \mapsto F(z) x$ is in $\mathcal{H}(K)$ for each $x \in \mathcal{X}$. A consequence of the closed-graph theorem is that the multiplication operator $M_F \colon x \mapsto F(\cdot) x$ is then bounded as an operator from $\mathcal{X}$ into $\mathcal{H}(K)$. With this notation in hand we can pose the following interpolation problem:

Problem 2.3. *Given a positive kernel K along with $F \in \mathcal{M}(\mathcal{X}, \mathcal{H}(K))$ and $\mathbf{x} \in \mathcal{X}$,*

(i) *Find all functions $f \in \mathcal{H}(K)$ satisfying*

$$M_F^* f = \mathbf{x}. \tag{2.7}$$

(ii) *Find all functions $f \in \mathcal{H}(K)$ satisfying (2.7) with $\|f\|_{\mathcal{H}(K)} \leq 1$.*

As a straightforward application of the general Hilbert space results in Lemma 2.1, Remark 2.2 and the preceding discussion, we have the following solution of Problem 2.3.

Proposition 2.4. *Problem 2.3 (ii) has a solution if and only if*

$$P \geq \mathbf{x}\mathbf{x}^*, \quad \text{where} \quad P := M_F^* M_F. \tag{2.8}$$

Problem 2.3 (i) has a solution if and only if $\mathbf{x} \in \operatorname{Ran} P^{\frac{1}{2}}$.

Proof. By specializing the Douglas lemma to the case where

$$A = M_F^* : \mathcal{H}(K) \to \mathcal{X} \quad \text{and} \quad B = \mathbf{x} \in \mathcal{X} \cong \mathcal{L}(\mathbb{C}, \mathcal{X}), \tag{2.9}$$

we see that solutions $X : \mathbb{C} \to \mathcal{H}(K)$ to problem (2.1) necessarily have the form of a multiplication operator M_f for some function $f \in \mathcal{H}(K)$. This observation leads us to (2.8). On the other hand, by the second equality in (2.8), $\operatorname{Ran} P^{\frac{1}{2}} = \operatorname{Ran} M_F^*$. Thus, $\mathbf{x}$ belongs to $\operatorname{Ran} P^{\frac{1}{2}}$ if and only if it belongs to $\operatorname{Ran} M_F^*$, that is, if and only if equality $M_F^* f = \mathbf{x}$ holds for some $f \in \mathcal{H}(K)$, which means that this f solves Problem 2.3. $\qquad \square$

Let us assume that the operator $P = M_F^* M_F$ is strictly positive definite. Then the operator $M_F P^{-\frac{1}{2}}$ is an isometry and the space

$$\mathcal{N} = \{F(z)x : x \in \mathcal{X}\} \quad \text{with norm} \quad \|Fx\|_{\mathcal{H}(S)} = \|P^{\frac{1}{2}}x\|_{\mathcal{X}} \tag{2.10}$$

is isometrically included in $\mathcal{H}(K)$. Moreover, the orthogonal complement of $\mathcal{N}$ in $\mathcal{H}(K)$ is the reproducing kernel Hilbert space $\mathcal{H}(\widetilde{K})$ with reproducing kernel

$$\widetilde{K}(z, \zeta) = K(z, \zeta) - F(z)P^{-1}F(\zeta)^*. \tag{2.11}$$

The following theorem is an adaptation of Lemma 2.1 to the special case (2.9).

Theorem 2.5. *Assume that condition (2.8) holds and that P is strictly positive definite. Let $\widetilde{K}$ be the kernel defined in (2.11).*

1. *A function $f : \Omega \to \mathcal{Y}$ solves Problem 2.3 (ii) if and only if it is of the form*

$$f(z) = F(z)P^{-1}\mathbf{x} + h(z) \tag{2.12}$$

   for some function $h \in \mathcal{H}(\widetilde{K})$ subject to $\|h\|_{\mathcal{H}(\widetilde{K})} \leq \sqrt{1 - \|P^{-\frac{1}{2}}\mathbf{x}\|^2}$.

2. *The representation (2.12) is orthogonal (in the metric of $\mathcal{H}(K)$) so that $F(z)P^{-1}\mathbf{x}$ is the minimal-norm solution of Problem 2.3.*

3. *Problem 2.3 (ii) has a unique solution if and only if*

$$\|P^{-\frac{1}{2}}\mathbf{x}\| = 1 \quad \text{or} \quad \widetilde{K}(z, \zeta) \equiv 0.$$

Proof. It is readily seen that

$$X_1 = P^{-\frac{1}{2}}\mathbf{x} \in \mathcal{X} \cong \mathcal{L}(\mathbb{C}, \mathcal{X}) \quad \text{and} \quad X_2 = P^{-\frac{1}{2}}M_F^* \in \mathcal{L}(\mathcal{H}(K_S), \mathcal{X})$$

are the operators X_1 and X_2 from (2.4) after specialization to the case (2.9). Statements (2) and (3) now follow from Remark 2.2, since $P^{-\frac{1}{2}}\mathbf{x} \in \mathcal{L}(\mathbb{C}, \mathcal{X})$ being isometric means that $\|P^{-\frac{1}{2}}\mathbf{x}\| = 1$ and, on the other hand, the isometric property for the operator $M_F P^{-\frac{1}{2}}$ means that the space $\mathcal{N}$ defined in (2.10) is equal to the whole space $\mathcal{H}(K)$. Thus $\mathcal{H}(\widetilde{K}) = \mathcal{H}(K) \ominus \mathcal{N} = \{0\}$ or $\widetilde{K} \equiv 0$.

In the present framework, the parametrization formula (2.4) takes the form

$$M_f = M_F P^{-\frac{1}{2}}\mathbf{x} + \sqrt{1 - \|P^{-\frac{1}{2}}\mathbf{x}\|^2} \cdot (I - M_F P^{-1}M_F^*)^{\frac{1}{2}}Q \tag{2.13}$$

where Q is equal to the operator of multiplication $M_k : \mathbb{C} \to \mathcal{H}(\widetilde{K})$ by a function $k \in \mathcal{H}(\widetilde{K})$ with $\|k\| \leq 1$. Since $M_F P^{-\frac{1}{2}}$ is an isometry, the second term on the right-hand side of (2.13) is equal to the operator M_h of multiplication by a function $h \in \mathcal{H}(\widetilde{K})$ such that $\|h\|_{\mathcal{H}(\widetilde{K})} = \|h\|_{\mathcal{H}(K)} \leq \sqrt{1 - \|P^{-\frac{1}{2}}\mathbf{x}\|^2}$. $\qquad\square$

Remark 2.6. The parametrization formula (2.12) can be obtained in a more analytic way (still originating from (2.2)) as follows. Specializing (2.2) to the case (2.9) we conclude that a function f solves Problem 2.3 (ii) if and only if

$$\mathbf{P} = \begin{bmatrix} 1 & \mathbf{x}^* & M_f^* \\ \mathbf{x} & P & M_F^* \\ M_f & M_F & I_{\mathcal{H}(K)} \end{bmatrix} \geq 0.$$

The latter condition is equivalent to the positivity on $\Omega \times \Omega$ of the following kernel:

$$\mathbf{K}(z, \zeta) = \begin{bmatrix} 1 & \mathbf{x}^* & f(\zeta)^* \\ \mathbf{x} & P & F(\zeta)^* \\ f(z) & F(z) & K(z, \zeta) \end{bmatrix} \succeq 0. \tag{2.14}$$

This equivalence is justified by the fact that the set of all vectors of the form

$$g(z) = \sum_{j=1}^{r} \begin{bmatrix} c_j \\ x_j \\ K(\cdot, z_j) y_j \end{bmatrix} \qquad (c_j \in \mathbb{C}, \ y_j \in \mathcal{Y}, \ x_j \in \mathcal{X}, \ z_j \in \Omega)$$

is dense in $\mathbb{C} \oplus \mathcal{X} \oplus \mathcal{H}(K)$ and since for every such vector,

$$\langle \mathbf{P}g, \ g \rangle_{\mathbb{C} \oplus \mathcal{X} \oplus \mathcal{H}(K_S)} = \sum_{j,\ell=1}^{r} \left\langle \mathbf{K}(z_j, z_\ell) \begin{bmatrix} c_\ell \\ x_\ell \\ y_\ell \end{bmatrix}, \ \begin{bmatrix} c_j \\ x_j \\ y_j \end{bmatrix} \right\rangle_{\mathbb{C} \oplus \mathcal{X} \oplus \mathcal{Y}}.$$

If in addition, P is strictly positive definite, one can take its Schur complement in $\mathbf{K}$ to get the inequality

$$\begin{bmatrix} 1 - \|P^{-\frac{1}{2}}\mathbf{x}\|^2 & f(\zeta)^* - \mathbf{x}^* P^{-1} F(\zeta)^* \\ f(z) - F(z)P^{-1}\mathbf{x} & K(z, \zeta) - F(z)P^{-1}F(\zeta)^* \end{bmatrix} \succeq 0$$

which is equivalent to (2.14). By [9, Theorem 2.2], the latter positivity is equivalent to the membership of the function $h(z) := f(z) - F(z)P^{-1}\mathbf{x}$ in the space $\mathcal{H}(\widetilde{K})$ together with the norm constraint $\|h\|_{\mathcal{H}(\widetilde{K})}^2 \leq 1 - \|P^{-\frac{1}{2}}\mathbf{x}\|$.

Remark 2.7. The function h on the right-hand side of (2.12) is in fact the general solution of the homogeneous interpolation problem $M_F^* h = 0$. Thus the first part of Theorem 2.5 states that the solution set of the corresponding homogeneous interpolation problem coincides with the reproducing kernel Hilbert space $\mathcal{H}(\widetilde{K})$. The results of this sort hold true even in a more general setting of Hilbert modules [1, 10]. The most interesting part, however, is to get a more detailed parametrization of $\mathcal{H}(\widetilde{K})$. Although Problem 2.3 is too general to get such a parametrization, in the context of Nevanlinna–Pick type Problem 1.1 in sub-Bergman spaces $\mathcal{H}(K_{S,n})$ we shall see that something can be done in this direction (see Theorem 3.8 below).

3. The main result

To apply the general results from Section 2 to sub-Bergman spaces $\mathcal{H}(K_{S,n})$ we must show that Problem 1.1 is a particular case of Problem 2.3. In other words (see formulas (1.9) and (2.7)), we need to find an $\mathcal{L}(\mathcal{X}, \mathcal{Y})$-valued function $F(z)$ in the multiplier space $\mathcal{M}(\mathcal{X}, \mathcal{H}(K))$ such that the adjoint of the multiplication operator M_F (in metric of $\mathcal{H}(K_{S,n})$) is equal to the adjoint of the observability operator $\mathcal{O}_{n,E,T}$ (in metric of $\mathcal{A}_n(\mathcal{Y})$). This function F will be constructed in Section 3.2 below. Then we will immediately get the parametrization formula (2.12) for the solution set of Problem 1.1 and then Theorem 3.8 (the main result of the paper) will give a more detailed description of the solution set of the associated homogeneous problem; the result involves the construction of multiplier functions $\Theta_n \in \mathcal{M}(\mathcal{A}_n(\mathcal{Y}))$ and $\Theta_k \in \mathcal{M}(\mathcal{A}_k(\mathcal{X}), \mathcal{A}_n(\mathcal{Y}))$ for $k = 1, 2, \ldots, n-1$. The additional ingredients needed to arrive at this result are introduced in Sections 3.1–3.4 below.

Before starting this program, we make one last observation. For an operator $A: \mathcal{X} \to \mathcal{H}(K_{S,k}) \subset \mathcal{A}_k(\mathcal{Y})$, the adjoint operator can be taken in the metric of $\mathcal{A}_k(\mathcal{Y})$ as well as in the metric of $\mathcal{H}(K_{S,k})$ and these two operations in general are not the same. To avoid confusion, in what follows we use the notation A^* for the adjoint of A in the metric of $\mathcal{A}_k(\mathcal{Y})$ and $A^{[*]}$ for the adjoint of A in the metric of $\mathcal{H}(K_{S,k})$. The precise value of k ($1 \le k \le n$) occurring here will be clear from the context.

3.1. Operators N and P_k

Recall that if the pair (E, T) is n-output stable, then the *n-observability gramian*

$$\mathcal{G}_{n,E,T} := (\mathcal{O}_{n,E,T})^* \mathcal{O}_{n,E,T} = \sum_{j=0}^{\infty} \mu_{n,j}^{-1} \cdot T^{*j} E^* E T^j \qquad (3.1)$$

is bounded on $\mathcal{X}$ and the strong convergence of the power series in its representation (3.1) follows from the definition of the inner product in $\mathcal{A}_n(\mathcal{Y})$. One may conclude that (E, T) is n-output stable if and only if the power series in (3.1) converges weakly (and therefore strongly, since all terms are positive semidefinite). The power series representation (3.1) suggests that $\mathcal{G}_{n,E,T}$ be defined for $n = 0$ by simply letting $\mathcal{G}_{0,E,T} := E^* E$.

Lemma 3.1. *If (E, T) is n-output stable, then it is k-output stable for all $k = 0, \ldots, n-1$ and the observability gramians satisfy the Stein identity*

$$\mathcal{G}_{k,E,T} - T^* \mathcal{G}_{k,E,T} T = \mathcal{G}_{k-1,E,T} \quad \text{for all} \quad k = 1, \ldots, n. \qquad (3.2)$$

Proof. Since $\mu_{n,j} \le \mu_{n-1,j}$ (see (1.2)), we conclude that if the power series in (3.1) converges for some integer n, it also converges for any positive integer $k < n$. Identity (3.2) follows from power series representations for $\mathcal{G}_{m,E,T}$ and $\mathcal{G}_{m-1,E,T}$, due to the binomial-coefficient identity $\binom{m}{k} = \binom{m-1}{k} + \binom{m-1}{k-1}$. $\square$

The evaluation map (1.7) extends to Schur-class functions $S \in \mathcal{S}(\mathcal{U}, \mathcal{Y})$ by

$$(E^*S)^{\wedge L}(T^*) = \mathcal{O}^*_{n,E,T} M_S|_{\mathcal{U}}.$$

Lemma 3.2. *Let (E,T) be n-output stable, let $S \in \mathcal{S}(\mathcal{U}, \mathcal{Y})$ be a Schur-class function and let $N \in \mathcal{L}(\mathcal{U}, \mathcal{X})$ be defined by*

$$N^* = (E^*S)^{\wedge L}(T^*) := \mathcal{O}^*_{n,E,T} M_S|_{\mathcal{U}}. \tag{3.3}$$

Then the pair (N,T) is n-output stable and the following equality holds:

$$\mathcal{O}^*_{n,E,T} M_S = \mathcal{O}^*_{n,N,T} : \mathcal{A}_n(\mathcal{U}) \to \mathcal{X}. \tag{3.4}$$

Furthermore, the operators

$$P_k := \mathcal{G}_{k,E,T} - \mathcal{G}_{k,N,T} : \mathcal{X} \to \mathcal{X} \tag{3.5}$$

satisfy the Stein identities

$$P_k - T^* P_k T = P_{k-1} \quad for \quad k = 1, \ldots, n, \tag{3.6}$$

as well as the inequalities

$$P_k \geq P_{k-1} \geq 0 \quad for \quad k = 2, \ldots, n. \tag{3.7}$$

Proof. Making use of power series representations

$$S(z) = \sum_{j \geq 0} S_j z^j \quad and \quad h(z) = \sum_{j \geq 0} h_j z^j$$

of a given $S \in \mathcal{S}(\mathcal{U}, \mathcal{Y})$ and of an arbitrary fixed function $h \in \mathcal{A}_n(\mathcal{U})$ we have

$(M_S h)(z) = \sum_{\ell=0}^{\infty} \left(\sum_{j=0}^{k} S_j h_{\ell-j} \right) z^\ell$ which together with (1.8) implies

$$\mathcal{O}^*_{n,E,T} M_S h = (E^*(Sh))^{\wedge L}(T^*) = \sum_{\ell=0}^{\infty} T^{*\ell} E^* \left(\sum_{j=0}^{\ell} S_j h_{\ell-j} \right), \tag{3.8}$$

Note that the latter series converges weakly since the pair (E,T) is n-output stable and $Sh \in \mathcal{A}_n(\mathcal{Y})$. If we regularize the series by replacing S_j by $r^j S_j$ and by replacing h_i by $r^i h_i$, we even get that the double series in (3.8), after taking the inner product against a fixed vector $x \in \mathcal{X}$, converges absolutely. We may then rearrange the series to have the form

$$\langle \mathcal{O}^*_{n,E,T} M_{S_r} h_r, x \rangle = \sum_{j,k=0}^{\infty} \langle r^{j+k}(T^*)^{j+k} E^* S_j h_k, x \rangle.$$

We may then invoke Abel's theorem to take the limit as $r \nearrow 1$ (justified by the facts that (E,T) is n-output stable and that $Sh \in \mathcal{A}_n(\mathcal{Y})$) to get

$$\mathcal{O}^*_{n,E,T} M_S h = (E^*Sh)^{\wedge L}(T^*) = \sum_{j,k=0}^{\infty} (T^*)^{j+k} E^* S_j h_k. \tag{3.9}$$

On the other hand, due to (3.3),

$$\mathcal{O}^*_{n,N,T} h = (N^* h)^{\wedge L}(T^*) = \sum_{k=0}^{\infty} T^{*k} N^* h_k = \sum_{k=0}^{\infty} T^{*k} \mathcal{O}^*_{n,E,T} S h_k$$

$$= \sum_{k=0}^{\infty} T^{*k} \left(\sum_{j=0}^{\infty} T^{*j} E^* S_j \right) h_k = \sum_{j,k=0}^{\infty} (T^*)^{j+k} E^* S_j h_k$$

where all the series converge weakly, since that in (3.9) does. Since h was picked arbitrarily in $\mathcal{A}_n(\mathcal{U})$, the last equality and (3.9) imply (3.4). Therefore, the operator $\mathcal{O}^*_{n,N,T} : \mathcal{A}_n(\mathcal{U}) \to \mathcal{X}$ is bounded and hence the pair (N,T) is n-output stable.

By Lemma 3.1, the pairs (E,T) and (N,T) are k-output stable for all $k = 1, \ldots, n$. By using the definition (3.5) of the operators P_k along with the identities (3.2), we get

$$P_k - T^* P_k T = \mathcal{G}_{k,E,T} - \mathcal{G}_{k,N,T} - T^* \left(\mathcal{G}_{k,E,T} - \mathcal{G}_{k,N,T} \right) T$$

$$= \left(\mathcal{G}_{k,E,T} - T^* \mathcal{G}_{k,E,T} T \right) - \left(\mathcal{G}_{k,N,T} - T^* \mathcal{G}_{k,N,T} T \right)$$

$$= \mathcal{G}_{k-1,E,T} - \mathcal{G}_{k-1,N,T} =: P_{k-1}$$

and we arrive at (3.6). Since M_S is a contraction on $\mathcal{A}_k(\mathcal{Y})$, we may replace n by k in the preceding proof to conclude that

$$\mathcal{O}^*_{k,E,T} M_S = \mathcal{O}^*_{k,N,T} : \mathcal{A}_k(\mathcal{U}) \to \mathcal{X} \quad \text{for} \quad k = 1, \ldots, n. \tag{3.10}$$

Therefore, from (3.5) and (3.10) we see that

$$P_k = \mathcal{O}^*_{k,E,T} \mathcal{O}_{k,E,T} - \mathcal{O}^*_{k,N,T} \mathcal{O}_{k,N,T} = \mathcal{O}^*_{k,E,T} \left(I - M_S M_S^* \right) \mathcal{O}_{k,E,T} \geq 0.$$

Finally, we invoke (3.6) to conclude that $P_k = T^* P_k T + P_{k-1} \geq P_{k-1}$ which completes the proof of (3.7). □

3.2. Functions F_k^S

We now assume that we are given the data set $(E, T, S(z), \mathbf{x})$ for an interpolation problem as in Problem 1.1. By making use of the operator N defined by (3.3), we now introduce $\mathcal{L}(\mathcal{X}, \mathcal{Y})$-valued functions

$$F_k^S(z) = (E - S(z)N)(I - zT)^{-k}, \quad k = 1, \ldots, n, \tag{3.11}$$

which are therefore completely specified by the data set $(E, N, T, S(z), \mathbf{x})$. For the multiplication operator $M_{F_k^S}$ we have

$$M_{F_k^S} = \mathcal{O}_{k,E,T} - M_S \mathcal{O}_{k,N,T} = (I - M_S M_S^*) \mathcal{O}_{k,E,T} \tag{3.12}$$

where the first equality follows from (3.11) and (1.6), while the second equality is a consequence of (3.10).

Lemma 3.3. *Let (E,T) be an n-output stable pair, let $S \in \mathcal{S}(\mathcal{U}, \mathcal{Y})$ be a Schur-class function and let N and P_k be defined as in (3.3) and (3.5) respectively. Then*

1. *The function F_k^S given by (3.11) is in the space of multipliers $\mathcal{M}(\mathcal{X}, \mathcal{H}(K_{S,k}))$, and moreover*

$$M_{F_k^S}^{[*]} M_{F_k^S} = P_k \quad and \quad M_{F_k^S}^{[*]} = \mathcal{O}_{k,E,T}^*. \tag{3.13}$$

2. *The kernel $R_k(z, \zeta) = \begin{bmatrix} P_k & F_k^S(\zeta)^* \\ F_k^S(z) & K_{S,k}(z,\zeta) \end{bmatrix}$ is positive on $\mathbb{D} \times \mathbb{D}$.*

Proof. Formula (3.12) and the range characterization of $\mathcal{H}(K_{S,k})$ imply that $M_{F_k^S}$ maps $\mathcal{X}$ into $\mathcal{H}(K_{S,k})$. Furthermore, it follows from (3.11), (1.5), (3.4), and (3.5) that

$$\|F_k^S x\|_{\mathcal{H}(K_{S,k})}^2 = \langle (I - M_S M_S^*) \mathcal{O}_{k,E,T} x, \mathcal{O}_{k,E,T} x \rangle_{A_k(\mathcal{Y})}$$
$$= \langle (\mathcal{O}_{k,E,T}^* \mathcal{O}_{k,E,T} - \mathcal{O}_{k,N,T}^* \mathcal{O}_{k,N,T}) x, x \rangle_{\mathcal{X}} = \langle P_k x, x \rangle_{\mathcal{X}}$$

for every $x \in \mathcal{X}$ which implies the first relation in (3.13). On the other hand, upon making subsequent use of (3.12) and (1.5), we see that

$$\langle x, M_{F_k^S}^{[*]} f \rangle_{\mathcal{X}} = \langle F_k^S x, f \rangle_{\mathcal{H}(K_{S,k})} = \langle (I - M_S M_S^*) \mathcal{O}_{k,E,T} x, f \rangle_{\mathcal{H}(K_{S,k})}$$
$$= \langle \mathcal{O}_{k,E,T} x, f \rangle_{A_k(\mathcal{Y})} = \langle x, \mathcal{O}_{k,E,T}^* f \rangle_{\mathcal{X}}$$

for every $f \in \mathcal{H}(K_{S,k})$ and $x \in \mathcal{X}$, which implies the second equality in (3.13). By the first equality in (3.13), the operator

$$\begin{bmatrix} P_k & M_{F_k^S}^{[*]} \\ M_{F_k^S} & I \end{bmatrix} = \begin{bmatrix} M_{F_k^S}^{[*]} M_{F_k^S} & M_{F_k^S}^{[*]} \\ M_{F_k^S} & I \end{bmatrix} = \begin{bmatrix} M_{F_k^S}^{[*]} \\ I \end{bmatrix} \begin{bmatrix} M_{F_k^S} & I \end{bmatrix} \in \mathcal{L}\left(\begin{bmatrix} \mathcal{X} \\ \mathcal{H}(K_{S,k}) \end{bmatrix} \right)$$

is positive semidefinite; as in Remark 2.6, this condition in turn is equivalent to positivity of the kernel $R_k(z, \zeta)$ on $\mathbb{D} \times \mathbb{D}$. $\qquad \square$

We now conclude from (3.13) that the interpolation condition (1.9) in Problem 1.1 can be written in the form $M_{F_n^S}^* f = \mathbf{x}$. Therefore Proposition 1.1 and Theorem 2.5 apply leading us to the following result.

Lemma 3.4. *Let N and P_n be defined as in (3.3) and (3.5) respectively.*

1. *Problem 1.1 (i) has a solution if and only if $\mathbf{x} \in \operatorname{Ran} P_n^{\frac{1}{2}}$.*
2. *If P_n is strictly positive definite, the function $f_{\min}(z) = F_n^S(z) P_n^{-1} \mathbf{x}$ solves Problem 1.1 (i) and has the minimal possible norm $\|f_{\min}\|_{\mathcal{H}(K_{S,n})} = \|P_n^{-\frac{1}{2}} \mathbf{x}\|$.*
3. *A function $h \in \mathcal{H}(K_{S,n})$ satisfies $(E^* h)^{\wedge L}(T^*) = 0$ if and only if h is in the reproducing kernel Hilbert space $\mathcal{H}(\widetilde{K}_{S,n})$ with reproducing kernel*

$$\widetilde{K}_{S,n}(z, \zeta) = \frac{I_{\mathcal{Y}} - S(z) S(\zeta)^*}{(1 - z\bar{\zeta})^n} - F_n^S(z) P_n^{-1} F_n^S(\zeta)^*. \tag{3.14}$$

3.3. J-inner function Θ and the Schur-class function $\mathcal{E}$

The assumption $P_n > 0$ allowed us to get a simple explicit formula for the minimal-norm solution in Lemma 3.4. The parametrization of the space $\mathcal{H}(\widetilde{K}_{S,n})$ will be established in Theorem 3.8 below under the stronger assumption $P_1 > 0$. By (3.7) this condition implies that $P_k > 0$ for all $k = 1, \ldots, n$.

Let us observe that if P_1 is boundedly invertible, then so is the observability gramian $\mathcal{G}_{1,E,T}$ (see formula (3.5)) which in turn implies (see, e.g., [8]) that the operator T is *strongly stable* in the sense that T^n converge to zero in the strong operator topology. Let J be the signature operator given by

$$J = \begin{bmatrix} I_{\mathcal{Y}} & 0 \\ 0 & -I_{\mathcal{U}} \end{bmatrix} \quad \text{and let} \quad \Theta(z) = \begin{bmatrix} a(z) & b(z) \\ c(z) & d(z) \end{bmatrix} \tag{3.15}$$

be a $\mathcal{L}(\mathcal{Y} \oplus \mathcal{U})$-valued function such that for all $z, \zeta \in \mathbb{D}$,

$$\frac{J - \Theta(z)J\Theta(\zeta)^*}{1 - z\bar\zeta} = \begin{bmatrix} E \\ N \end{bmatrix} (I - zT)^{-1} P_1^{-1} (I - \bar\zeta T^*)^{-1} \begin{bmatrix} E^* & N^* \end{bmatrix}. \tag{3.16}$$

The function Θ is determined by equality (3.16) uniquely up to a constant J-unitary factor on the right. One possible choice of Θ satisfying (3.16) is

$$\Theta(z) = D + z \begin{bmatrix} E \\ N \end{bmatrix} (I - zT)^{-1} B$$

where the operator $\begin{bmatrix} B \\ D \end{bmatrix} : \mathcal{Y} \oplus \mathcal{U} \to \begin{bmatrix} \mathcal{X} \\ \mathcal{Y} \oplus \mathcal{U} \end{bmatrix}$ is an injective solution to the J-Cholesky factorization problem

$$\begin{bmatrix} B \\ D \end{bmatrix} J \begin{bmatrix} B^* & D^* \end{bmatrix} = \begin{bmatrix} P_1^{-1} & 0 \\ 0 & J \end{bmatrix} - \begin{bmatrix} T \\ E \\ N \end{bmatrix} P_1^{-1} \begin{bmatrix} T^* & E^* & N^* \end{bmatrix}.$$

Such a solution exists due to identity (3.2) for $k = 1$, that is,

$$P_1 - T^* P_1 T = E^* E - N^* N. \tag{3.17}$$

If $\operatorname{spec}(T) \cap \mathbb{T} \neq \mathbb{T}$ (which is the case if, e.g., $\dim \mathcal{X} < \infty$), then a function Θ satisfying (3.16) can be taken in the form

$$\Theta(z) = I + (z - \mu) \begin{bmatrix} E \\ N \end{bmatrix} (I_{\mathcal{X}} - zT)^{-1} P_1^{-1} (\mu I_{\mathcal{X}} - T^*)^{-1} \begin{bmatrix} E^* & -N^* \end{bmatrix} \tag{3.18}$$

where μ is an arbitrary point in $\mathbb{T} \setminus \operatorname{spec}(T)$ (see [7]). For Θ of the form (3.18), the verification of identity (3.16) is straightforward and relies on the Stein identity (3.17) only. It follows from (3.16) that Θ is J-contractive on $\mathbb{D}$, i.e., that $\Theta(z)J\Theta(z)^* \leq J$ for all $z \in \mathbb{D}$. A less trivial fact is that due to the strong stability of T, the function Θ is J-inner; that is, the nontangential boundary values $\Theta(t)$ exist for almost all $t \in \mathbb{T}$ and are J-unitary: $\Theta(t)J\Theta(t)^* = J$. Every J-inner function $\Theta = \begin{bmatrix} a & b \\ c & d \end{bmatrix}$ gives rise to the one-to-one linear fractional transform

$$\mathcal{E} \mapsto \mathbf{T}_\Theta[\mathcal{E}] := (a\mathcal{E} + b)(c\mathcal{E} + d)^{-1} \tag{3.19}$$

mapping the Schur class $\mathcal{S}(\mathcal{U}, \mathcal{Y})$ into itself. In case Θ is a J-inner function satisfying identity (3.16), the transform (3.19) establishes a one-to-one correspondence between $\mathcal{S}(\mathcal{U}, \mathcal{Y})$ and the set of all Schur class functions G such that $(E^*G)^{\wedge L}(T^*) = N^*$. Since the given function S satisfies the latter condition by definition (3.3) of N, it follows that $S = \mathbf{T}_\Theta[\mathcal{E}]$ for some (uniquely determined) function $\mathcal{E} \in \mathcal{S}(\mathcal{U}, \mathcal{Y})$ which is recovered from S by

$$\mathcal{E} = (a - Sc)^{-1}(Sd - b). \qquad (3.20)$$

Since Θ is J-inner, it follows that $d(z)$ is boundedly invertible for all $z \in \mathbb{D}$ and also $\|d^{-1}(z)c(z)\| < 1$ for all $z \in \mathbb{D}$.

The following result gives the construction of the multiplier Φ_n needed for the Main Result (Theorem 3.8 below); the remaining multipliers Φ_k ($k = 1, \ldots, n-1$) are obtained in Lemma 3.7 below (formula (3.30)).

Lemma 3.5. *Let us assume that Schur-class functions S and $\mathcal{E}$ are related as in (3.20). Then the following identity holds for all $z \in \mathbb{D}$:*

$$\frac{I_\mathcal{Y} - S(z)S(\zeta)^*}{1 - z\bar{\zeta}} - F_1^S(z)P_1^{-1}F_1^S(\zeta)^* = \Phi_n(z)\frac{I_\mathcal{Y} - \mathcal{E}(z)\mathcal{E}(\zeta)^*}{1 - z\bar{\zeta}}\Phi_n(\zeta)^*, \qquad (3.21)$$

where $F_1^S(z) = (E - S(z)N)(I - zT)^{-1}$ (according to formula (3.11)) and where

$$\Phi_n = (a - bd^{-1}c)(I + \mathcal{E}d^{-1}c)^{-1}. \qquad (3.22)$$

Proof. Substituting (3.20) into (3.22) gives

$$\Phi_n = (a - bd^{-1}c)(I + (a - Sc)^{-1}(Sd - b)d^{-1}c)^{-1}$$

$$= (a - bd^{-1}c)\left(a - Sc + (Sd - b)d^{-1}c\right)^{-1}(a - Sc) = a - Sc.$$

Recalling the matrix representation for Θ in (3.15) and the expression (3.19) for $\mathcal{E}$, we therefore have

$$\begin{bmatrix} I_\mathcal{Y} & -S \end{bmatrix} \Theta = \begin{bmatrix} a - Sc & b - Sd \end{bmatrix} = \Phi_n \begin{bmatrix} I_\mathcal{Y} & -\mathcal{E} \end{bmatrix}. \qquad (3.23)$$

Multiplying both sides of (3.16) by $\begin{bmatrix} I_\mathcal{Y} & -S(z) \end{bmatrix}$ on the left and by $\begin{bmatrix} I_\mathcal{Y} \\ -S(\zeta)^* \end{bmatrix}$ on the right we get

$$\begin{bmatrix} I_\mathcal{Y} & -S(z) \end{bmatrix} \frac{J - \Theta(z)J\Theta(\zeta)^*}{1 - z\bar{\zeta}} \begin{bmatrix} I_\mathcal{Y} \\ -S(\zeta)^* \end{bmatrix} = F_1^S(z)P_1^{-1}F_1^S(\zeta)^*$$

which can be rearranged (see the formula (3.15) for J) to

$$\frac{I_\mathcal{Y} - S(z)S(\zeta)^*}{1 - z\bar{\zeta}} - F_1^S(z)P_1^{-1}F_1^S(\zeta)^* = \begin{bmatrix} I_\mathcal{Y} & -S(z) \end{bmatrix} \frac{\Theta(z)J\Theta(\zeta)^*}{1 - z\bar{\zeta}} \begin{bmatrix} I_\mathcal{Y} \\ -S(\zeta)^* \end{bmatrix}$$

which together with (3.23) implies (3.21). $\qquad \square$

In case $\Theta = \begin{bmatrix} a & b \\ c & d \end{bmatrix}$ is taken in the form (3.18), the formula (3.22) takes the form

$$\Phi_n(z) = (I + (z - \mu)E\Delta(z)^{-1}E^*)(I + (z - \mu)\mathcal{E}(z)N\Delta(z)^{-1}E^*)^{-1} \qquad (3.24)$$

where
$$\Delta(z) = (\mu I - T^*)P_1(I - zT) - (z - \mu)N^*N.$$
Indeed, let us first note that
$$\left(I - (z - \mu)N(I - zT)^{-1}P_1^{-1}(\mu I - T^*)^{-1}N^*\right)^{-1}$$
$$= I + (z - \mu)\left(I - (z - \mu)N(I - zT)^{-1}P_1^{-1}(\mu I - T^*)^{-1}N^*\right)^{-1}$$
$$\times N(I - zT)^{-1}P_1^{-1}(\mu I - T^*)^{-1}N^*$$
$$= I + (z - \mu)N\left(I - (z - \mu)(I - zT)^{-1}P_1^{-1}(\mu I - T^*)^{-1}N^*N\right)^{-1}$$
$$\times (I - zT)^{-1}P_1^{-1}(\mu I - T^*)^{-1}N^*$$
$$= I + (z - \mu)N\Delta(z)^{-1}N^*.$$
With this result in hand we see next that
$$d(z)^{-1}c(z) = (z - \mu)\left(I - (z - \mu)N(I - zT)^{-1}P_1^{-1}(\mu I - T^*)^{-1}N^*\right)^{-1}$$
$$\times N(I - zT)^{-1}P_1^{-1}(\mu I - T^*)^{-1}E^*$$
$$= (z - \mu)\left(I + (z - \mu)N\Delta(z)^{-1}N^*\right)$$
$$\times N(I - zT)^{-1}P_1^{-1}(\mu I - T^*)^{-1}E^*$$
$$= (z - \mu)N\Delta(z)^{-1}\left(\Delta(z) + (z - \mu)N^*N\right)$$
$$\times (I - zT)^{-1}P_1^{-1}(\mu I - T^*)^{-1}E^*$$
$$= (z - \mu)N\Delta(z)^{-1}E^*.$$
Therefore we get
$$a(z) - b(z)d(z)^{-1}c(z) = I + (z - \mu)E(I - zT)^{-1}P_1^{-1}(\mu I - T^*)^{-1}$$
$$\times \left(E^* + (z - \mu)N^*N\Delta(z)^{-1}E^*\right)$$
$$= I + (z - \mu)E(I - zT)^{-1}P_1^{-1}(\mu I - T^*)^{-1}$$
$$\times \left(\Delta(z) + (z - \mu)N^*N\right)\Delta(z)^{-1}E^*$$
$$= I + (z - \mu)E\Delta(z)^{-1}E^*$$
and (3.24) follows from the two latter equalities and (3.22).

3.4. Inner functions Ψ_k

Our next construction is similar to that in the previous section. Due to identities (3.6) and (strict) inequalities (3.7), we can find operators $\mathbf{B}_j, \mathbf{D}_j \colon \mathcal{X} \to \mathcal{X}$ so that
$$\begin{bmatrix} T & \mathbf{B}_j \\ P_j^{\frac{1}{2}} & \mathbf{D}_j \end{bmatrix}\begin{bmatrix} P_{j+1}^{-1} & 0 \\ 0 & I_\mathcal{X} \end{bmatrix}\begin{bmatrix} T^* & P_j^{\frac{1}{2}} \\ \mathbf{B}_j^* & \mathbf{D}_j^* \end{bmatrix} = \begin{bmatrix} P_{j+1}^{-1} & 0 \\ 0 & I_\mathcal{X} \end{bmatrix}$$
and
$$\begin{bmatrix} T^* & P_j^{\frac{1}{2}} \\ \mathbf{B}_j^* & \mathbf{D}_j^* \end{bmatrix}\begin{bmatrix} P_{j+1} & 0 \\ 0 & I_\mathcal{X} \end{bmatrix}\begin{bmatrix} T & \mathbf{B}_j \\ P_j^{\frac{1}{2}} & \mathbf{D}_j \end{bmatrix} = \begin{bmatrix} P_{j+1} & 0 \\ 0 & I_\mathcal{X} \end{bmatrix}.$$

In fact, the latter equalities determine $\mathbf{B}_j$ and $\mathbf{D}_j$ uniquely up to a common unitary factor on the right:

$$\mathbf{B}_j = (P_{j+1}^{-1} - AP_{j+1}^{-1}A^*)^{\frac{1}{2}}, \quad \mathbf{D}_j = -P_j^{-\frac{1}{2}}A^*P_j\mathbf{B}_j.$$

We now define the functions

$$\Psi_j(z) = \mathbf{D}_j + zP_j^{\frac{1}{2}}(I - zT)^{-1}\mathbf{B}_j \tag{3.25}$$

for $j = 1, \ldots, n-1$, which are inner and satisfy the identities

$$\frac{I_{\mathcal{X}} - \Psi_j(z)\Psi_j(\zeta)^*}{1 - z\bar{\zeta}} = P_j^{\frac{1}{2}}(I - zT)^{-1}P_{j+1}^{-1}(I - \bar{\zeta}T^*)^{-1}P_j^{\frac{1}{2}}. \tag{3.26}$$

As was shown in [4, Proposition 7.2], the function Ψ_j in (3.25) for $j = 1, \ldots, n-1$ can alternatively be given by

$$\Psi_j(z) = P_j^{\frac{1}{2}}(I - zT)^{-1}P_{j+1}^{-1}(zI - T^*)\mathbf{B}_j^{-1}.$$

If $\mathrm{spec}(T) \cap \mathbb{T} \neq \mathbb{T}$, then a function Ψ_j satisfying (3.26) can be taken in the form

$$\Psi_j(z) = I + (z - \mu)P_j^{\frac{1}{2}}(I_{\mathcal{X}} - zT)^{-1}P_{j+1}^{-1}(\mu I_{\mathcal{X}} - T^*)^{-1}P_j^{\frac{1}{2}} \tag{3.27}$$

where μ is an arbitrary point in $\mathbb{T} \setminus \mathrm{spec}(T)$. For this choice, an alternative formula for Ψ_j is the following

$$\Psi_j(z) = P_j^{\frac{1}{2}}(I - zT)^{-1}P_{j+1}^{-1}(zI - T^*)(\mu I - T^*)^{-1}P_{j+1}(I - \mu T)^{-1}P_j^{-\frac{1}{2}}.$$

Lemma 3.6. *Let* $\Psi_1, \ldots, \Psi_{n-1}$ *be the inner functions defined in* (3.25). *Then*

$$\sum_{k=1}^{n-1}(I - zT)^{-k+1}P_k^{-\frac{1}{2}}\frac{\Psi_k(z)\Psi_k(\zeta)^*}{(1 - z\bar{\zeta})^{n-k}}P_k^{-\frac{1}{2}}(I - \bar{\zeta}T^*)^{-k+1}$$

$$= \frac{P_1^{-1}}{(1 - z\bar{\zeta})^{n-1}} - (I - zT)^{-n+1}P_n^{-1}(I - \bar{\zeta}T^*)^{-n+1}. \tag{3.28}$$

Proof. Equality

$$(I - zT)^{-k+1}P_k^{-\frac{1}{2}}\frac{I_{\mathcal{X}} - \Psi_k(z)\Psi_k(\zeta)^*}{1 - z\bar{\zeta}}P_k^{-\frac{1}{2}}(I - \bar{\zeta}T^*)^{-k+1}$$

$$= (I - zT)^{-k}P_{k+1}^{-1}(I - \bar{\zeta}T^*)^{-k}$$

follows immediately from (3.26) and can be written equivalently as

$$(I - zT)^{-k+1}P_k^{-\frac{1}{2}}\frac{\Psi_k(z)\Psi_k(\zeta)^*}{(1 - z\bar{\zeta})^{n-k}}P_j^{-\frac{1}{2}}(I - \bar{\zeta}T^*)^{-j+1}$$

$$= (I - zT)^{-k+1}\frac{P_k^{-1}}{(1 - z\bar{\zeta})^{n-k}}(I - \bar{\zeta}T^*)^{-k+1}$$

$$- (I - zT)^{-k}\frac{P_{k+1}^{-1}}{(1 - z\bar{\zeta})^{n-k-1}}(I - \bar{\zeta}T^*)^{-k}.$$

Summing up the latter equalities for $k = 1, \ldots, n-1$ we get (3.28).   □

3.5. The main result

Now we are in a position to represent the kernel (3.14) as the sum of n positive kernels.

Lemma 3.7. *The kernel $\widetilde{K}_{S,n}(z,\zeta)$ defined in (3.14) can be represented as*

$$\widetilde{K}_{S,n}(z,\zeta) = \Phi_n(z)K_{\mathcal{E},n}(z,\zeta)\Phi_n(\zeta)^* + \sum_{k=1}^{n-1} \frac{\Phi_k(z)\Phi_k(\zeta)^*}{(1-z\bar{\zeta})^k} \qquad (3.29)$$

where Φ_n is defined in (3.24) and where

$$\Phi_k(z) = F_{n-k}^S(z)P_{n-k}^{-\frac{1}{2}}\Psi_{n-k}(z) \qquad (k=1,\ldots,n-1). \qquad (3.30)$$

Proof. We first divide both sides in (3.21) by $(1-z\bar{\zeta})^{n-1}$ and combine the obtained identity with (3.14) to get

$$\widetilde{K}_{S,n}(z,\zeta) = \Phi_n(z)\frac{I_{\mathcal{Y}} - \mathcal{E}(z)\mathcal{E}(\zeta)^*}{(1-z\bar{\zeta})^n}\Phi_n(\zeta)^*$$

$$+ \frac{F_1^S(z)P_1^{-1}F_1^S(\zeta)^*}{(1-z\bar{\zeta})^{n-1}} - F_n^S(z)P_n^{-1}F_n^S(\zeta)^*. \qquad (3.31)$$

Multiplying both sides in (3.28) by $F_1^S(z) = (E - S(z)N)(I - zT)^{-1}$ on the left and by $F_1^S(\zeta)^*$ on the right we get

$$\sum_{k=1}^{n-1} F_k^S(z)P_k^{-\frac{1}{2}}\frac{\Psi_k(z)\Psi_k(\zeta)^*}{(1-z\bar{\zeta})^{n-k}}P_k^{-\frac{1}{2}}F_k^S(\zeta)^*$$

$$= \frac{F_1^S(z)P_1^{-1}F_1^S(\zeta)^*}{(1-z\bar{\zeta})^{n-1}} - F_n^S(z)P_n^{-1}F_n^S(\zeta)^*$$

which being plugged into (3.31) gives the desired representation (3.29). $\square$

Identity (3.29) means that the map

$$X\colon \widetilde{K}_{S,n}(\cdot,\zeta)y \mapsto \begin{bmatrix} K_{\mathcal{E},n}(\cdot,\zeta)\Phi_n(\zeta)^*y \\ \frac{1}{(1-\cdot\bar{\zeta})^{n-1}}\Phi_{n-1}(\zeta)^*y \\ \vdots \\ \frac{1}{(1-\cdot\bar{\zeta})^1}\Phi_1(\zeta)^*y \end{bmatrix}$$

extends by linearity and continuity to an isometry from $\mathcal{H}(\widetilde{K}_{S,n})$ to $\mathcal{H}(K_{\mathcal{E},n}) \oplus \mathcal{A}_{n-1}(\mathcal{X}) \oplus \cdots \oplus \mathcal{A}_1(\mathcal{X})$. Furthermore, standard reproducing-kernel arguments show that X^* is given by multiplication by the function $\Phi := \begin{bmatrix} \Phi_n & \cdots & \Phi_1 \end{bmatrix}$. We conclude that the meaning of the identity (3.29) is that the function Φ is an coisometric multiplier from $\mathcal{H}(K_{\mathcal{E},n}) \oplus \mathcal{A}_{n-1}(\mathcal{X}) \oplus \cdots \oplus \mathcal{A}_1(\mathcal{X})$ onto $\mathcal{H}(\widetilde{K}_{S,n})$. Combining this observation with statement (3) in Lemma 3.4 we arrive at the following result.

Theorem 3.8. *Let us assume that the data set of Problem 1.1 is such that the operator P_1 defined in (3.5) is strictly positive definite. Let Φ_k be defined as in (3.22), (3.30). Then:*

1. *A function f is a solution of Problem 1.1 (i) if and only if it is of the form*

$$f = f_{\min} + \Phi_n h_n + \sum_{j=1}^{n-1} \Phi_j h_j, \qquad f_{\min}(z) = F_n^S(z) P_n^{-1} \mathbf{x} \qquad (3.32)$$

for some choice of $h_n \in \mathcal{H}(K_{\mathcal{E},n})$ and $h_j \in \mathcal{A}_j(\mathcal{X})$ for $j = 1, \ldots, n-1$.

2. *A function f is a solution of Problem 1.1 (ii) if and only if it is of the form (3.32) for some $h_n \in \mathcal{H}(K_{\mathcal{E},n})$ and $h_j \in \mathcal{A}_j(\mathcal{X})$ for $j = 1, \ldots, n-1$ such that*

$$\|h_n\|_{\mathcal{H}(K_{S,n})}^2 + \sum_{j=1}^{n-1} \|h_j\|_{\mathcal{A}_j(\mathcal{X})}^2 \le 1 - \|P^{-\frac{1}{2}} \mathbf{x}\|^2.$$

In particular, such solutions exist if and only if $\|P^{-\frac{1}{2}} \mathbf{x}\|^2 \le 1$.

If $S \equiv 0$, then the space $\mathcal{H}(K_{S,n})$ amounts to the Bergman space $\mathcal{A}_n(\mathcal{Y})$. In this case, we get from (3.3) that $N = 0$ and subsequently, $P_k = \mathcal{G}_{k,E,T}$ for $k = 1, \ldots, n$. We then have equality (3.23) with $\mathcal{E} \equiv 0$ and Φ_n and inner $\mathcal{L}(\mathcal{Y})$-valued function subject to

$$\frac{I_{\mathcal{Y}} - \Psi_n(z) \Psi_n(\zeta)^*}{1 - z\bar{\zeta}} = E(I - zT)^{-1} \mathcal{G}_{1,E,T}^{-1} (I - \bar{\zeta}T^*)^{-1} E^*.$$

In the present context, Theorem 3.8 parametrizes the solution set of the interpolation problem with operator argument in $\mathcal{A}_n(\mathcal{Y})$. The homogeneous version of this problem can be interpreted as a Beurling-type theorem for $\mathcal{A}_n(\mathcal{Y})$.

Another particular case where $n = 1$ partly recovers results on interpolation in de Branges–Rovnyak spaces presented in [6].

4. Examples

Parametrization (3.32) is especially transparent in case $\dim \mathcal{U} = \dim \mathcal{Y} = 1$ and $\dim \mathcal{X} < \infty$. Besides, in this case (as we will see below), the matrices P_k are all positive definite for $k = 2, \ldots, n$. The matrix P_1 may be singular but in the scalar-valued case we are able to handle this option as well.

If $\dim \mathcal{X} < \infty$ and $\dim \mathcal{U} = \dim \mathcal{Y} = 1$, then with respect to an appropriate basis of $\mathcal{X}$ the output stable observable pair (E, T) has the following form: T is a block diagonal matrix $T = \text{diag}\{T_1, \ldots, T_k\}$ with the diagonal block T_i equal to the upper triangular $n_i \times n_i$ Jordan block with the number $\bar{z}_i \in \mathbb{D}$ on the main diagonal and E is the row vector

$$E = \begin{bmatrix} E_1 & \cdots & E_k \end{bmatrix}, \quad \text{where} \quad E_i = \begin{bmatrix} 1 & 0 & \ldots 0 \end{bmatrix} \in \mathbb{C}^{1 \times n_i}.$$

It is not hard to show that for (E, T) as above and for every function f analytic at $z_1, \ldots, z_k$, evaluation (1.7) amounts to

$$(E^* f)^{\wedge L}(T^*) = \text{Col}_{1 \le i \le k} \text{Col}_{0 \le j < n_i} \frac{f^{(j)}(z_i)}{j!}. \qquad (4.1)$$

If we specify the entries of the column $\mathbf{x}^*$ by letting

$$\mathbf{x}^* = \mathrm{Col}_{1 \le i \le k} \, \mathrm{Col}_{0 \le j < n_i} \, x_{ij},$$

then it is readily seen that Problem 1.1 amounts to the following Lagrange–Sylvester interpolation problem:

LSP: *Given a scalar Schur-class function $S \in \mathcal{S}$, distinct points $z_1, \ldots, z_k \in \mathbb{D}$ and a collection $\{x_{ij}\}$ of complex numbers, find all functions $f \in \mathcal{H}(K_{S,n})$ such that*

$$f^{(j)}(z_i)/j! = x_{ij} \quad for \quad j = 0, \ldots, n_i - 1; \ i = 1, \ldots, k.$$

The auxiliary column N^* is now defined from the derivatives of the given function S via formula (4.1), and we define the matrix P_1 as the unique solution of the Stein equation (3.17). This matrix P_1 turns out to be equal to the Schwarz–Pick matrix

$$P_1 = \left[\left[\frac{1}{\ell! r!} \frac{\partial^{\ell+r}}{\partial z^\ell \partial \bar{\zeta}^r} \left. \frac{1 - S(z)\overline{S(\zeta)}}{1 - z\bar{\zeta}} \right|_{\substack{z=z_i \\ \zeta=z_j}} \right]_{\substack{r=0,\ldots,n_j-1 \\ \ell=0,\ldots,n_i-1}} \right]_{i,j=1}^{k},$$

which in turn is known to be positive definite unless S is a Blaschke product of degree $d < \mathbf{n} := n_1 + \cdots + n_k$, in which case P_1 is positive semidefinite and rank $P_1 = d$. It is not hard to show that the matrices P_k defined in (3.5) are equal to higher-order Schwarz–Pick matrices

$$P_k = \left[\left[\frac{1}{\ell! r!} \frac{\partial^{\ell+r}}{\partial z^\ell \partial \bar{\zeta}^r} \left. \frac{1 - S(z)\overline{S(\zeta)}}{(1 - z\bar{\zeta})^k} \right|_{\substack{z=z_i \\ \zeta=z_j}} \right]_{\substack{r=0,\ldots,n_j-1 \\ \ell=0,\ldots,n_i-1}} \right]_{i,j=1}^{k}.$$

Lemma 4.1. *If a Schur-class function is not a unimodular constant, then the matrices P_k are positive definite for $k \ge 2$.*

Proof. By (3.7), it suffices to prove the statement for $k = 2$. Observe that if $C_i \in \mathbb{C}^{1 \times n_i}$ is a row-vector with a non-zero left-most entry and $T_i \in \mathbb{C}^{n_i \times n_i}$ is the upper triangular Jordan block, then the pair (C_i, T_i) is observable in the sense that the gramian $\mathcal{G}_{1,C_i,T_i}$ is positive definite. Consequently, if $C = \begin{bmatrix} C_1 & \cdots & C_k \end{bmatrix}$ is a block-row vector with all blocks C_i having nonzero left-most entries and if $T = \mathrm{diag}\{T_1, \ldots, T_k\}$ is the conformally decomposed block-diagonal matrix as above, then the pair (C, T) is observable. In particular we may take C equal to the top row in the matrix P_1. The left-most entries in its blocks are equal to $\frac{1 - S(z_1)\overline{S(z_i)}}{1 - z_1 \bar{z}_i}$ and are non-zero unless S is a unimodular constant. Thus, for this choice of C we have

$$0 < \mathcal{G}_{1,C,T} = \sum_{j=0}^{\infty} T^{*j} C^* C T^k \le \sum_{j=0}^{\infty} T^{*j} P_1 T^j. \tag{4.2}$$

Since the spectral radius of T is strictly less than one, it follows from the Stein identity $P_2 - T^* P_2 T = P_1$ that P_2 can be represented via converging series

$$P_2 = \sum_{j=0}^{\infty} T^{*j} P_1 T^j$$

and now we conclude from (4.2) that $P_2 > 0$ regardless of whether P_1 is invertible or not. □

We now proceed to three different cases.

Case 1: S is not a finite Blaschke product or it is a finite Blaschke product of degree $\deg S > \mathbf{n}$. In this case $P_1 > 0$ and all the solutions f to the problem **LSP** are given by formula (3.32), where now all the ingredients are not only explicit but also computable.

Case 2: S be a Blaschke product of degree $\deg S = \mathbf{n}$. Then the matrix P_1 is invertible, but the associated function $\mathcal{E}$ defined by (3.19) is a unimodular constant, so that the corresponding sub-Bergman space $\mathcal{H}(K_{\mathcal{E},n})$ is trivial. Observe, that in this case, the formula (3.21) takes the form

$$\frac{I_y - S(z)S(\zeta)^*}{1 - z\bar{\zeta}} = F_1^S(z) P_1^{-1} F_1^S(\zeta)^*. \tag{4.3}$$

Furthermore, the parametrizing formula (3.32) does not contain the second term on the right. In particular, if $n = 1$, then Problem 1.1(i) has a unique solution.

Case 3: $\deg S < \mathbf{n}$. Since the matrices P_k are invertible for $k = 2, \ldots, n$, the formula (3.32) for $f_{\min}$ as well as formulas (3.30) for Φ_k for $k = 1, \ldots, n-2$ still make sense and by the preceding analysis,

$$\frac{F_2^S(z) P_2^{-1} F_2^S(\zeta)^*}{(1 - z\bar{\zeta})^{n-1}} - F_n^S(z) P_n^{-1} F_n^S(\zeta)^* = \sum_{k=1}^{n-2} \frac{\Phi_k(z) \Phi_k(\zeta)^*}{(1 - z\bar{\zeta})^k}. \tag{4.4}$$

The only remaining question is to modify appropriately the function Φ_{n-1}.

Let us consider conformal block decompositions

$$P_1 = \begin{bmatrix} \widetilde{P}_1 & X \\ X^* & Y \end{bmatrix}, \quad T = \begin{bmatrix} \widetilde{T} & \widetilde{T}_1 \\ 0 & \widetilde{T}_2 \end{bmatrix}, \quad E = \begin{bmatrix} \widetilde{E} & \widetilde{E}_1 \end{bmatrix}, \quad N = \begin{bmatrix} \widetilde{N} & \widetilde{N}_1 \end{bmatrix} \tag{4.5}$$

with $\widetilde{P}_1, \widetilde{T} \in \mathbb{C}^{d \times d}$ and $\widetilde{E}, \widetilde{N} \in \mathbb{C}^{1 \times d}$ where $d := \deg S$. Making use of (4.5) and taking the advantage of the upper triangular structure of T, we also decompose the function F_1^S as

$$F_1^S(z) = \begin{bmatrix} \widetilde{F}_1^S(z) & \widehat{F}_1^S(z) \end{bmatrix}, \quad \text{where} \quad \widetilde{F}_1^S(z) = (\widetilde{E} - S(z)\widetilde{N})(I - z\widetilde{T})^{-1}. \tag{4.6}$$

The block $\widetilde{P}_1$ is the Schwarz–Pick matrix of a Blaschke product of degree d based on d points (counted with multiplicities) and therefore, $\widetilde{P}_1$ is invertible. On the other hand the subproblem of **LSP** based on these points is of the type considered in Case 2 and thus, by virtue of (4.3) we have

$$\frac{I_{\mathcal{Y}} - S(z)S(\zeta)^*}{1 - z\bar{\zeta}} = \tilde{F}_1^S(z)\tilde{P}_1^{-1}\tilde{F}_1^S(\zeta)^*. \qquad (4.7)$$

Since $\operatorname{rank} P_1 = \operatorname{rank} \tilde{P}_1 = d$ it follows that $Y = X^* \tilde{P}_1^{-1} X$ so that P_1 can be represented as

$$P_1 = \begin{bmatrix} \tilde{P}_1^{\frac{1}{2}} \\ X^* \tilde{P}_1^{-\frac{1}{2}} \end{bmatrix} \begin{bmatrix} \tilde{P}_1^{\frac{1}{2}} & \tilde{P}_1^{-\frac{1}{2}} X \end{bmatrix}.$$

From this representation and from the fact that the kernel $R_1(z, \zeta)$ is positive on $\mathbb{D} \times \mathbb{D}$ (see statement (2) in Lemma 3.3) we conclude that

$$F_1^S(z) \begin{bmatrix} \tilde{P}_1^{-1} X \\ -I \end{bmatrix} \equiv 0$$

which implies that the entries $\tilde{F}_1^S$ and $\hat{F}_1^S$ in (4.6) are related by $\hat{F}_1^S = \tilde{F}_1^S \tilde{P}_1^{-1} X$ so that $F_1^S(z)$ can be written as

$$F_1^S(z) = \tilde{F}_1^S(z) \begin{bmatrix} I & \tilde{P}_1^{-1} X \end{bmatrix}. \qquad (4.8)$$

We now define the function

$$\Phi_{n-1}(z) := \tilde{F}_1^S(z) \tilde{P}_1^{-\frac{1}{2}} \tilde{\Psi}_1(z) \qquad (4.9)$$

where $\tilde{\Psi}_1$ is the inner $\mathbb{C}^{d \times d}$-valued function given by

$$\tilde{\Psi}_1(z) = I + (z - \mu)\tilde{P}_1^{\frac{1}{2}} \begin{bmatrix} I & \tilde{P}_1^{-1} X \end{bmatrix} (I - zT)^{-1} P_2^{-1} (\mu I - T^*)^{-1} \begin{bmatrix} I \\ X^* \tilde{P}_1^{-1} \end{bmatrix} \tilde{P}_1^{\frac{1}{2}}$$

(compare with (3.27)) and satisfying the identity

$$\frac{I_d - \tilde{\Psi}_1(z)\tilde{\Psi}_1(\zeta)^*}{1 - z\bar{\zeta}} = \tilde{P}_1^{\frac{1}{2}} \begin{bmatrix} I & \tilde{P}_1^{-1} X \end{bmatrix} (I - zT)^{-1} P_2^{-1} (I - \bar{\zeta}T^*)^{-1} \begin{bmatrix} I \\ X^* \tilde{P}_1^{-1} \end{bmatrix} \tilde{P}_1^{\frac{1}{2}}$$

similar to that in (3.26). By (4.8),

$$F_2^S(z) = F_1^S(z)(I - zT)^{-1} = \tilde{F}_1^S(z) \begin{bmatrix} I & \tilde{P}_1^{-1} X \end{bmatrix} (I - zT)^{-1},$$

which together with (4.7), (4.9) and the previous identity implies

$$F_2^S(z)P_2^{-1}F_2^S(\zeta)^* = \tilde{F}_1^S(z)\tilde{P}_1^{-\frac{1}{2}} \frac{I_d - \tilde{\Psi}_1(z)\tilde{\Psi}_1(\zeta)^*}{1 - z\bar{\zeta}} \tilde{P}_1^{-\frac{1}{2}} \tilde{F}_1^S(\zeta)^*$$

$$= \frac{I_{\mathcal{Y}} - S(z)S(\zeta)^*}{1 - z\bar{\zeta}} - \frac{\Phi_{n-1}(z)\Phi_{n-1}(\zeta)^*}{1 - z\bar{\zeta}}.$$

Therefore the kernel (3.14) can be written as

$$\tilde{K}_{S,n}(z, \zeta) = \frac{\Phi_{n-1}(z)\Phi_{n-1}(\zeta)^*}{(1 - z\bar{\zeta})^n} + \frac{F_2^S(z)P_2^{-1}F_2^S(\zeta)^*}{(1 - z\bar{\zeta})^{n-1}} - F_n^S(z)P_n^{-1}F_n^S(\zeta)^*$$

which together with (4.4) implies

$$\widetilde{K}_{S,n}(z,\zeta) = \sum_{k=1}^{n-1} \frac{\Phi_k(z)\Phi_k(\zeta)^*}{(1 - z\bar{\zeta})^k}.$$

Thus in the present case we have the same parametrization of the solution set but the parameter h_{n-1} is taken in $\mathcal{A}_{n-1}(\mathbb{C}^d)$ rather than in $\mathcal{A}_{n-1}(\mathbb{C}^{\mathbf{n}})$.

5. Some open questions and directions for future work

5.1. Stein equations, inertia theorems, and associated orthogonal polynomials

In the classical setting, Stein equations and associated inertial theorems (where the solution of the Stein equation may be indefinite) are closely associated with orthogonal polynomials with respect to an appropriate weight on the unit circle and location of the zeros of these polynomials (inside or outside the unit circle). Here instead of a single Stein equation we have a nested family of Stein equations (3.6), (3.7) conceivably associated with a family of orthogonal polynomials with respect to a weight on the unit disk rather than on the unit circle. It would be of interest to extend the classical theory to this nested/Bergman-space setting. This would follow up on one of the interests of Leonia Lerer (see, e.g., [15, 16]) to whom this paper is dedicated.

5.2. Overlapping spaces

In the classical theory of de Branges–Rovnyak spaces (see in particular Sarason's book [18]), a prominent role is played by the overlapping space

$$\mathcal{H}\left(\frac{1 - S(z)S(\zeta^*)}{(1 - z\bar{\zeta})}\right) \cap \mathcal{H}\left(\frac{S(z)S(\zeta)^*}{(1 - z\bar{\zeta})}\right).$$

It would be of interest to determine if something significant can be said about the analogous overlapping spaces $\mathcal{H}\left(\frac{1-S(z)S(\zeta)^*}{(1-z\zeta)^n}\right) \cap \mathcal{H}\left(\frac{S(z)S(\zeta)^*}{(1-z\zeta)^n}\right)$ for $n > 1$; some results in this direction appear in [19].

5.3. Characterization of de Branges–Rovnyak spaces via backward-shift operator identities

In the classical setting ($n = 1$) the de Branges–Rovnyak reproducing kernel Hilbert spaces $\mathcal{H}(K_{S,1})$, and more generally reproducing kernel Hilbert spaces with kernel of the form $\frac{J - \Theta(z)J\Theta(w)^*}{1 - z\bar{\zeta}}$, can be characterized via invariance under the backward-shift operator combined with an appropriate functional identity (see Section 5.3 of [2] for a very general formulation and additional references and history). A natural question for future investigation is whether such a characterization can be given for reproducing kernel Hilbert spaces with kernel of the form $\frac{J - \Theta(z)J\Theta(w)^*}{(1 - z\bar{\zeta})^n}$.

Acknowledgement

We would like to thank our colleague Sanne ter Horst for useful comments on an early draft of this manuscript.

References

[1] D. Alpay and V. Bolotnikov, *On tangential interpolation in reproducing kernel Hilbert modules and applications*, in: Topics in interpolation theory (eds. H. Dym, B. Fritzsche, V. Katsnelson and B. Kirstein), pp. 37–68, **OT 95**, Birkhäuser, Basel, 1997.

[2] D. Arov and H. Dym, *J-Contractive Matrix Valued Functions and Related Topics*, Encyclopedia of Mathematics and its Applications **116**, Cambridge, 2008.

[3] J.A. Ball and V. Bolotnikov, *Contractive multipliers from Hardy space to weighted Hardy space*, Preprint, arXiv:1209.3690

[4] J.A. Ball and V. Bolotnikov, *Weighted Bergman spaces: shift-invariant subspaces and input/state/output linear systems*, Integral Equations Operator Theory **76** (2013) no. 3, 301–356.

[5] J.A. Ball, V. Bolotnikov and S. ter Horst *Interpolation in de Branges–Rovnyak spaces*, Proc. Amer. Math. Soc. **139** (2011), no. 2, 609–618.

[6] J.A. Ball, V. Bolotnikov and S. ter Horst, *Abstract interpolation in vector-valued de Branges–Rovnyak spaces*, Integral Equations Operator Theory **70** (2011), no. 2, 227–268.

[7] J.A. Ball, I. Gohberg, and L. Rodman. *Interpolation of Rational Matrix Functions*, OT45, Birkhäuser Verlag, 1990.

[8] J.A. Ball and M.W. Raney, *Discrete-time dichotomous well-posed linear systems and generalized Schur–Nevanlinna–Pick interpolation*, Complex Anal. Oper. Theory **1** (2007), 1–54.

[9] F. Beatrous and J. Burbea, *Positive-definiteness and its applications to interpolation problems for holomorphic functions*, Trans. Amer. Math. Soc., **284** (1984), no. 1, 247–270.

[10] V. Bolotnikov and L. Rodman, *Remarks on interpolation in reproducing kernel Hilbert spaces*, Houston J. Math. **30** (2004), no. 2, 559–576.

[11] L. de Branges and J. Rovnyak, *Canonical models in quantum scattering theory*, Perturbation Theory and its Applications in Quantum Mechanics (ed. C. Wilcox), pp. 295–392, Wiley, New York, 1966.

[12] L. de Branges and J. Rovnyak, *Square Summable Power Series*, Holt, Rinehart and Winston, New–York, 1966.

[13] R.G. Douglas, *On majorization, factorization, and range inclusion of operators on Hilbert space*, Proc. Amer. Math. Soc. **17** (1966), 413–415.

[14] O. Giselsson and A. Olofsson, *On some Bergman shift operators*, Complex Anal. Oper. Theory **6** (2012), 829–842.

[15] L. Lerer and A.C.M. Ran, *A new inertia theorem for Stein equations, inertia of invertible Hermitian block Toeplitz matrices and matrix orthogonal polynomials*, Integral Equations and Operator Theory **47** (2003) no. 3, 339–360

[16] L. Lerer, I. Margulis, and A.C.M. Ran, *Inertia theorems based on operator Lyapunov equations*, Oper. Matrices **2** (2008) no. 2, 153–166.

[17] M. Rosenblum and J. Rovnyak, *Hardy Classes and Operator Theory*, Oxford University Press, 1985.

[18] D.E. Sarason, *Sub-Hardy Hilbert Spaces in the Unit Disk*, John Wiley & Sons, Inc., New York, 1994.

[19] S. Sultanic, *Sub-Bergman Hilbert spaces*, J. Math. Anal. Appl. **324** (2006), no. 1, 639–649.

[20] K. Zhu, *Sub-Bergman Hilbert spaces on the unit disk.* Indiana Univ. Math. J. **45** (1996), no. 1, 165–176.

[21] K. Zhu, *Sub-Bergman Hilbert spaces on the unit disk. II*, J. Funct. Anal. **202** (2003), no. 2, 327–341

Joseph A. Ball
Department of Mathematics
Virginia Tech
Blacksburg, VA 24061-0123, USA
e-mail: joball@math.vt.edu

Vladimir Bolotnikov
Department of Mathematics
The College of William and Mary
Williamsburg VA 23187-8795, USA
e-mail: vladi@math.wm.edu

Operator Theory:
Advances and Applications, Vol. 237, 41–78

Zero Sums of Idempotents and Banach Algebras Failing to be Spectrally Regular

H. Bart, T. Ehrhardt and B. Silbermann

Dedicated to Leonia Lerer, in celebration of his seventieth birthday

Abstract. A large class of Banach algebras is identified allowing for non-trivial zero sums of idempotents, hence failing to be spectrally regular. Belonging to it are the C^*-algebras known under the name Cuntz algebras. Other Banach algebras lying in the class are those of the form $\mathcal{L}(X)$ with X a (non-trivial) Banach space isomorphic to a (finite) direct sum of at least two copies of X. There do exist (somewhat exotic) Banach spaces for which $\mathcal{L}(X)$ is spectrally regular.

Mathematics Subject Classification (2010). Primary 46H99, 47C15; Secondary 30G30, 46E15.

Keywords. Logarithmic residue, spectral regularity, (zero) sum of idempotents, Cuntz algebra, space of (bounded) continuous functions, Cantor type set.

1. Introduction

All algebras considered in this paper are assumed to be associative. A logarithmic residue is a contour integral of the type

$$\frac{1}{2\pi i}\int_{\partial\Delta} f'(\lambda)f(\lambda)^{-1}d\lambda, \tag{1}$$

where the analytic function f has its values in a unital complex Banach algebra $\mathcal{B}$ and $\partial\Delta$ is a suitable contour in the complex plane $\mathbb{C}$, in fact the positively oriented boundary of a Cauchy domain Δ. In the scalar case $\mathcal{B} = \mathbb{C}$, the expression (1) is equal to the number of zeros of f in Δ, multiplicities of course taken into account. Thus, in that situation, the integral (1) vanishes if and only if f takes non-zero values, not only on $\partial\Delta$ (which has been implicitly assumed in order to let (1) make sense) but on all of Δ. This state of affairs leads to the following question: *if for a Banach algebra-valued analytic function f the integral (1) vanishes, can one conclude that f takes invertible values on all of Δ?*

There are many situations where the answer to this question is positive (see
[6], [8], [9], [11], [12], [13] and [15]); in general it is negative, however. The Banach
algebra $\mathcal{L}(\ell_2)$ of all bounded linear operators on ℓ_2 is a counterexample (see [5] and
[6]). This comes about from the fact that $\mathcal{L}(\ell_2)$ allows for non-trivial zero sums
of idempotents, i.e., a finite collection of non-zero idempotents adding up to zero.
Indeed, for Banach algebras featuring that phenomenon the answer to the above
question is always negative.

For a long time, $\mathcal{L}(\ell_2)$ was basically the only known counterexample in connection with the issue stated above. In this paper, the existence of non-trivial zero
sums of idempotents will be established for a large class of Banach algebras. Belonging to it are the C^*-algebras known under the name Cuntz algebras. Other
Banach algebras lying in the class are those of the form $\mathcal{L}(X)$ with X a non-trivial
Banach space isomorphic to a (finite) direct sum of at least two copies of X. Here
$\mathcal{L}(X)$ stands for the Banach algebra of bounded linear operators on X.

This brings us to a description of the contents of the different sections to be
found below. Apart from the introduction contained in Section 1 and the list of
references, the paper consists of seven sections. Section 2 is of a preliminary nature
and introduces the main concepts. Among them is that of spectral regularity of
a Banach algebra, meaning basically that for the algebra in question the answer
to the question formulated above is always positive. It is also recalled from [6]
that a spectrally regular Banach algebra does not allow for non-trivial zero sums
of idempotents. Section 3 contains information about zero sums of idempotents
in general algebras. Special attention is paid to sums involving a small number of
terms. Section 4 gives a criterion for a Banach algebra to have a non-trivial zero
sum of idempotents, hence failing to be spectrally regular. The pertinent condition
is phrased in terms reminiscent of the defining characteristic for the C^*-algebras
called Cuntz algebras, but there is no restriction here to the C^*-context (as is the
case in [15]). Sections 5 and 6 deal with the situation where the Banach algebra
under consideration is of the form $\mathcal{L}(X)$ with X an infinite-dimensional Banach
space. The idempotents in $\mathcal{L}(X)$ then are the projections in X. There do exist
infinite-dimensional Banach spaces X for which $\mathcal{L}(X)$ is spectrally regular so that
non-trivial zero sums of projections in X do not exist. This follows by combining
results from [11] with the quite remarkable examples given in [2], [22] and [1].
Section 5 is concerned with the situation where a finite number of projections
add up to an operator which is quasinilpotent or compact. For non-trivial sums
of that type, the conclusion is drawn that they involve at least five projections
with infinite rank and co-rank. At the end of the section, the following embedding
issue comes up: for Banach spaces X and Y, when can $\mathcal{L}(Y)$ be viewed as a
continuously embedded subalgebra of $\mathcal{L}(X)$? One simple case in which there is a
positive answer is identified. Generally speaking however, the issue seems to be
rather non-trivial. In Section 6, the criterion exhibited in Section 3 leads to the
conclusion that $\mathcal{L}(X)$ allows for non-trivial zero sums of projections, hence is not
spectrally regular, whenever X is non-trivial and isomorphic to a (finite) direct
sum of at least two copies of itself. In Sections 7 and 8, this is used to identify new

examples of Banach algebras lacking the property of being spectrally regular. Some of these relate to deep problems in general topology and the geometry of Banach spaces. This is especially true for Banach algebras of the form $\mathcal{L}(X)$ with X taken to be a Banach space of bounded continuous functions. For instance, using a truly remarkable theorem of A.A. Miljutin [28], the following result is obtained: If S is an uncountable compact metrizable topological space and $\mathcal{C}(S; \mathbb{C})$ stands for the Banach space of all complex continuous functions on S (endowed with the max-norm), then the operator algebra $\mathcal{L}(\mathcal{C}(S; \mathbb{C}))$ allows for non-trivial zero sums of projections, hence is not spectrally regular. In two examples given towards the end of the paper, a (generalizing) modification of the well-known Cantor construction plays an important role.

One final remark to close the introduction. The expression (1) defines the *left* logarithmic residue of f. There is also a *right* version, obtained by replacing the left logarithmic derivative $f'(\lambda)f(\lambda)^{-1}$ by the right logarithmic derivative $f(\lambda)^{-1}f'(\lambda)$. For some special cases, the relationship between left logarithmic residues and right logarithmic residues has been investigated: see [7], [8], [9], and [11]. As far as the issues considered in the present paper are concerned, the results that can be obtained for the left and the right version of the logarithmic residue are analogous to each other. Therefore the qualifiers left and right will be suppressed in what follows.

2. Preliminaries

A *spectral configuration* is a triple $(\mathcal{B}, \Delta, f)$ where $\mathcal{B}$ is a non-trivial unital complex Banach algebra, Δ is a bounded Cauchy domain in $\mathbb{C}$ (see [34] or [20]) and f is a $\mathcal{B}$-valued analytic function defined on an open neighborhood of the closure of Δ and having invertible values on all of the boundary $\partial\Delta$ of Δ. With such a spectral configuration, taking $\partial\Delta$ to be positively oriented, one can associate the contour integral

$$LR(f; \Delta) = \frac{1}{2\pi i} \int_{\partial\Delta} f'(\lambda)f(\lambda)^{-1} d\lambda.$$

We call it the *logarithmic residue associated with* $(\mathcal{B}, \Delta, f)$; sometimes the term *logarithmic residue of f with respect to Δ* is used as well.

The spectral configuration $(\mathcal{B}, \Delta, f)$ is called *winding free* when the logarithmic residue $LR(f; \Delta) = 0$, *spectrally winding free* if $LR(f; \Delta)$ is quasinilpotent, and *spectrally trivial* in case f takes invertible values on all of Δ. This terminology is taken from [13].

In [13] a unital Banach algebra $\mathcal{B}$ is said to be spectrally regular if a spectral configuration having $\mathcal{B}$ as the underlying Banach algebra is spectrally trivial whenever it is spectrally winding free. Here we shall work with a (possibly) somewhat weaker notion. We shall call $\mathcal{B}$ *spectrally regular* if a spectral configuration having $\mathcal{B}$ as the underlying Banach algebra is spectrally trivial whenever it is winding free. Whether this notion is strictly weaker than the one employed (under the same name) in [13] is not known. It makes sense to adopt the weaker form of spectral regularity here because in this paper we are (mainly) interested in "negative results", i.e., in results having as conclusion that the Banach algebra under consideration fails to be spectrally regular. In that case we call it *spectrally irregular*. So the unital Banach algebra $\mathcal{B}$ is spectrally irregular if and only if there exists a spectral configuration having $\mathcal{B}$ as the underlying Banach algebra which is winding free but not spectrally trivial.

Closely connected to the issue of spectral (ir)regularity is that of zero sums of idempotents. Let $\mathcal{A}$ be an algebra (possibly without norm). A (finite) number of idempotents $p_1, \ldots, p_n$ in $\mathcal{A}$ are said to form a *zero sum* if they add up to the zero element in $\mathcal{A}$, i.e., if $p_1 + \cdots + p_n = 0$. The zero sum is called *trivial* when $p_j = 0$, $j = 1, \ldots, n$. Non-triviality of a zero sum of idempotents then means that at least one among the idempotents involved does not vanish. By leaving out the zero terms, such a non-trivial zero sum can be transformed into a *genuine* zero sum, that is one where all terms are non-zero.

In line with what has been the case up to now, here too spectral irregularity will be brought to light via the construction of non-trivial zero sums of idempotents. The background for this is the following basic result taken from [5].

Theorem 2.1. *Let $\mathcal{B}$ be a unital Banach and let $p_1, \ldots, p_n$ be idempotents in $\mathcal{B}$. If $\mathcal{B}$ is spectrally regular and $p_1 + \cdots + p_n = 0$, then $p_j = 0$, $j = 1, \ldots, n$.*

We say that Banach algebra $\mathcal{B}$ has the *non-trivial zero sum property* if there exist a positive integer m and non-zero idempotents $p_1, \ldots, p_m$ in $\mathcal{B}$ such that $p_1 + \cdots + p_m = 0$. The above theorem can then be read as follows: *if $\mathcal{B}$ has the non-trivial zero sum property, then $\mathcal{B}$ is spectrally irregular*; schematically:

$$\text{non-trivial zero sum property} \Rightarrow \text{spectral irregularity}.$$

It is an open problem whether or not the reverse implication is valid too. No example is known of a spectrally irregular Banach algebra which fails to have the non-trivial zero sum property, so of a spectrally irregular Banach algebra allowing for trivial zeros sums of idempotents only.

We close this section by mentioning that the issue of spectral (ir)regularity is closely related to non-commutative Gelfand theory (cf. [31]). In this connection families of matrix representations play an important role: the existence of certain specific families for a Banach algebra $\mathcal{B}$ implies spectral regularity of $\mathcal{B}$ (see [13] and [14]). The strongest result of this type is that $\mathcal{B}$ is spectrally regular when $\mathcal{B}$ allows for a radical-separating family of matrix representations. Here a family $\{\phi_\omega : \mathcal{B} \to \mathbb{C}^{m_\omega \times m_\omega}\}_{\omega \in \Omega}$ is called *radical-separating* if it separates the points of $\mathcal{B}$ modulo the radical of $\mathcal{B}$. Thus spectral irregularity brings with it that $\mathcal{B}$ does not allow for a radical-separating family of matrix representations; schematically:

$$\text{non-trivial zero sum property} \Rightarrow \text{spectral irregularity}$$

$$\Rightarrow \text{absence of radical-separating families.}$$

The absence of radical-separating families (so a-fortiori having the non-trivial zero sum property) implies the non-existence of the families of other types considered in [13] and [14].

3. Sums of small numbers of idempotents

In this section we focus on (zero) sums of a small number of idempotents (not counting the possibly repeated occurrence of the unit element). To put the (new) material to be presented into proper perspective, we first recall some known facts (see [5]).

If $\mathcal{A}$ is any algebra (normable or not), then a zero sum of three idempotents in $\mathcal{A}$ is necessarily trivial. There is an example of an algebra which allows for a genuine zero sum of four idempotents. The algebra in question is, however, non-normable. It must be because, as Theorem 4.3 from [5] asserts, zero sums of four idempotents in a normed algebra are always trivial. We shall return to this (non-trivial) result, in a moment. In [30] it is shown that every bounded linear operator on the separable Hilbert space ℓ_2 can be written as the sum of five idempotents in the Banach algebra $\mathcal{L}(\ell_2)$ of all bounded linear operators on ℓ_2. An immediate consequence of this is that $\mathcal{L}(\ell_2)$ allows for non-trivial zero sums of six idempotents. In fact one can make do with one less: the five idempotents constructed in [5], Example 3.1 yield a genuine zero sum in $\mathcal{L}(\ell_2)$.

We now return to Theorem 4.3 from [5]. As was already indicated, the theorem says that a zero sum of four idempotents in a Banach algebra (or, more generally a normed algebra) is always trivial. Here is an extension of this result.

Theorem 3.1. *Let p_1, p_2, p_3 and p_4 be idempotents in a non-trivial Banach algebra $\mathcal{B}$ with unit element $e_\mathcal{B}$, and let ν be a non-negative integer. If*

$$p_1 + p_2 + p_3 + p_4 + \nu e_\mathcal{B} = 0,$$

then $\nu = 0$ and $p_1 = p_2 = p_3 = p_4 = 0$.

Theorem 4.3 in [5] corresponds to the case where the integer ν is a priori assumed to be zero. As to the proof of Theorem 3.1, we shall first show that $\nu = 0$ which brings us in the situation considered in [5]. Then we shall extend the argument to cover the case $\nu = 0$ too. In this way, a new proof of Theorem 4.3 in [5] is obtained which is more transparent than the original one. The argument given in [5], although conceptually elementary, is technically quite complicated. The reasoning presented below is suggested by the material in [19], Section 3. The spectrum of an element $x \in \mathcal{B}$ is denoted by $\sigma(x)$.

Proof. As already indicated we first show that $\nu = 0$. Put $x_1 = p_1 + p_2 - e_\mathcal{B}$ and $x_2 = p_3 + p_4 - e_\mathcal{B}$. Then, by Lemma 3 in [19],

$$\lambda \in \sigma(x_j) \setminus \{-1, +1\} \quad \Leftrightarrow \quad -\lambda \in \sigma(x_j) \setminus \{-1, +1\}, \qquad j = 1, 2.$$

As $x_1 + x_2 + (\nu + 2)e_\mathcal{B} = 0$, we also have

$$\lambda \in \sigma(x_1) \;\Leftrightarrow\; -(\lambda + \nu + 2) \in \sigma(x_2).$$

Introduce the set of (negative) integers

$$N = \{-\nu - 1, -2\nu - 3, -3\nu - 5, \ldots\} \cup \{-\nu - 3, -2\nu - 5, -3\nu - 7, \ldots\},$$

and assume $\sigma(x_1)$ contains an element μ which is not in N. Then $-\mu - \nu - 2$ belongs to $\sigma(x_2)$ and $\mu + \nu + 2 \in \sigma(x_2)$ provided that $-\mu - \nu - 2 \notin \{-1, +1\}$. The latter comes down to $\mu \neq -\nu - 1, -\nu - 3$ which certainly holds because μ is not in N. So $\mu + \nu + 2 \in \sigma(x_2)$. But then $-\mu - 2\nu - 4 \in \sigma(x_1)$, and we get $\mu + 2\nu + 4 \in \sigma(x_1)$ because $-\mu - 2\nu - 4 \notin \{-1, +1\}$. Proceeding in this way (formally by induction of course), we obtain $\mu + 2k(\nu + 2) \in \sigma(x_1), k = 0, 1, 2, \ldots$. As $\nu + 2$ is positive, this conflicts with the boundedness of $\sigma(x_1)$ implied by the non-triviality of $\mathcal{B}$. The conclusion is that $\sigma(x_1) \subset N$.

Next assume that there is an element $\mu \in \sigma(x_1)$ which does not belong to $-N \cup \{-1, +1\}$. As μ differs from -1 and $+1$, we have $-\mu \in \sigma(x_1)$, and this implies $-\mu \in N$. But then $\mu \in -N$ and we have a contradiction. The conclusion is that $\sigma(x_1) \subset -N \cup \{-1, +1\}$.

The upshot of these arguments is that $\sigma(x_1) \subset N \cap [-N \cup \{-1, +1\}]$. Now N consists of negative integers, so N and $-N$ are disjoint. Hence $\sigma(x_1)$ is contained in $N \cap \{-1, +1\}$, and it follows that the latter is non-empty. Thus either -1 or $+1$ must belong to N, and this is the case only when $\nu = 0$, as desired.

In this way we have arrived at the situation considered in [5], Theorem 4.3: we have four idempotents p_1, p_2, p_3 and p_4 in a Banach algebra adding up to the zero element. The following argument provides a new (and more transparent) proof for the conclusion of [5], namely that $p_1 = p_2 = p_3 = p_4 = 0$.

With $x_1 = p_1 + p_2 - e_\mathcal{B}$ and $x_2 = p_3 + p_4 - e_\mathcal{B}$ as before, we have $x_1 + x_2 + 2e_\mathcal{B} = 0$, and so $\lambda \in \sigma(x_1) \Leftrightarrow -(\lambda + 2) \in \sigma(x_2)$. Also with $\nu = 0$, the set N introduced above becomes $N = \{-1, -3, -5, \ldots\}$. As we have seen above, $\sigma(x_1) \subset N \cap \{-1, +1\}$, and it follows that $\sigma(x_1) = \{-1\}$. We shall now investigate the behavior of the resolvent $(\lambda e - x_1)^{-1}$ which is defined and analytic on $\mathbb{C} \setminus \{-1\}$.

Take $\lambda \neq -1, +1$, and put $y = p_1 - p_2$. As has been observed in [19], and can be easily verified, $x_1^2 + y^2 = e_\mathcal{B}$ and $yx_1 = -x_1y$. With the help of these identities, one easily obtains the identities $y = (\lambda e_\mathcal{B} - x_1)y(\lambda e_\mathcal{B} + x_1)^{-1}$ and

$$(\lambda e_\mathcal{B} - x_1)(\lambda e_\mathcal{B} + x_1 - y(\lambda e_\mathcal{B} + x_1)^{-1}y)$$

$$= \lambda^2 e_\mathcal{B} - x_1^2 - (\lambda e_\mathcal{B} - x_1)y(\lambda e_\mathcal{B} + x_1)^{-1}y$$

$$= \lambda^2 e_\mathcal{B} - x_1^2 - y^2$$

$$= (\lambda^2 - 1)e_\mathcal{B}.$$

Dividing by $\lambda^2 - 1$ we get

$$(\lambda e_{\mathcal{B}} - x_1)\left(\frac{1}{\lambda^2 - 1}(\lambda e_{\mathcal{B}} + x_1 - y(\lambda e_{\mathcal{B}} + x_1)^{-1}y)\right) = e_{\mathcal{B}}. \qquad (2)$$

In the same way one proves that, again for $\lambda \neq -1, +1$,

$$\left(\frac{1}{\lambda^2 - 1}(\lambda e_{\mathcal{B}} + x_1 - y(\lambda e_{\mathcal{B}} + x_1)^{-1}y)\right)(\lambda e_{\mathcal{B}} - x_1) = e_{\mathcal{B}},$$

and, in combination with (2), this leads to

$$(\lambda e_{\mathcal{B}} - x_1)^{-1} = \left(\frac{1}{\lambda^2 - 1}(\lambda e_{\mathcal{B}} + x_1 - y(\lambda e_{\mathcal{B}} + x_1)^{-1}y)\right).$$

Now $1 \notin \sigma(x_1)$, so $(\lambda e_{\mathcal{B}} + x_1)^{-1}$ is analytic in a neighborhood of -1, and it follows that $(\lambda e_{\mathcal{B}} - x_1)^{-1}$ has a simple pole at -1. But then $\left(\lambda e_{\mathcal{B}} - (e_{\mathcal{B}} + x_1)\right)^{-1}$ has a simple pole at the origin. From $\sigma(e_{\mathcal{B}} + x_1) = \{0\}$, we see that $e_{\mathcal{B}} + x_1$ is quasinilpotent. Hence, by standard spectral theory,

$$\left(\lambda e_{\mathcal{B}} - (e_{\mathcal{B}} + x_1)\right)^{-1} = \frac{1}{\lambda} e_{\mathcal{B}}, \qquad \lambda \neq 0.$$

Thus $\lambda e_{\mathcal{B}} - (e_{\mathcal{B}} + x_1) = \lambda e_{\mathcal{B}}$ for all complex λ, and so $e_{\mathcal{B}} + x_1 = 0$. In other words $p_1 + p_2 = 0$. From $p_1 = p_1^2 = (-p_2)^2 = p_2^2 = p_2 = -p_1$ it is now clear that $p_1 = p_2 = 0$. In the same way one shows that p_3 and p_4 vanish too. $\qquad \square$

Corollary 3.2. *Let p_1, p_2, p_3 and p_4 be idempotents in a Banach algebra $\mathcal{B}$ with unit element $e_{\mathcal{B}}$, let ν be a non-negative integer, and let $\mathcal{J}$ be a proper closed two-sided ideal in $\mathcal{B}$. If*

$$p_1 + p_2 + p_3 + p_4 + \nu e_{\mathcal{B}} \in \mathcal{J},$$

then $\nu = 0$ and $p_1, p_2, p_3, p_4 \in \mathcal{J}$.

Proof. Pass to the quotient algebra $\mathcal{B}/\mathcal{J}$ (which is non-trivial because $\mathcal{J}$ is proper), and apply Theorem 3.1 with $\mathcal{B}$ replaced by $\mathcal{B}/\mathcal{J}$. $\qquad \square$

4. Banach algebras of Cuntz type

The material in the next paragraph is presented by way of motivation for a definition that will given below. The symbol $\delta_{k,l}$ stands for the Kronecker delta.

The *Cuntz algebra* $\mathcal{O}_n$ is the universal unital C^*-algebra generated by n isometries $v_1, \ldots, v_n \in \mathcal{O}_n$ satisfying the identities

$$\sum_{j=1}^{n} v_j v_j^* = e, \qquad v_k^* v_l = \delta_{k,l} e, \quad k, l = 1, \ldots, n, \qquad (3)$$

where e is the unit element in $\mathcal{O}_n$. Here n is an integer larger than or equal to 2. The first to consider this algebra was J. Cuntz [17]. The Cuntz algebras are universal in the sense that for fixed n, any two concrete realization generated by isometries $v_1, \ldots, v_n$ and $\tilde{v}_1, \ldots, \tilde{v}_n$, respectively, are *-isomorphic to each other (cf. [17], [18]). For completeness, and to make the proper connection with [17], we note that the relations (3) come down to the same as

$$\sum_{j=1}^{n} v_j v_j^* = e, \qquad v_k^* v_k = e, \quad k = 1, \ldots, n.$$

To see this multiply the first part of (3) from the left with v_k^* and from the right with v_k, and recall that a sum of nonnegative elements in a C^*-algebra can only vanish when so do all its terms.

Returning to general Banach algebras (not necessarily C^*), we stipulate that a non-trivial unital Banach algebra $\mathcal{B}$ will be said to have the *Cuntz n-property* if n is an integer larger than one and there exist elements $v_1, \ldots, v_n, w_1, \ldots, w_n$ in $\mathcal{B}$ such that

$$\sum_{j=1}^{n} v_j w_j = e_{\mathcal{B}}, \qquad w_k v_l = \delta_{k,l} e_{\mathcal{B}}, \quad k, l = 1, \ldots, n. \tag{4}$$

We emphasize that this definition does not imply that the algebra is generated by the elements $v_1, \ldots, v_n, w_1, \ldots, w_n$. The following statement can be easily verified. If, for $s = 1, 2$, the Banach algebra $\mathcal{B}$ has the Cuntz n_s-property, then $\mathcal{B}$ has the Cuntz $(n_1 + n_2 - 1)$-property; hence, if $\mathcal{B}$ has the Cuntz n-property, then $\mathcal{B}$ has the Cuntz $(kn - k + 1)$-property for each positive integer k. The argument is as follows. Suppose $v_{s,1}, \ldots, v_{s,n_s}, w_{s,1}, \ldots, w_{s,n_s} \in \mathcal{B}$ satisfy the identities

$$\sum_{j=1}^{n_s} v_{s,j} w_{s,j} = e_{\mathcal{B}}, \qquad w_{s,k} v_{s,l} = \delta_{k,l} e_{\mathcal{B}}, \quad k, l = 1, \ldots, n_s.$$

For $j = 1, \ldots, n_1$, write $v_j = v_{2,1} v_{1,j}$, $w_j = w_{1,j} w_{2,1}$. Also, for the values of j ranging from $n_1 + 1$, up to $n_1 + n_2 - 1$, put $v_j = v_{2,j-n_1+1}$ and $w_j = w_{2,j-n_1+1}$. Then (4) holds with $n = n_1 + n_2 - 1$.

A non-trivial unital Banach algebra $\mathcal{B}$ is said to be of *Cuntz type* if $\mathcal{B}$ has the Cuntz n-property for some integer n larger than one. Such an algebra necessarily is non-commutative.

Theorem 4.1. *Let m be an integer larger than or equal to five, let $\mathcal{B}$ be a non-trivial unital Banach algebra, and suppose $\mathcal{B}$ is of Cuntz type. Then $\mathcal{B}$ allows for a zero sum of m non-zero idempotents.*

Thus Banach algebras of Cuntz type have the non-trivial zero sum property introduced in Section 2 and are (therefore) spectrally irregular. As has been mentioned in Section 3, in a Banach algebra non-trivial zero sums of idempotents involving less than five idempotents cannot exist.

Proof. We shall break up the reasoning into seven steps. In the first, some preparatory action is taken.

Step 1. If $A_1, \ldots, A_l$ are matrices, their direct sum will be denoted by $A_1 \oplus \cdots \oplus A_l$ or $\oplus_{j=1}^{l} A_j$. So $\oplus_{j=1}^{l} A_j$ is the block diagonal matrix with $A_1, \ldots, A_l$ on the diagonal (in this order). In case all the matrices A_j coincide with a single matrix A, we use $A^{\oplus l}$ for $\oplus_{j=1}^{l} A_j = A \oplus \cdots \oplus A$ (with l terms in the direct sum). If A is the sum of k non-zero idempotents, then so is $A^{\oplus l}$. Indeed, if $A = \sum_{j=1}^{k} P_j$, then $A^{\oplus l} = \sum_{j=1}^{k} P_j^{\oplus l}$.

Whenever convenient complex matrices will be identified with matrices having entries in $\mathcal{B}$. The identification goes via "tensorizing" with $e_{\mathcal{B}}$, i.e., via replacing each scalar entry by the corresponding multiple of the unit element in $\mathcal{B}$.

Step 2. Consider the matrices

$$M_1 = \frac{1}{2} \begin{bmatrix} -5 & 35 \\ -1 & 7 \end{bmatrix}, \qquad M_2 = \frac{1}{2} \begin{bmatrix} -5 & -35 \\ 1 & 7 \end{bmatrix}.$$

$$N_1 = \frac{1}{6} \begin{bmatrix} 5 & 15 & -1 \\ 5 & 15 & -1 \\ 70 & 210 & -14 \end{bmatrix}, \quad N_2 = \frac{1}{6} \begin{bmatrix} 5 & -15 & -1 \\ -5 & 15 & 1 \\ 70 & -210 & -14 \end{bmatrix}, \quad N_3 = \frac{1}{3} \begin{bmatrix} 10 & 0 & 1 \\ 0 & 0 & 0 \\ -70 & 0 & -7 \end{bmatrix}.$$

These are idempotents, regardless of whether they are considered as complex matrices (in which case they have rational entries and rank 1), or as matrices with entries in $\mathcal{B}$. Also

$$M_1 + M_2 = \begin{bmatrix} -5 & 0 \\ 0 & 7 \end{bmatrix}, \qquad N_1 + N_2 + N_3 = \begin{bmatrix} 5 & 0 & 0 \\ 0 & 5 & 0 \\ 0 & 0 & -7 \end{bmatrix}.$$

Thus $[-5] \oplus [7]$ is the sum of the two non-zero idempotents M_1 and M_2, and $[5] \oplus [5] \oplus [-7]$ is the sum of the three non-zero idempotents N_1, N_2 and N_3.

Step 3. For simplification of notation, put $d = n - 1$. Then $[[-5] \oplus [7]]^{\oplus d}$ is a sum of two non-zero idempotents, and $[[5] \oplus [5] \oplus [-7]]^{\oplus d}$ is a sum of three non-zero idempotents. In fact

$$[[-5] \oplus [7]]^{\oplus d} = M_1^{\oplus d} + M_2^{\oplus d},$$

$$[[5] \oplus [5] \oplus [-7]]^{\oplus d} = N_1^{\oplus d} + N_2^{\oplus d} + N_3^{\oplus d}.$$

Now choose permutation similarities Π_M and Π_N (having the effect of interchanging rows and columns) such that $\Pi_M^{-1} = \Pi_M$ and $\Pi_N^{-1} = \Pi_N$ while, moreover,

$$\Pi_M \left([[-5] \oplus [7]]^{\oplus d} \right) \Pi_M = [-5]^{\oplus d} \oplus [7]^{\oplus d},$$

$$\Pi_N \left([[5] \oplus [5] \oplus [-7]]^{\oplus d} \right) \Pi_N = [5]^{\oplus (d+1)} \oplus [5]^{\oplus (d-1)} \oplus [-7]^{\oplus d}.$$

It follows that the right-hand sides of these expressions are the sum of two and three idempotents, respectively:

$$[-5]^{\oplus d} \oplus [7]^{\oplus d} = \sum_{k=1}^{2} \Pi_M \left(M_k^{\oplus d} \right) \Pi_M,$$

$$[5]^{\oplus(d+1)} \oplus [5]^{\oplus(d-1)} \oplus [-7]^{\oplus d} = \sum_{k=1}^{3} \Pi_N \left(N_k^{\oplus d} \right) \Pi_N.$$

Step 4. Before proceeding with the main line of the argument, we make an auxiliary remark. Let $v \in \mathcal{B}^{j \times k}$, $w \in \mathcal{B}^{k \times j}$ and suppose wv is the identity element in $\mathcal{B}^{k \times k}$. Further assume $a \in \mathcal{B}^{k \times k}$ is a sum of l non-zero idempotents in $\mathcal{B}^{k \times k}$. Then the product $vaw \in \mathcal{B}^{j \times j}$ is a sum of l non-zero idempotents in $\mathcal{B}^{j \times j}$. To see this, write $a = r_1 + \cdots + r_l$ with $r_1, \ldots, r_l$ non-zero idempotents in $\mathcal{B}^{k \times k}$. Then we have $vaw = vr_1 w + \cdots + vr_l w$ with $(vr_i w)^2 = vr_i wvr_i w = vr_i^2 w = vr_i w$. From $wvr_i wv = r_i$ one sees that $vr_i w$ cannot be zero.

Step 5. From now on $v_1, \ldots, v_n, w_1, \ldots, w_n$ will be elements in $\mathcal{B}$ satisfying (4). Put $V = [v_1 \ \ldots \ v_{d+1}]$ and $W = [w_1 \ \ldots \ w_{d+1}]^{\top}$, where $\top$ signals the operation of taking the transpose. Then $V \in \mathcal{B}^{1 \times (d+1)}$, $W \in \mathcal{B}^{(d+1) \times 1}$, and (4) can be rephrased by saying that VW is the identity element in $\mathcal{B}$ and WV is the identity element in $\mathcal{B}^{(d+1) \times (d+1)}$. Write

$$V_1 = V \oplus [e_{\mathcal{B}}]^{\oplus(2d-1)} = V \oplus [e_{\mathcal{B}}]^{\oplus(d-1)} \oplus [e_{\mathcal{B}}]^{\oplus d},$$

$$W_1 = W \oplus [e_{\mathcal{B}}]^{\oplus(2d-1)} = W \oplus [e_{\mathcal{B}}]^{\oplus(d-1)} \oplus [e_{\mathcal{B}}]^{\oplus d}.$$

These matrices belong to $\mathcal{B}^{2d \times 3d}$ and $\mathcal{B}^{3d \times 2d}$, respectively. Also $W_1 V_1$ is the identity element in $\mathcal{B}^{3d \times 3d}$. In combination with what we obtained in Step 4, this gives that $V_1 \left([5]^{\oplus(d+1)} \oplus [5]^{\oplus(d-1)} \oplus [-7]^{\oplus d} \right) W_1$ is a sum of three non-zero idempotents in $\mathcal{B}^{2d \times 2d}$:

$$V_1 \left([5]^{\oplus(d+1)} \oplus [5]^{\oplus(d-1)} \oplus [-7]^{\oplus d} \right) W_1 = \sum_{k=1}^{3} V_1 \Pi_N \left(N_k^{\oplus d} \right) \Pi_N W_1.$$

Using the defining expressions for V_1 and W_1, we can rewrite the left-hand side of this identity as $5VW \oplus [5]^{\oplus(d-1)} \oplus [-7]^{\oplus d}$ and this, in view of the fact that VW is the identity element in $\mathcal{B}$, is equal to $[5]^{\oplus d} \oplus [-7]^{\oplus d}$. So the latter is the sum of three non-zero idempotents in $\mathcal{B}^{2d \times 2d}$:

$$[5]^{\oplus d} \oplus [-7]^{\oplus d} = \sum_{k=1}^{3} V_1 \Pi_N \left(N_k^{\oplus d} \right) \Pi_N W_1.$$

Again referring to Step 2, we recall that $[-5]^{\oplus d} \oplus [7]^{\oplus d}$ is the sum of the two non-zero idempotents in $\mathcal{B}^{2d \times 2d}$, namely $\Pi_M \left(M_1^{\oplus d} \right) \Pi_M$ and $\Pi_M \left(M_2^{\oplus d} \right) \Pi_M$. Thus the zero element in $\mathcal{B}^{2d \times 2d}$ appears as the sum of five non-zero idempotents:

$$\sum_{k=1}^{2} \Pi_M \left(M_k^{\oplus d} \right) \Pi_M + \sum_{k=1}^{3} V_1 \Pi_N \left(N_k^{\oplus d} \right) \Pi_N W_1 = 0.$$

Step 6. Next we make a reduction from $\mathcal{B}^{2d \times 2d}$ to $\mathcal{B}$. Put

$$V_2 = \begin{bmatrix} v_1 & \cdots & v_d & v_{d+1}v_1 & \cdots & v_{d+1}v_d \end{bmatrix},$$

$$W_2 = \begin{bmatrix} w_1 & \cdots & w_d & w_1 w_{d+1} & \cdots & w_d w_{d+1} \end{bmatrix}^\top.$$

These matrices belong to $\mathcal{B}^{1 \times 2d}$ and $\mathcal{B}^{2d \times 1}$, respectively. Also $W_2 V_2$ is the identity element in $\mathcal{B}^{2d \times 2d}$, and so (again on account of Step 4) the zero element in $\mathcal{B}$ appears as the sum of five non-zero idempotents in $\mathcal{B}$:

$$\sum_{k=1}^{2} V_2 \Pi_M \left(M_k^{\oplus d} \right) \Pi_M W_2 + \sum_{k=1}^{3} V_2 V_1 \Pi_N \left(N_k^{\oplus d} \right) \Pi_N W_1 W_2 = 0. \qquad (5)$$

Step 7. We have proved the statement in the theorem now for $m = 5$: there exist five non-zero idempotents $p(5)_1, \ldots, p(5)_5 \in \mathcal{B}$ with $p(5)_1 + \cdots + p(5)_5 = 0$. To get the result for arbitrary m larger than or equal to five, it suffices to deal with the cases $m = 6$ up to and including $m = 9$. Indeed, we then have that for $l = 5, 6, 7, 8, 9$ there are non-zero idempotents $p(l)_1, \ldots, p(l)_l \in \mathcal{B}$ such that $p(l)_1 + \cdots + p(l)_l = 0$. Given any integer m larger than or equal to five, we write m in the form $5k + l$ with k a non-negative integer and $l \in \{5, 6, 7, 8, 9\}$. Then

$$kp_1 + \cdots + kp_5 + p(l)_1 + \cdots + p(l)_l = 0$$

is a zero sum involving m non-zero idempotents in $\mathcal{B}$.

We finish the proof by establishing the existence of the idempotents $p(l)_1, \ldots, p(l)_l$ featuring in the previous paragraph. Take $l \in \{6, 7, 8, 9\}$, and put

$$q_j = \begin{cases} \begin{bmatrix} p_j & 0 \\ 0 & 0 \end{bmatrix}, & j = 1, \ldots, (l-5), \\[2mm] \begin{bmatrix} p_j & 0 \\ 0 & p_{5+j-l} \end{bmatrix}, & j = (l-4), \ldots, 5, \\[2mm] \begin{bmatrix} 0 & 0 \\ 0 & p_{5+j-l} \end{bmatrix}, & j = 6, \ldots, l. \end{cases}$$

Then $q_1, \ldots, q_l$ are l non-zero idempotents in $\mathcal{B}^{2 \times 2}$ adding up to the zero element in $\mathcal{B}^{2 \times 2}$. For $p(l)_j$ we can now take $[v_1 \, v_2] q_j [w_1 \, w_2]^\top$, $j = 1, \ldots, l$. Indeed, $[w_1 \, w_2]^\top [v_1 \, v_2]$ is the unit element in $\mathcal{B}^{2 \times 2}$. $\qquad \square$

Elaborating on the above proof, we rewrite the identity (5) as

$$\sum_{k=1}^{2} V_M \left(M_k^{\oplus(n-1)} \right) W_M + \sum_{k=1}^{3} V_N \left(N_k^{\oplus(n-1)} \right) W_N = 0,$$

where $V_M \in \mathcal{B}^{1 \times 2(n-1)}$ and $W_M \in \mathcal{B}^{2(n-1) \times 1}$ are given by

$$V_M = \begin{bmatrix} v_1 & \cdots & v_{n-1} & v_n v_1 & \cdots & v_n v_{n-1} \end{bmatrix} \Pi_M$$

$$W_M = \Pi_M \begin{bmatrix} w_1 & \cdots & w_{n-1} & w_1 w_n & \cdots & w_{n-1} w_n \end{bmatrix}^\top ,$$

and $V_N \in \mathcal{B}^{1 \times 3(n-1)}$ and $W_N \in \mathcal{B}^{3(n-1) \times 1}$ by

$$V_N = \begin{bmatrix} v_1^2 & v_1 v_2 & \cdots & v_1 v_n & v_2 & \cdots & v_{n-1} & v_n v_1 & \cdots & v_n v_{n-1} \end{bmatrix} \Pi_N$$

$$W_N = \Pi_N \begin{bmatrix} w_1^2 & w_2 w_1 & \cdots & w_n w_1 & w_2 & \cdots & w_{n-1} & w_1 w_n & \cdots & w_{n-1} w_n \end{bmatrix}^\top .$$

Anticipating on what we shall need in the proof of Theorem 6.3 below, we note that $W_M V_M$ and $W_N V_N$ are the identity elements in $\mathcal{B}^{2(n-1) \times 2(n-1)}$ and $\mathcal{B}^{3(n-1) \times 3(n-1)}$, respectively.

In our reasoning, the idempotents $p_1, \ldots, p_m$ meant in Theorem 4.1 come up as linear combinations involving rational coefficients of monomials in the elements $v_1, \ldots, v_n, w_1, \ldots, w_n$ satisfying (4). It is possible to give additional information, for instance on the number of the monomials involved (maximally $9(n-1)^2$) and their degree (at most 4), but we refrain from giving further details here. Instead we say something more about the cases $n = 2$, $m = 5$ and $n = 3$, $m = 5$.

When $n = 2$, the matrix Π_M can be chosen to be the 2×2 identity matrix, and one obtains

$$V_M = \begin{bmatrix} v_1 & v_2 v_1 \end{bmatrix}, \qquad W_M = \begin{bmatrix} w_1 & w_1 w_2 \end{bmatrix}^\top .$$

Also, for Π_N one can take the 3×3 identity matrix which leads to

$$V_N = \begin{bmatrix} v_1^2 & v_1 v_2 & v_2 v_1 \end{bmatrix}, \qquad W_N = \begin{bmatrix} w_1^2 & w_2 w_1 & w_1 w_2 \end{bmatrix}^\top .$$

The idempotents in $\mathcal{B}$ associated with the 2×2 matrices M_k, $k = 1, 2$, from Step 2 in the proof of Theorem 4.1 are now $\begin{bmatrix} v_1 & v_2 v_1 \end{bmatrix} M_k \begin{bmatrix} w_1 & w_1 w_2 \end{bmatrix}^\top$. Similarly, those corresponding to the 3×3 matrices N_k, $k = 1, 2, 3$ (see again Step 2), can be written as $\begin{bmatrix} v_1^2 & v_1 v_2 & v_2 v_1 \end{bmatrix} N_k \begin{bmatrix} w_1 & w_1 w_2 & w_2 w_1 \end{bmatrix}^\top$. These five non-zero idempotents, involving degree four polynomials in the elements v_1, v_2, w_1 and w_2, add up to the zero element in $\mathcal{B}$.

In case $n = 3$, the matrix Π_M can be chosen to be the 4×4 permutation similarity corresponding to the exchange of the second and the third row (column), and one gets

$$V_M = \begin{bmatrix} v_1 & v_3 v_1 & v_2 & v_3 v_2 \end{bmatrix},$$

$$W_M = \begin{bmatrix} w_1 & w_1 w_3 & w_2 & w_2 w_3 \end{bmatrix}^\top .$$

Also, for Π_N one can take the 6×6 permutation similarity corresponding to the exchange of the third and the fifth row (column), which leads to

$$V_N = \begin{bmatrix} v_1^2 & v_1 v_2 & v_3 v_1 & v_2 & v_1 v_3 & v_3 v_2 \end{bmatrix},$$

$$W_N = \begin{bmatrix} w_1^2 & w_2 w_1 & w_1 w_3 & w_2 & w_3 w_1 & w_2 w_3 \end{bmatrix}^\top .$$

The idempotents in $\mathcal{B}$ associated with the 2×2 matrices M_k, $k = 1, 2$, from Step 2 in the proof of Theorem 4.1 are now

$$\begin{bmatrix} v_1 & v_3 v_1 & v_2 & v_3 v_2 \end{bmatrix} \begin{bmatrix} M_k & 0 \\ 0 & M_k \end{bmatrix} \begin{bmatrix} w_1 & w_1 w_3 & w_2 & w_2 w_3 \end{bmatrix}^{\top}.$$

Similarly, those corresponding to the 3×3 matrices N_k, $k = 1, 2, 3$ (see once more Step 2), can be written as

$$\begin{bmatrix} v_1^2 & v_1 v_2 & v_3 v_1 & v_2 & v_1 v_3 & v_3 v_2 \end{bmatrix} \begin{bmatrix} N_k & 0 \\ 0 & N_k \end{bmatrix}$$
$$\times \begin{bmatrix} w_1^2 & w_2 w_1 & w_1 w_3 & w_2 & w_3 w_1 & w_2 w_3 \end{bmatrix}^{\top}.$$

These five non-zero idempotents, involving degree four polynomials in the elements v_1, v_2, v_3, w_1, w_2 and w_3, add up to the zero element in $\mathcal{B}$.

For these small values of n (2 and 3) and m (= 5), the five polynomials that came up above can of course be computed explicitly. Once this is done, it also possible to prove directly that they constitute non-zero idempotents in $\mathcal{B}$ which add up to the zero element of $\mathcal{B}$. For other low values of n and m such an approach might work too; for higher values it becomes practically unmanageable though.

A Cuntz algebra obviously is a Banach algebra (actually a C^*-algebra) of Cuntz type. Thus we have the following direct consequence of Theorem 4.1.

Corollary 4.2. *Given an integer m larger than or equal to five, a Cuntz algebra allows for a zero sum of m non-zero idempotents.*

In particular Cuntz algebras have the non-trivial zero sum property and are (consequently) spectrally irregular.

5. Sums of idempotents in Banach algebras of the form $\mathcal{L}(X)$

Throughout this section, X will denote a (non-trivial) Banach space. Adopting standard terminology, by a *projection in X* we mean an idempotent in the Banach algebra $\mathcal{L}(X)$ of bounded linear operators on X. If P is a projection in X, the (possibly infinite) dimension of the range of P will be called the *rank of P*, written rank P. The following result generalizes Proposition 2.1 in [5] which states that non-trivial zero sums of finite rank projections cannot exist.

Proposition 5.1. *Let $P_1, \ldots, P_m$ be finite rank projections in X and assume their sum $P_1 + \cdots + P_m$ is quasinilpotent. Then $P_k = 0$, $k = 1, \ldots, m$.*

So, in fact, the sum $P_1 + \cdots + P_m$ vanishes.

Proof. Put $S = P_1 + \cdots + P_n$. Then S is quasinilpotent and has finite rank. Hence S is nilpotent and the trace of S, written trace S, vanishes. As $P_1, \ldots, P_m$ are projections, we have trace $P_k = \operatorname{rank} P_k$, $k = 1, \ldots m$. It follows that

$$\sum_{k=1}^{n} \operatorname{rank} P_k = \sum_{k=1}^{n} \operatorname{trace} P_k = \operatorname{trace}\left(\sum_{k=1}^{n} P_k\right) = \operatorname{trace} S = 0,$$

and we get $P_k = 0$, $k = 1, \ldots, n$, as desired. $\square$

If P is a projection in X, the (possibly infinite) dimension of the null space of P is called the *co-rank of* P, written co-rank P. Note that co-rank P coincides with rank $(I_X - P)$, where I_X is the identity operator on X.

Proposition 5.2. *Let m be a positive integer and let $P_1, \ldots, P_m$ be projections in X. Assume these projections have all finite co-rank and $P_1 + \cdots + P_m$ is quasinilpotent. Then X is finite dimensional and $P_k = 0$, $k = 1, \ldots, m$.*

So, actually, the sum $P_1 + \cdots + P_m$ vanishes.

Proof. Taking into account Proposition 5.1, it suffices to show that X is finite dimensional. Put $S = P_1 + \cdots + P_m$. Then S is quasinilpotent and, consequently, $mI_X - S$ is invertible. On the other hand

$$mI_X - S = \sum_{k=1}^{m}(I_X - P_k)$$

is of finite rank. But then so is $I_X = (mI_X - S)^{-1}(mI_X - S)$, and the finite dimensionality of X follows. $\square$

Theorem 5.3. *Let m be a positive integer, let $P_1, \ldots, P_m$ be projections in X, and suppose the sum $P_1 + \cdots + P_m$ is compact. Then (precisely) one of the following statements holds:*

(a) *$P_1, \ldots, P_m$ are all of finite rank (hence so is their sum),*
(b) *$m \geq 5$ and at least five among the projections $P_1, \ldots, P_m$ have both infinite rank and co-rank.*

It is worthwhile to say a few words about the situation when all the projections $P_1, \ldots, P_m$ have finite co-rank. If that is the case and, in addition, $P_1 + \cdots + P_m$ is compact, then (a) holds, i.e., $P_1, \ldots, P_m$ are all of finite rank as well. It follows that $I_X = (I_X - P_1) + P_1$ has finite rank, and we arrive at one of the conclusions also appearing in Proposition 5.2, namely that X is finite dimensional. So a compact sum of finite co-rank projections can only occur when the underlying space is finite dimensional.

Proof. First assume that each of the projections $P_1, \ldots, P_m$ is either of finite rank or of finite co-rank. Write k for the number of projections among $P_1, \ldots, P_m$ that are of finite co-rank. If $k = 0$, we have (a). So suppose k is at least one. Renumbering (if necessary), we can achieve the situation where $P_1, \ldots, P_k$ have finite co-rank and $P_{k+1}, \ldots, P_m$ are of finite rank. Now

$$\sum_{j=1}^{k} P_k = \sum_{j=1}^{n} P_k - \sum_{j=k+1}^{n} P_k,$$

where the first sum in the right-hand side is compact (by hypothesis) and the second of finite rank. So $P_1 + \cdots + P_k$ is compact. The projections $(I_X - P_1), \ldots, (I_X - P_k)$ are of finite rank. Further

$$I_X = \frac{1}{k}\left(\sum_{j=1}^{k} P_k + \sum_{j=1}^{k}(I_X - P_k) \right).$$

It follows that I_X is compact and, consequently, X is finite dimensional. Under these circumstances, the validity of (a) is a triviality.

Next consider the case when among $P_1, \ldots, P_m$, there are idempotents which are neither of finite rank nor of finite co-rank. Let there be n of those. We may assume (renumbering if necessary) that $P_1, \ldots, P_n$ are of this type and (hence) $P_{n+1}, \ldots, P_m$ are not, i.e., they are of finite rank or finite co-rank. Let ν be the number of idempotents among $P_{n+1}, \ldots, P_m$ that have finite co-rank, and suppose (without loss of generality) that $P_{n+1}, \ldots, P_{n+\nu}$ are of that kind. Then $P_{n+\nu+1}, \ldots, P_n$ have finite rank.

The same is true for the projections $(I_X - P_{n+1}), \ldots, (I_X - P_{n+\nu})$, and it follows that $P_1 + \cdots + P_n + \nu I_X$ is compact. Now apply Corollary 3.2 with $\mathcal{B} = \mathcal{L}(X)$ and taking for $\mathcal{J}$ the ideal of compact operators on X. This gives that all the projections $P_1, \ldots, P_n$ are compact. But then they are of finite rank, which is impossible in view of how the number n has been introduced. So n (and a fortiori m) must be at least five, as claimed in (b). $\qquad\square$

In the situation where the sum of idempotents in Theorem 5.3 is both compact and quasinilpotent (for instance because it vanishes), the conclusion of the theorem can be sharpened.

Theorem 5.4. *Let m be a positive integer and let $P_1, \ldots, P_m$ be projections in X. Suppose the sum $P_1 + \cdots + P_m$ is compact and quasinilpotent. Then (precisely) one of the following statements holds:*

(a) $P_k = 0, \; k = 1, \ldots, m$ *(so, in fact, the sum $P_1 + \cdots + P_m$ vanishes),*
(b) $m \geq 5$ *and at least five among the idempotents $P_1, \ldots, P_n$ have both infinite rank and co-rank.*

Proof. Combine Theorem 5.3 and Proposition 5.1. $\qquad\square$

As we have seen, in dealing with non-trivial zero sums of idempotents, the number five plays a special role. This fact is underlined by the following result on zero sums of five projections.

Corollary 5.5. *Let P_1, P_2, P_3, P_4 and P_5 be projections in X, not all equal to the zero operator on X, and assume $P_1 + P_2 + P_3 + P_4 + P_5 = 0$. Then all five projections P_1, P_2, P_3, P_4 and P_5 have both infinite rank and co-rank.*

Before we proceed with some additional observations, we make a connection with [15]. Notions like finite rank and compactness can be introduced in a meaningful way for elements in C^*-algebras (see [4] and [24]). It turns out that the results obtained so far in this section have analogues in this C^*-context. For details we refer to [15], Section 3 in particular. The proofs presented there (of Propositions 3.9 and 3.10, of Theorems 3.12 and 3.13, and of Corollary 3.14) are modifications of those given here.

Now let us return to zero sums of projections in the given Banach space X. Suppose we have such a sum:

$$P_1 + \cdots + P_m = 0, \tag{6}$$

with m a positive integer and $P_1, \ldots, P_m$ non-zero projections in X. Then necessarily $m \geq 5$, and at least five among the idempotents $P_1, \ldots, P_n$ have both infinite rank and co-rank (Theorem 5.4). Without loss of generality we may assume that P_m is of this type. Thus both the image and the null space of P_m have infinite dimension. Let n be any positive integer larger than m, and let $r_1, \ldots, r_{n-m}$ be positive integers too. Then a routine argument shows that P_m can be written as $P_m = Q_m + Q_{m+1} + \cdots + Q_n$ with Q_m a projection of both infinite rank and co-rank, and Q_j a projection of finite rank r_{j-m}, $j = m+1, \ldots, n$. In this way we arrive at the zero sum

$$P_1 + \cdots + P_{m-1} + Q_m + Q_{m+1} + \cdots + Q_n = 0, \tag{7}$$

involving a total of n non-zero idempotents, featuring just as many projections of both infinite rank and co-rank as there are in the original zero sum (6), and compared to that one having $n-m$ additional projections of prescribed finite rank. One may ask whether it is also possible to transform (6) into a zero sum (7) with n terms by writing P_m as a sum $P_m = Q_m + Q_{m+1} + \cdots + Q_n$ involving only projections of both infinite rank and co-rank. This is problematic because, as has been shown in [22], there are Banach spaces lacking complemented subspaces with both infinite dimension and codimension; see however Theorem 6.3 below.

Suppose that besides X another (non-trivial) Banach space Y is given. If there exists an injective continuous Banach algebra homomorphism $\Phi : \mathcal{L}(Y) \to \mathcal{L}(X)$ (possibly non-unital) and if $\mathcal{L}(Y)$ has the non-trivial zero sum property, then obviously so does $\mathcal{L}(X)$. Indeed, if $P_1 + \cdots + P_m = 0$ is a non-trivial zero sum of projections in X, then $\Phi(P_1) + \cdots + \Phi(P_m) = 0$ is a non-trivial zero sum of projections in Y. This straightforward observation lead to the following embedding issue: *when can $\mathcal{L}(Y)$ be viewed as a continuously embedded subalgebra of $\mathcal{L}(X)$?* This is a non-trivial question indeed. Even first attempts to deal with it touch on deep problems in the geometry of Banach spaces. Here are some observations.

There is one very simple case in which the answer is positive.

Proposition 5.6. *Assume Y is isomorphic to a complemented subspace of X. Then $\mathcal{L}(Y)$ can be continuously embedded into $\mathcal{L}(X)$, i.e., there exists an injective continuous Banach algebra homomorphism from $\mathcal{L}(Y)$ into $\mathcal{L}(X)$.*

Proof. We may assume that Y is a complemented subspace of X. Let J be the natural embedding of the (closed) subspace Y into X and let P be the projection of X onto Y, viewed as an operator from X into Y. Then $J : Y \to X$ and $P : X \to Y$ are bounded linear operators, injective and surjective, respectively. Also PJ is the identity operator on Y. Define $\Phi : \mathcal{L}(Y) \to \mathcal{L}(X)$ by $\Phi(T) = JTP$, $T \in \mathcal{L}(Y)$. Then Φ is an injective continuous Banach algebra homomorphism. $\qquad\square$

The complementedness assumption in Proposition 5.6 is essential: when Y is a non-complemented closed subspace Y of X it may happen that an injective continuous Banach algebra homomorphism from $\mathcal{L}(Y)$ into $\mathcal{L}(X)$ does not exist. For an example we need to delve into the geometry of Banach spaces.

For quite some time it was an open problem whether there exists an infinite-dimensional Banach space Z that solves the so-called scalar-plus-compact problem. This means that Z has only very few operators in the sense that each bounded linear operator on Z is the sum of a scalar multiple of I_Z and a compact operator on Z. It was in [2] that a first example was given. In the recent paper [1], another space E with the property in question has been produced, this time having the additional feature that E contains a closed subspace F which is isomorphic to the separable Hilbert space ℓ_2. As is clear from Corollary 2.2 in [10], a Banach space that solves the scalar-plus-compact problem does not allow for non-trivial zero sums of projections (see also [11], Corollary 3.4). So this is the case for E. However, as has been established in [5], Example 3.1, such a non-trivial zero sum of projections does exist for the Banach space ℓ_2. By isomorphy, this carries over to F. Hence there is no injective continuous Banach algebra homomorphism (possibly non-unital) of $\mathcal{L}(F)$ into $\mathcal{L}(E)$. This conclusion is corroborated by the fact that $\mathcal{L}(E)$ is spectrally regular (see Corollary 4.3 in [11]) and $\mathcal{L}(F)$ is not (cf. [13], Section 4 in particular).

6. Cuntz type Banach algebras of the form $\mathcal{L}(X)$

As was mentioned in Section 1 (Introduction), the first example found of a unital Banach algebra failing to be spectrally regular was the operator algebra $\mathcal{L}(\ell_2)$. Actually, this algebra allows for non-trivial zero sums of idempotents which implies that it lacks the property of being spectrally regular. Example 3.1 in [5], exhibiting this, fits in the framework of Cuntz type algebras developed in Section 4. In the present section, the set up in question will be considered for Banach algebras of the form $\mathcal{L}(X)$. But first we make some introductory remarks.

As we have seen already at the end of the previous section, it can happen that $\mathcal{L}(X)$ is spectrally regular even when X has infinite dimension. The examples from [2] and [1] mentioned there are rather spectacular and connected to deep problems from Banach space geometry. Another similarly remarkable instance where $\mathcal{L}(X)$ is spectrally regular can be found in [22]. Indeed, an example is given there of an infinite-dimensional Banach space X such that each bounded linear operator on X is the sum of a scalar multiple of I_X and a strictly singular operator on X. In that case the arguments given in the first part of [11], Section 4 apply upon slight modification.

Next we turn to the investigation of the situation where $\mathcal{L}(X)$ is of Cuntz type (hence spectrally irregular because of the occurrence of non-trivial zero sums of idempotents). The following observation is straightforward.

Proposition 6.1. *If the Banach algebra $\mathcal{B}$ has the Cuntz n-property, then so has the operator algebra $\mathcal{L}(\mathcal{B})$.*

Proof. Let $v_1, \ldots, v_n, w_1, \ldots, w_n$ be elements in $\mathcal{B}$ satisfying (4). Replacing these elements by their left regular representations on $\mathcal{B}$ (considered as a Banach space), we obtain operators $V_1, \ldots, V_n, W_1, \ldots, W_n$ in $\mathcal{L}(\mathcal{B})$ such that (4) holds with the lower case letters replaced by the corresponding upper case ones, and with I_X replaced by the identity operator $I_\mathcal{B}$ on $\mathcal{B}$. □

From Proposition 6.1 we see that with every Banach algebra of Cuntz type there comes one of the form $\mathcal{L}(X)$ with X a Banach space. This in itself is already reason enough to pay special attention to the case of such operator algebras. Also it is an elementary fact that a unital Banach algebra $\mathcal{B}$ can be identified (for instance via the use of left regular representations) with a Banach subalgebra of $\mathcal{L}(\mathcal{B})$. Proposition 6.1 and its proof now tell us that each Banach algebra having the Cuntz n-property can be viewed as a Banach subalgebra of a Banach algebra having the Cuntz n-property too and being of the type $\mathcal{L}(X)$ for some Banach space X. This is an additional reason for now looking at Banach algebras of the form $\mathcal{L}(X)$.

Theorem 6.2. *Let X be a non-trivial Banach space, and let n be an integer larger than one. Then the operator algebra $\mathcal{L}(X)$ has the Cuntz n-property if and only if X is isomorphic to X^n, where X^n denotes the direct sum of n copies of X.*

For the arguments below to work, the norm on X^n needs to have an appropriate relationship with the given norm on X. What matters is that the following linear operators are continuous: the embeddings

$$X \ni x \mapsto (0, \ldots, 0, x, 0, \ldots, 0)^\top \in X^n,$$

with x in the kth position, and the projections

$$X^n \ni (x_1, \ldots, x_n)^\top \mapsto x_k \in X.$$

Here k is allowed to take the values $1, \ldots, n$. Such norms are mutually equivalent, and we will settle here on the norm $||| \cdot ||| : X^n \to [0, \infty)$ given by

$$\||(x_1, \ldots, x_n)^\top\|| = \max_{k=1,\ldots,n} \|x_k\|, \qquad (x_1, \ldots, x_n)^\top \in X^n,$$

where $\| \cdot \|$ is the norm on X. One could also take any norm on X^n induced by a monotone norm on $\mathbb{C}^n$ (see [16] for the definition). For N such a monotone norm on $\mathbb{C}^n$, the norm $||| \cdot |||_N$ on X induced by N has the form $|||(x_1, \ldots, x_n)^\top|||_N = N(\|x_1\|, \ldots, \|x_n\|)$, $(x_1, \ldots, x_n)^\top \in X^n$.

Proof. First let us deal with the "only if part" of the theorem. So assume the existence of $V_1, \ldots, V_n, W_1, \ldots, W_n$ in $\mathcal{L}(X)$ satisfying

$$\sum_{j=1}^{n} V_j W_j = I_X, \qquad W_k V_l = \delta_{k,l} I_X, \quad k, l = 1, \ldots, n, \tag{8}$$

where I_X stands for the identity operator on X. Now introduce the bounded linear operators $V : X^n \to X$ and $W : X \to X^n$ by

$$V(x_1, \ldots, x_n)^\top = V_1 x_1 + \cdots + V_n x_n, \qquad (x_1, \ldots, x_n)^\top \in X^n, \tag{9}$$

$$W x = (W_1 x, \ldots, W_n x)^\top, \qquad x \in X. \tag{10}$$

By (8) these are each others inverse. Thus X is isomorphic to X^n.

Next we turn to the "if part" of the theorem. Let the bounded linear operators $V : X^n \to X$ and $W : X \to X^n$ be each others inverse. The choice made for the norm on X^n implies that the expressions (9) and (10) determine bounded linear operators $V_1, \ldots, V_n, W_1, \ldots, W_n$ on X. The identities (8) follow from $VW = I_X$ and $WV = I_{X^n}$. $\qquad\square$

Combining Proposition 6.1 and Theorem 6.2, one immediately gets that the following is true. If the Banach algebra $\mathcal{B}$ has the Cuntz n-property, then $\mathcal{B}$, viewed as Banach space, is isomorphic to $\mathcal{B}^n$. The converse is not true. For an example, take $\mathcal{B} = \ell_\infty$ with the coordinatewise product. Clearly ℓ_∞ and ℓ_∞^2 are isomorphic. However ℓ_∞, being commutative, is not of Cuntz type. Note also that ℓ_∞ provides an example of a unital Banach algebra $\mathcal{B}$ which (being commutative) is spectrally regular while $\mathcal{L}(\mathcal{B})$, being of Cuntz type, lacks this property. Contrasting with this we have that $\mathcal{B}$ is spectrally regular whenever $\mathcal{L}(\mathcal{B})$ is. Indeed, $\mathcal{B}$ can be viewed as a Banach subalgebra of $\mathcal{L}(\mathcal{B})$ (for instance via left regular representations) and Corollary 4.1 in [13] applies.

As an immediate consequence of Theorems 6.2 and 4.1 we have the following result. If X is a non-trivial Banach space, n is an integer larger than one, and the Banach spaces X^n and X are isomorphic, then the operator algebra $\mathcal{L}(X)$ allows for non-trivial zero sums of idempotents (hence it is spectrally irregular). In fact we can say a bit more (cf. the second paragraph after Corollary 5.5).

Theorem 6.3. *Let X be a non-trivial Banach space, let n be an integer larger than one, and assume X is isomorphic to X^n. Then, given an integer m larger than or equal to five, there exist m projections $P_1, \ldots, P_m$ in X, all of infinite rank and infinite co-rank, and such that $P_1 + \cdots + P_m = 0$.*

Proof. For this we return to the proof of Theorem 4.1 and the remark following it. The Banach algebra $\mathcal{B}$ featuring there is now taken to be $\mathcal{L}(X)$ and, whenever convenient, complex matrices will be identified with operator matrices having entries in $\mathcal{L}(X)$, in this case via "tensorizing" with I_X.

With the 2×2 matrices M_1 and M_2 and the 3×3 matrices N_1, N_2 and N_3 as is Step 2 of the proof of Theorem 4.1, we have a non-trivial zero sum of five projections in X, namely

$$\sum_{k=1}^{2} V_M \left(M_k^{\oplus(n-1)} \right) W_M + \sum_{k=1}^{3} V_N \left(N_k^{\oplus(n-1)} \right) W_N = 0,$$

where the bounded linear operators V_M, W_M, V_N and W_N act as follows

$$V_M : X^{2(n-1)} \to X, \qquad W_M : X \to X^{2(n-1)},$$

$$V_N : X^{3(n-1)} \to X, \qquad W_N : X \to X^{3(n-1)},$$

and the products $W_M V_M$ and $W_N V_N$ yield the identity operators on $X^{2(n-1)}$ and $X^{3(n-1)}$, respectively. Recall that, as the scalar matrices, M_1, M_2, N_1, N_2 and N_3 are rank one idempotents. Also note that the operators V_M and V_N are injective, and that W_M and W_N are surjective. It now suffices to establish the following auxiliary result (valid because X is infinite dimensional): if R is a rank one $k \times k$ idempotent matrix, $V : X^k \to X$ is an injective linear operator, and $W : X \to X^k$ is a surjective linear operator, then both the dimension of the range space of the operator $V R W : X \to X$ and that of its null space are infinite.

To see this, we reason as follows. Modulo similarity it may be assumed that R has one in the $(1,1)$th position and zeros everywhere else. Viewed as an operator matrix, R then has I_X in the $(1,1)$th position and the zero operator on X everywhere else. From this (using the infinite dimensionality of X) it follows that the range of R, written $\operatorname{Im} R$, and the null space of R, denoted by $\operatorname{Ker} R$, both have infinite dimension. Of course R is viewed here as an operator on X^k. As W is surjective, $\operatorname{Im} V R W = V[\operatorname{Im} R]$, and the injectivity of V gives that $\operatorname{Im} V R W$ and $\operatorname{Im} R$ have the same dimension. Again using the injectivity of V, we get $\operatorname{Ker} V R W = W^{-1}[\operatorname{Ker} R]$. The surjectivity of W now implies that $W[\operatorname{Ker} V R W] = \operatorname{Ker} R$ and we see that the dimension of $\operatorname{Ker} R$ does not exceed that of $\operatorname{Ker} V R W$. Thus both the dimension of $\operatorname{Im} V R W$ and that of $\operatorname{Ker} V R W$ are infinite.

This covers the case $m = 5$. For $m > 5$, use a similar argument and the construction described in Step 7 of the proof of Theorem 4.1. $\qquad\square$

The following corollary will be used at several places in the next section (for instance in the proof of Theorems 7.2 and 7.4).

Corollary 6.4. *Let X and Y be non-trivial Banach spaces, let Y be isomorphic to a complemented subspace of X, and suppose Y is isomorphic to Y^n where n is an integer larger than one. Then, given an integer m larger than or equal to five, there exist m projections $P_1, \ldots, P_m$ in X, all of infinite rank and infinite co-rank, and such that $P_1 + \cdots + P_m = 0$.*

In particular $\mathcal{L}(X)$ is spectrally irregular.

Proof. We may assume that Y is a complemented subspace of X so that the situation is as in the proof of Proposition 5.6. Let $\Phi : \mathcal{L}(Y) \to \mathcal{L}(X)$ be the injective continuous Banach algebra homomorphism constructed there. Clearly maps projections in Y of infinite rank into projections of infinite rank in X, and the analogue of this for co-ranks is valid too. By Theorem 6.3 there exist m projections $Q_1, \ldots, Q_m$ in Y, all of infinite rank and infinite co-rank, and such that $Q_1 + \cdots + Q_m = 0$. For $j = 1, \ldots, m$, put $P_j = \Phi(Q_j)$. Then the projections $P_1, \ldots, P_m$ have the desired properties. $\qquad \square$

7. Applications to specific Banach spaces

We will now use the material presented in the previous section to identify certain Banach spaces X for which the operator algebra $\mathcal{L}(X)$ allows for non-trivial zero sums, hence is spectrally irregular. In most cases this will be done by showing that $\mathcal{L}(X)$ is of Cuntz type so that Theorem 4.1 applies; occasionally we will need to refer to Corollary 6.4.

For Σ a non-empty set, $1 \leq p \leq \infty$, and Z a Banach space, let $\ell_p(\Sigma; Z)$ denote the Banach space of all ℓ_p-functions from Σ into Z, i.e., the functions $f : \Sigma \to Z$ for which the following quantities are finite: in case $p = \infty$,

$$\|f\|_\infty = \sup_{\sigma \in \Sigma} \|f(\sigma)\|_Z,$$

in case $1 \leq p < \infty$,

$$\|f\|_p = \left(\sup_{F \text{ finite subset of } \Sigma} \sum_{\sigma \in F} \|f(\sigma)\|_Z^p \right)^{\frac{1}{p}}.$$

Here $\| \cdot \|_Z$ stands for the (given) norm on Z. With the usual algebraic operations and the norm given by the expressions above, $\ell_p(\Sigma; Z)$ is a Banach space.

Theorem 7.1. *Let Σ be an infinite set, let $1 \leq p \leq \infty$, and let Z be a Banach space. Then the Banach space $\ell_p(\Sigma; Z)$ is isomorphic to its square $\ell_p(\Sigma; Z)^2$. Also, when Z is non-trivial, the operator algebra $\mathcal{L}(\ell_p(\Sigma; Z))$ has the Cuntz 2-property.*

Proof. By Theorem 6.2 it is sufficient to prove the first part of the theorem. Write Σ as the disjoint union of Σ_1 and Σ_2 of two sets both sets having the same cardinality as Σ. This is possible by virtue of the basic set theoretical result saying that the sum of two infinite cardinalities is that same cardinality again. Let $\phi_1 : \Sigma \to \Sigma_1$ and $\phi_2 : \Sigma \to \Sigma_2$ be bijections. Now define $W : \ell_p(\Sigma; Z) \to \left(\ell_p(\Sigma; Z) \right)^2$ by stipulating that $Wf = (W_1 f, W_2 f)^\top$ where $W_j : \ell_p(\Sigma; Z) \to \ell_p(\Sigma; Z)$ is given by

$$W_j f = f|_{\Sigma_j} \circ \phi_j : \Sigma \to Z, \qquad j = 1, 2.$$

Also introduce $V : \left(\ell_p(\Sigma; Z) \right)^2 \to \ell_p(\Sigma; Z))$ by $V(f, g)^\top = V_1 f + V_2 g$ with

$$V_1 f|_{\Sigma_1} = f \circ \phi_1^{-1}, \quad V_1 f|_{\Sigma_2} = 0, \quad V_2 g|_{\Sigma_2} = g \circ \phi_2^{-1}, \quad V_2 g|_{\Sigma_1} = 0.$$

Then V and W are each others inverse. $\qquad \square$

As a particular case of Theorem 7.1 we have that, for $1 \leq p \leq \infty$, the operator algebras $\mathcal{L}(\ell_p)$ are of Cuntz type. Hence they allow for non-trivial zero sums of idempotents and are spectrally irregular. The case $p = 2$ was already covered in [6]; see also [5].

In the same vein, by pasting together two copies of the real line $\mathbb{R}$, it becomes clear that $L_p(\mathbb{R})$ is isomorphic to a direct sum of two copies of itself. Thus $\mathcal{L}(L_p(\mathbb{R}))$ has the Cuntz 2-property. Possible generalizations of this result involve measure spaces (Σ, μ) such that two or more copies of (Σ, μ) can be combined into a measure space which is equivalent to one copy of (Σ, μ), in the sense that there exists a bijective measurable function whose inverse is measurable too.

As is usual, the subspaces of ℓ_∞ consisting of those complex sequences having a limit, respectively limit zero, are denote by c, respectively c_0. Clearly c_0 is isomorphic to the direct sum c_0^2 of two copies of itself. Thus Theorem 6.2 guarantees that $\mathcal{L}(c_0)$ has the Cuntz 2-property. But then $\mathcal{L}(c_0)$ allows for non-trivial zero sums, hence is spectrally irregular. The same conclusion holds for $\mathcal{L}(c)$. This can be derived from Corollary 6.4 upon noting that c_0 is a complemented subspace of c (having codimension one).

The above observations concerning c_0 and $\mathcal{L}(c_0)$ can be brought into a more general context. Let $c_0(\Sigma; Z)$ be the (closed) subspace of $\ell_\infty(\Sigma; Z)$ consisting of the functions f from Σ into Z having the following property: for each $\varepsilon > 0$ there exist a finite subset $F \subset \Sigma$ (depending on ε) such that $\|f(t)\|_Z < \varepsilon$ for all $t \in \Sigma \setminus F$. Here, as before, $\| \cdot \|_Z$ denotes the norm on the Banach space Z.

Theorem 7.2. *Let Σ be an infinite set and let Z be a Banach space. Then the Banach space $c_0(\Sigma; Z)$ isomorphic to its square $c_0(\Sigma; Z)^2$. Also, when Z is non-trivial, the operator algebra $\mathcal{L}(c_0(\Sigma; Z))$ has the Cuntz 2-property.*

The proof of Theorem 7.2 follows the same line of thought as the argument given for Theorem 7.1.

We can also obtain an analogue of c by stipulating that $c(\Sigma; Z)$ is the (closed) subspace of $\ell_\infty(\Sigma; Z)$ consisting of the functions $f : \Sigma \to Z$ having the following property: there exists $z \in Z$ (depending on f) such that for each $\varepsilon > 0$ there exists a finite subset $F \subset \Sigma$ (depending on ε) with $\|f(\sigma) - z\|_Z < \varepsilon$ for all $\sigma \in \Sigma \setminus F$. Clearly $c_0(\Sigma; Z)$ is a closed subspace of $c(\Sigma; Z)$. Another closed subspace of $c(\Sigma; Z)$ is the set $\widehat{Z}$ of all constant functions on Σ (which can be viewed as a copy of Z). As $c(\Sigma; Z) = c_0(\Sigma; Z) \dotplus \widehat{Z}$ we have that $c_0(\Sigma; Z)$ is complemented in $c(\Sigma; Z)$. Thus by Corollary 6.4, for each integer m larger than or equal to five, the operator algebra $\mathcal{L}(c(\Sigma; Z))$ allows for zero sums involving m non-zero idempotents, hence is spectrally irregular.

To further demonstrate the applicability of Theorem 4.1 and Corollary 6.4, we present a couple of results involving functions that are continuous on the real line $\mathbb{R}$ with the possible exception of the points in a certain infinite subset of $\mathbb{R}$ where jumps occur.

Let Γ be an infinite subset of $\mathbb{R}$ having no accumulation point in $\mathbb{R}$. By $C_\Gamma(\mathbb{R}; Z)$ we denote the subspace of $\ell_\infty(\mathbb{R}; Z)$ consisting of all functions $f : \mathbb{R} \to Z$

that are continuous on $\mathbb{R}$ except possibly in the points of Γ where jumps may occur. The latter means that for each $a \in \Gamma$ both $\lim_{t \uparrow a} f(t)$ and $\lim_{t \downarrow a} f(t)$ exist, one or both of them possibly different from the value of f at a. It is not hard to see that $C_\Gamma(\mathbb{R}; Z)$ is closed in $\ell_\infty(\mathbb{R}; Z)$.

A few preliminary remarks are in order.

As Γ has no accumulation point in $\mathbb{R}$, every compact subset of $\mathbb{R}$ contains only a finite number of points of Γ. Suppose Γ is bounded below. Then we can find a monotonically increasing sequence $a_1, a_2, a_3, \ldots$ of real numbers such that $\lim_{k \to \infty} a_k = \infty$ and $\Gamma = \{a_1, a_2, a_3, \ldots\}$. It is now easy to see that $C_\Gamma(\mathbb{R}; Z)$ is isomorphic to $C_\mathbb{N}(\mathbb{R}; Z)$, where $\mathbb{N}$ stands for the set of non-negative integers.

Next assume that Γ is bounded above. Then there exists a monotonically decreasing sequence $b_1, b_2, b_3, \ldots$ of real numbers with $\lim_{k \to \infty} b_k = -\infty$ and $\Gamma = \{\ldots, b_3, b_2, b_1\}$, and again it follows that $C_\Gamma(\mathbb{R}; Z)$ is isomorphic to $C_\mathbb{N}(\mathbb{R}; Z)$.

Finally, in case Γ is neither bounded below nor bounded above, there are a monotonically increasing sequence $a_1, a_2, a_3, \ldots$ and a monotonically decreasing sequence $b_1, b_2, b_3, \ldots$ of real numbers for which $b_1 < a_1$,

$$\lim_{k \to \infty} a_k = \infty, \qquad \lim_{k \to \infty} b_k = -\infty,$$

and $\Gamma = \{\ldots, b_3, b_2, b_1, a_1, a_2, a_3, \ldots\}$. The conclusion is now that $C_\Gamma(\mathbb{R}; Z)$ is isomorphic to $C_\mathbb{Z}(\mathbb{R}; Z)$, where $\mathbb{Z}$ denotes the set of all integers.

Theorem 7.3. *Let Γ be an infinite subset of the real line $\mathbb{R}$ having no accumulation point in $\mathbb{R}$. Suppose Γ is neither bounded below nor above. Then the Banach space $C_\Gamma(\mathbb{R}; Z)$ is isomorphic to its square $C_\Gamma(\mathbb{R}; Z)^2$. Also, when the Banach space Z is non-trivial, the operator algebra $\mathcal{L}(C_\Gamma(\mathbb{R}; Z))$ has the Cuntz 2-property.*

Proof. It is enough to show an isomorphy between $C_\Gamma(\mathbb{R}; Z)$ and $C_\Gamma(\mathbb{R}; Z)^2$. As $C_\Gamma(\mathbb{R}; Z)$ is isomorphic to $C_\mathbb{Z}(\mathbb{R}; Z)$, it suffices to consider the $\Gamma = \mathbb{Z}$. Write $\mathbb{R}$ as the disjoint union of Σ_1 and Σ_2:

$$\Sigma_1 = \bigcup_{k \in \mathbb{Z}} (2k, 2k+1], \qquad \Sigma_2 = \bigcup_{k \in \mathbb{Z}} (2k+1, 2k+2].$$

Also define ϕ_1 and ϕ_2 on $\mathbb{R}$ by

$$\phi_1(t) = k + t, \quad \phi_2(t) = k + 1 + t, \qquad t \in (k, k+1], \ k \in \mathbb{Z}.$$

Then $\phi_1 : \mathbb{R} \to \Sigma_1$ and $\phi_2 : \mathbb{R} \to \Sigma_2$ are bijective with inverses given by

$$\phi_1^{-1}(s) = -k + s, \quad s \in (2k, 2k+1], \qquad k = 0, 1, 2, \ldots,$$

$$\phi_2^{-1}(s) = -k - 1 + s, \quad s \in (2k+1, 2k+2], \quad k = 0, 1, 2, \ldots.$$

Take f in $C_\mathbb{Z}(\mathbb{R}; Z)$. Then, along with f, the function $f|_{\Sigma_1} \circ \phi_1 : \mathbb{R} \to Z$ is continuous on $\mathbb{R}$ except maybe in the points of $\mathbb{Z}$ where jumps occur. Thus it belongs to $C_\mathbb{Z}(\mathbb{R}; Z)$, and the same is true for $f|_{\Sigma_2} \circ \phi_2 : \mathbb{R} \to Z$. Now introduce $W : C_\mathbb{Z}(\mathbb{R}; Z) \to (C_\mathbb{Z}(\mathbb{R}; Z))^2$ by stipulating that $Wf = (W_1 f, W_2 f)^\top$ with

$W_j f = f|_{\Sigma_j} \circ \phi_j$. Then W is bijective. For its inverse V from $\left(C_{\mathbb{Z}}(\mathbb{R}; Z)\right)^2$ into $C_{\mathbb{Z}}(\mathbb{R}; Z)$ we have $V(f,g)^{\top} = V_1 f + V_2 g$ with

$$V_1 f|_{\Sigma_1} = f \circ \phi_1^{-1}, \quad V_1 f|_{\Sigma_2} = 0, \quad V_2 g|_{\Sigma_2} = g \circ \phi_2^{-1}, \quad V_2 g|_{\Sigma_1} = 0,$$

and with this the argument is complete. □

Theorem 7.4. *Let Γ be a an infinite subset of the real line $\mathbb{R}$ having no accumulation point in $\mathbb{R}$. Suppose Γ is bounded below or above. Also assume Z to be non-trivial. Then, given an integer m larger than or equal to five, there exist m projections $P_1, \ldots, P_m$ in $C_{\Gamma}(\mathbb{R}; Z)$, all of finite rank and co-rank, and such that $P_1 + \cdots + P_m = 0$.*

In particular $\mathcal{L}\left(C_{\Gamma}(\mathbb{R}; Z)\right)$ is spectrally irregular.

Proof. As $C_{\Gamma}(\mathbb{R}; Z)$ is isomorphic to $C_{\mathbb{N}}(\mathbb{R}; Z)$, it suffices to consider the case $\Gamma = \mathbb{N}$. Write C_- for the (closed) subspace of $C_{\mathbb{N}}(\mathbb{R}; Z)$ consisting of all functions $f \in C_{\mathbb{N}}(\mathbb{R}; Z)$ for which f vanishes on $(0, \infty)$. Further let C_+ be the (closed) subspace of $C_{\mathbb{N}}(\mathbb{R}; Z)$ having as elements the functions $f \in C_{\mathbb{N}}(\mathbb{R}; Z)$ such that f vanishes on $(-\infty, 0]$. Then $C_{\mathbb{N}}(\mathbb{R}; Z) = C_- \dotplus C_+$. Also C_+ is isomorphic to C_+^2, as can be seen via an argument analogous to the proof of Theorem 7.3. Now apply Corollary 6.4. □

In the case where Γ is bounded below, we have the possibility to introduce the subalgebra $C_{\Gamma,-}(\mathbb{R}; Z)$ of $C_{\Gamma}(\mathbb{R}; Z)$ consisting of all $f \in C_{\Gamma}(\mathbb{R}; Z)$ for which $\lim_{t \to -\infty} f(t)$ exists in Z. This subalgebra is again closed, so we have a Banach subalgebra of $C_{\Gamma}(\mathbb{R}; Z)$ here. We do not know whether for Γ bounded below the Banach space $C_{\Gamma}(\mathbb{R}; Z)$ is isomorphic to $C_{\mathbb{Z}}(\mathbb{R}; Z)$, the space $C_{\Gamma,-}(\mathbb{R}; Z)$ is however.

Lemma 7.5. *Let Γ be a an infinite subset of the real line $\mathbb{R}$ having no accumulation point in $\mathbb{R}$. Suppose Γ is bounded below. Then $C_{\Gamma,-}(\mathbb{R}; Z)$ is isomorphic to $C_{\mathbb{Z}}(\mathbb{R}; Z)$.*

Proof. As in the proof of Theorem 7.4 it suffices to consider the case $\Gamma = \mathbb{N}$. Define the function ψ on $\mathbb{R}$ by

$$\psi(t) = \begin{cases} \log t, & t \in (0, 1], \\ 3k + 1 + t, & t \in (-k-1, -k], \quad k = 0, 1, 2, \ldots, \\ k + t, & t \in (k+1, k+2], \quad k = 0, 1, 2, \ldots. \end{cases}$$

Then $\psi : \mathbb{R} \to \mathbb{R}$ is bijective with inverse $\psi^{-1} : \mathbb{R} \to \mathbb{R}$ given by

$$\psi^{-1}(s) = \begin{cases} e^s, & s \in (-\infty, 0], \\ -3k - 1 + s, & s \in (2k, 2k+1], \quad k = 0, 1, 2, \ldots, \\ -k + s, & s \in (2k+1, 2k+2], \quad k = 0, 1, 2, \ldots. \end{cases}$$

Take f in $C_{\mathbb{N},-}(\mathbb{R}; Z)$. Then the function $f \circ \psi : \mathbb{R} \to Z$ belongs to $C_{\mathbb{Z}}(\mathbb{R}; Z)$. Introduce $W : C_{\mathbb{N},-}(\mathbb{R}; Z) \to C_{\mathbb{Z}}(\mathbb{R}; Z)$ by stipulating that $Wf = f \circ \phi$. Then W is bijective. For its inverse $V : C_{\mathbb{Z}}(\mathbb{R}; Z) \to C_{\mathbb{N},-}(\mathbb{R}; Z)$ we have $Vg = g \circ \psi^{-1}$.

Take f in $C_{\mathbb{Z}}(\mathbb{R}; Z)$. Then the function $f \circ \phi : \mathbb{R} \to Z$ belongs to $C_{\mathbb{N},-}(\mathbb{R}; Z)$. Introduce $W : C_{\mathbb{Z}}(\mathbb{R}; Z) \to C_{\mathbb{N},-}(\mathbb{R}; Z)$ by stipulating that $Wf = f \circ \phi$. Then W is bijective. For its inverse $V : C_{\mathbb{N},-}(\mathbb{R}; Z) \to C_{\mathbb{Z}}(\mathbb{R}; Z)$ we have $Vg = g \circ \phi^{-1}$.   □

The next result is now immediate from combining Lemma 7.5 and Theorem 7.3.

Theorem 7.6. *Let Γ be an infinite subset of the real line $\mathbb{R}$ having no accumulation point in $\mathbb{R}$. Suppose Γ is bounded below. Then the Banach space $C_{\Gamma,-}(\mathbb{R}; Z)$ is isomorphic to its square $C_{\Gamma,-}(\mathbb{R}; Z)^2$. Also, when the Banach space Z is non-trivial, the operator algebra $\mathcal{L}\big(C_{\Gamma,-}(\mathbb{R}; Z)\big)$ has the Cuntz 2-property.*

Under the assumption that Γ is bounded above, Lemma 7.5 and Theorem 7.6 hold with $C_{\Gamma,-}(\mathbb{R}; Z)$ replaced by the Banach subalgebra $C_{\Gamma,+}(\mathbb{R}; Z)$ of $C_{\Gamma}(\mathbb{R}; Z)$ consisting of all $f \in C_{\Gamma}(\mathbb{R}; Z)$ for which $\lim\limits_{t \to +\infty} f(t)$ exists in Z.

Theorems 7.3, 7.4 and 7.6 were concerned with functions on $\mathbb{R}$ being continuous with the possible exception of jump discontinuities only. Now we turn to Banach spaces consisting of continuous functions. For S a topological space and Z a (complex) Banach space, the expression $\mathcal{C}_{\infty}(S; Z)$ will denote the Banach space of all bounded continuous functions from S to Z, endowed with the usual algebraic operations (defined pointwise) and the sup-norm. Clearly $\mathcal{C}_{\infty}(S; Z)$ is a subspace of $\ell_{\infty}(S; Z)$. In case S is compact, $\mathcal{C}_{\infty}(S; Z)$ coincides with the Banach space $\mathcal{C}(S; Z)$ of all continuous functions from S to Z, again endowed with the usual algebraic operations (defined pointwise) and the max-norm.

We begin with some auxiliary observations.

Proposition 7.7. *Let S be a topological space, let Z be a Banach space, and let n be a positive integer larger than or equal to two. Suppose S is homeomorphic to the topological direct sum of n copies of itself. Then the Banach space $\mathcal{C}_{\infty}(S; Z)$ is isomorphic to $\mathcal{C}_{\infty}(S; Z)^n$. Also, when Z is non-trivial, the operator algebra $\mathcal{L}\big(\mathcal{C}_{\infty}(S; Z)\big)$ has the Cuntz n-property.*

The condition on S is met if and only if S can be written as the disjoint union of n clopen (which by definition means: open and closed) sets $S_1, \ldots, S_n$, each of which as a topological subspace of S (i.e., provided with the relative topology with respect to S) is homeomorphic to S. Note that we have a fractal type structure here. Each S_j is again the disjoint union of n clopen sets homeomorphic to S, and for these this is true again. And so on, indefinitely. We shall come back to this point after the proof. See also Examples A, B and C below.

Proof. Let $S_1, \ldots, S_n$ be clopen sets as above. For $j = 1, \ldots, n$, let ϕ_j be a homeomorphism of S onto S_j. Define $W : \mathcal{C}_{\infty}(S; Z) \to \mathcal{C}_{\infty}(S; Z)^n$ by

$$Wf = (W_1 f, \ldots, W_n f)^{\top}, \qquad W_j f = f|_{S_j} \circ \phi_j, \quad j = 1, \ldots, n.$$

Then W is linear and bounded. In fact, with the choice made for the norm on $\mathcal{C}_{\infty}(S; Z)^n$ in the paragraph directly following Theorem 6.2, the operator W is norm preserving. It is also bijective. This can be seen as follows. For $j = 1, \ldots, n$, let $V_j : \mathcal{C}_\infty(S; Z) \to \mathcal{C}_\infty(S; Z)$ be given by

$$V_j f|_{S_j} = f \circ \phi_j^{-1}, \qquad V_j f|_{S_k} = 0, \quad k = 1, \ldots, n, \ k \neq j.$$

Note that $V_j f \in \mathcal{C}_\infty(S; Z)$ because the sets $S_1, \ldots, S_n$ are clopen in S. The inverse $V : \mathcal{C}_\infty(S; Z)^n \to \mathcal{C}_\infty(S; Z)$ of the operator W is now given by $V(f_1, \ldots, f_n)^\top = V_1 f_1 + \cdots + V_n f_n$. This proves the first statement in the proposition; the second is immediate from Theorem 6.2. $\qquad\square$

Returning to the fractal type structure mentioned prior to the above proof, we recall that S_j is again the disjoint union of n clopen sets homeomorphic to S, and for these clopen sets this is true again. By induction, it follows that for any positive integer k, the space S is the disjoint union of kn clopen subsets of S, each homeomorphic to S. In particular, there exists a (countably) infinite collection of clopen subsets of S, each homeomorphic to S, possibly not mutually disjoint however.

In general it is not clear whether, under the conditions of Proposition 7.7, there is an infinite collection of mutually disjoint clopen subsets of S. A fortiori, it is not clear whether S can be written as the disjoint union of an infinite collection of clopen sets homeomorphic to S. When this is possible, we have the following result.

Proposition 7.8. *Let S be a topological space, let Z be a Banach space, and suppose S is the disjoint union of an infinite collection of clopen sets homeomorphic to S. Then the Banach space $\mathcal{C}_\infty(S; Z)$ is isomorphic to its square $\mathcal{C}_\infty(S; Z)^2$. Also, when Z is non-trivial, the operator algebra $\mathcal{L}\big(\mathcal{C}_\infty(S; Z)\big)$ has the Cuntz 2-property.*

Proof. Let Σ be an infinite index set, and assume S is the disjoint union of the family $\{S_\sigma\}_{\sigma \in \Sigma}$ of clopen subsets of S, each homeomorphic to S. In other words, S is the topological direct sum of the family $\{S_\sigma\}_{\sigma \in \Sigma}$. Using the type of argument featuring in the proof of Proposition 7.7, one finds that $\mathcal{C}_\infty(S; Z)$ is isomorphic to $\ell_\infty\big(\Sigma; \mathcal{C}_\infty(S; Z)\big)$. By Theorem 7.1, the latter space is isomorphic to its square, but then so is $\mathcal{C}_\infty(S; Z)$. This establishes the first statement in the proposition; the second now comes from Theorem 6.2. $\qquad\square$

By K we denote the familiar ternary Cantor set, also called the Cantor middle-third set. Recall that K is compact, so that $\mathcal{C}_\infty(K; Z)$ coincides with $\mathcal{C}(K; Z)$.

Corollary 7.9. *Let Z be a Banach space. Then the Banach space $\mathcal{C}(K; Z)$ is isomorphic to its square $\mathcal{C}(K; Z)^2$. Also, when Z is non-trivial, the operator algebra $\mathcal{L}\big(\mathcal{C}(K; Z)\big)$ has the Cuntz 2-property.*

Proof. Let K_- and K_+ be the intersection of K with the closed interval $[0, 1/3]$ and $[2/3, 1]$, respectively. Then K_- and K_+ are open sets in K. Also K is the disjoint union of K_- and K_+. So K is the topological direct sum of K_- and K_+. Now note that $K = 3K_-$ and $K = -2 + 3K_+$. It follows that both K_- and K_+

are homeomorphic to K. Now apply Proposition 7.7 to get the first part of the theorem, and Theorem 6.2 to get the second. □

The next result is a mild generalization of a truly remarkable result of A.A. Miljutin [28].

Theorem 7.10. *Let S be an uncountable compact metrizable topological space and suppose Z is a finite-dimensional Banach space. Then the Banach spaces $\mathcal{C}(S;Z)$ and $\mathcal{C}(K;Z)$ are isomorphic.*

Proof. Consider $\mathcal{C}_r(S;\mathbb{R})$, the Banach space of real-valued continuous functions on S, which is the real analogue of the complex Banach space $\mathcal{C}(S;\mathbb{C})$. By a celebrated result of A.A. Miljutin [28], the real Banach spaces $\mathcal{C}_r(S;\mathbb{R})$ and $\mathcal{C}_r(K;\mathbb{R})$ are isomorphic (see also [29], the remark below Theorem 6.2.5). A simple complexification argument now shows that $\mathcal{C}(S;\mathbb{C})$ and $\mathcal{C}(K;\mathbb{C})$ are isomorphic too. To make the step from complex-valued functions to those having values in the (complex) finite-dimensional Banach space Z, we argue as follows. Write n for the dimension of Z. As each n-dimensional complex Banach space is isomorphic to $\mathbb{C}^n$, we may assume that Z actually coincides with $\mathbb{C}^n$. For $j = 1, \ldots, n$, let $R_j : \mathbb{C}^n \to \mathbb{C}$ be the jth coordinate function. Take f in $\mathcal{C}(S;\mathbb{C}^n)$ and put $f_j = R_j \circ f$. Then $f_j \in \mathcal{C}(S;\mathbb{C})$. As we have already seen, there exists a bijective bounded linear operator from $\mathcal{C}(S;\mathbb{C})$ onto $\mathcal{C}(K;\mathbb{C})$. Let us denote it by B. For $k \in K$, now write $Tf(k) = \bigl(Bf_1(k), \ldots, Bf_n(k)\bigr)^\top$. Then $T(f)$ belongs to $\mathcal{C}(K;\mathbb{C}^n)$ and we have a mapping $T : \mathcal{C}(S;\mathbb{C}^n) \to \mathcal{C}(K;\mathbb{C}^n)$. This mapping is easily seen to be a bijective bounded linear operator. □

Theorem 7.11. *Let S be an uncountable compact metrizable topological space and suppose Z is a non-trivial finite-dimensional Banach space. Then the operator algebra $\mathcal{L}\bigl(\mathcal{C}(S;Z)\bigr)$ has the Cuntz 2-property.*

Proof. The Banach spaces $\mathcal{C}(S;Z)$ and $\mathcal{C}(K;Z)$ are isomorphic. Hence the same is true for $\mathcal{L}\bigl(\mathcal{C}(S;Z)\bigr)$ and $\mathcal{L}\bigl(\mathcal{C}(K;Z)\bigr)$. The desired result is now immediate from Corollary 7.9. □

In the next result, certain notions from general topology play a role. Here are the pertinent definitions taken from [29]. A non-empty topological space S is called *topologically complete* if there exists a complete metric on S which generates the topology of S. The space is said to be *nowhere locally compact* if no point of S has a neighborhood with compact closure. Finally, S is called *zero-dimensional* if S has a base consisting of clopen sets.

Theorem 7.12. *Let S be a non-empty topological space which is topologically complete, nowhere locally compact and zero-dimensional. Further let Z be a Banach space. Then the Banach space $\mathcal{C}_\infty(S;Z)$ is isomorphic to its square $\mathcal{C}_\infty(S;Z)^2$. Also, in case Z is non-trivial, the operator algebra $\mathcal{L}\bigl(\mathcal{C}(S;Z)\bigr)$ has the Cuntz 2-property.*

Proof. By a result of P. Alexandroff and P. Urysohn [3], cited as Theorem 1.9.8 in [29], the space S is homeomorphic to the space $\mathbb{P}$ of all irrational numbers. Let $\mathbb{P}_1$ be the set of negative irrational numbers, and let $\mathbb{P}_2$ be the set of the positive ones. Then $\mathbb{P}_1$ and $\mathbb{P}_2$ are clopen sets in $\mathbb{P}$ and $\mathbb{P}$ is the disjoint union of $\mathbb{P}_1$ and $\mathbb{P}_2$. Also both $\mathbb{P}_1$ and $\mathbb{P}_2$ are homeomorphic to $\mathbb{P}$. This is clear from the fact that these spaces satisfy the conditions mentioned in the theorem, but it can also be seen in a more direct way by constructing concrete homeomorphisms ϕ_1 from $\mathbb{P}$ onto $\mathbb{P}_1$ and ϕ_2 from $\mathbb{P}$ onto $\mathbb{P}_2$. For the latter on can take for instance $\phi_2 : \mathbb{P} \to \mathbb{P}_2$ given by

$$\phi_2(t) = \begin{cases} \dfrac{1}{1-t}, & t \in \mathbb{P},\ t < 0, \\ t+1, & t \in \mathbb{P},\ t > 0. \end{cases}$$

Now apply Proposition 7.7. $\qquad\square$

Theorem 7.13. *Let S be a non-empty topological space with a countable number of points and no isolated points. Further let Z be a Banach space. Then the Banach spaces $\mathcal{C}_\infty(S; Z)$ and $\mathcal{C}_\infty(S; Z)^2$ are isomorphic. Also, when Z is non-trivial, the operator algebra $\mathcal{L}\big(\mathcal{C}(S; Z)\big)$ has the Cuntz 2-property.*

Proof. By a result of W. Sierpiński [33] (see Theorem 1.9.6 in [29]), the space S is homeomorphic to $\mathbb{Q}$, the space of rational numbers. The argument for dealing with $\mathbb{Q}$ is similar to that given in the proof of the preceding result for $\mathbb{P}$; use, for instance, $\sqrt{2}$ as a "division point" instead of 0. $\qquad\square$

In each of the above specializations to concrete spaces, we have isomorphy of a Banach space with its square (hence with all its powers). In order to prove this, we needed some rather non-trivial results from general topology. In some cases one can avoid this by settling for something less, for instance isomorphy with the cube instead of the square. An example is given below. More on Banach spaces isomorphic to their cubes can be found in the next section.

Example A. Let S be the subset of the open interval $(-1, 1)$ consisting of the rational numbers

$$\sum_{k=1}^{\infty} 3\varepsilon_k \left(\frac{1}{4}\right)^k, \tag{11}$$

where the ε_k are allowed to take the numerical values $-1, 0, 1$ (and no others), while only a finite number among $\varepsilon_1, \varepsilon_2, \varepsilon_3, \dots$ may differ from zero, so that (11) is actually a finite sum. To get an idea of what is going on, let us look at a few cases.

The first is where all ε_k vanish. Then (11) only gives the number 0. Next consider the situation where $\varepsilon_k = 0$ for all $k \geq 2$. This leads to the three numbers $-\frac{3}{4}$, 0 and $\frac{3}{4}$. When $\varepsilon_k = 0$ for $k \geq 3$, the sum (11) reduces to $\frac{3}{4}\varepsilon_1 + \frac{3}{16}\varepsilon_2$ with $\varepsilon_1, \varepsilon_2 \in \{-1, 0, 1\}$, and so we arrive at the nine numbers

$$-\frac{15}{16}, \ -\frac{3}{4}, \ -\frac{9}{16}, \qquad -\frac{3}{16}, \ 0, \ \frac{3}{16}, \qquad \frac{9}{16}, \ \frac{3}{4}, \ \frac{15}{16}.$$

These include the three outcomes we already had in the previous stage. In case ε_k vanishes for all $k \geq 4$, the sum (11) becomes $\frac{3}{4}\varepsilon_1 + \frac{3}{16}\varepsilon_2 + \frac{3}{64}\varepsilon_3$ with the restrictions stipulated above on ε_1, ε_2 and ε_3. Thus, besides the nine numbers indicated above, we get eighteen additional ones, making a total of twentyseven:

$$
\begin{array}{ccccccccc}
-\dfrac{63}{64}, & -\dfrac{15}{16}, & \dfrac{57}{64}, & -\dfrac{51}{64}, & -\dfrac{3}{4}, & -\dfrac{45}{64}, & -\dfrac{39}{64}, & -\dfrac{9}{16}, & -\dfrac{33}{64}, \\[2mm]
-\dfrac{15}{64}, & -\dfrac{3}{16}, & -\dfrac{9}{64}, & -\dfrac{3}{64}, & 0, & \dfrac{3}{64}, & \dfrac{9}{64}, & \dfrac{3}{16}, & \dfrac{15}{64}, \\[2mm]
\dfrac{33}{64}, & \dfrac{9}{16}, & \dfrac{39}{64}, & \dfrac{45}{64}, & \dfrac{3}{4}, & \dfrac{51}{64}, & \dfrac{57}{64}, & \dfrac{15}{16}, & \dfrac{63}{64}.
\end{array}
$$

And so on.

The number of points in S is countably infinite and S, being a subspace of the real line, has a countable base. Also it is a straightforward matter to prove that S has no isolated points. For Z a non-trivial Banach space, Theorem 7.13 now gives that the operator algebra $\mathcal{L}(\mathcal{C}(S;Z))$ has the Cuntz 2-property. The proof given above employs Sierpinsky's (rather non-trivial) characterization of $\mathbb{Q}$ as the unique non-empty countable space without isolated points. Being content with the Cuntz 3-property, we can avoid the use of heavy machinery from general topology. Here is the argument.

Split S in three parts S_-, S_0 and S_+:

$$
S_- = S \cap \left(-1, -\frac{1}{2}\right), \qquad S_0 = S \cap \left(-\frac{1}{4}, \frac{1}{4}\right), \qquad S_+ = S \cap \left(\frac{1}{2}, 1\right).
$$

These parts correspond to $\varepsilon_1 = -1$, $\varepsilon_1 = 0$ and $\varepsilon_1 = 1$, respectively. Clearly S_-, S_0 and S_0 are open subsets of S which itself is the dsjoint union of these sets. But then S_-, S_0 and S_0 are closed in S too. Now note that $S = 3{+}4S_- = 4S_0 = -3{+}4S_+$. Hence S is homeomorphic to the topological direct sum of three copies of itself. Applying Proposition 7.7 we get that the operator algebra $\mathcal{L}(\mathcal{C}_\infty(S;Z))$ has the Cuntz 3-property. $\qquad\square$

The material presented above, is concerned with non-trivial Banach spaces that are isomorphic to their squares and, consequently, allow for non-trivial zero sums of projections. For such a Banach space X there exist "Cuntz operators" V_1, V_2, W_1 and W_2 in $\mathcal{L}(X)$ satisfying the identities

$$
V_1 W_1 + V_2 W_2 = I_X, \qquad W_j V_k = \delta_{j,k} I_X, \quad j,k = 1,2.
$$

For several of the above concrete instances of Banach spaces isomorphic to their square or a higher power we gave (or would be able to give) explicit descriptions of these Cuntz operators (cf. Theorems 7.1 and 7.6; see also Proposition 7.7 and Example A). In combination with the material presented in Section 4, such descriptions, when available, can be used to obtain explicit expressions for the projections forming a non-trivial zero sum. The expressions in question are complicated and not very illuminating; we refrain from giving further details here.

8. Banach spaces isomorphic to their cubes

The applications we gave in the previous section were concerned with Banach spaces having the Cuntz 2-property. However, this does not cover all possible cases. Indeed, Theorem 10 in [21] provides an example of a Banach space which is isomorphic to its cube while it is not isomorphic to its square. More generally it is shown in [23] that for every integer $k \geq 2$, there is a Banach space E such that E^n is isomorphic to E^m if and only if $m = n \,(\mathrm{mod}\ k)$. In particular the space E is then not isomorphic to E^k, but it is isomorphic to E^{k+1}, and the latter implies that $\mathcal{L}(E)$ is of Cuntz type, hence spectrally irregular because of the occurrence of non-trivial zero sums of idempotents. The conclusion is that one needs the full force of Theorem 6.2.

The examples of the type meant above are complicated. In this section we embark on a somewhat less ambitious endeavor: *to construct Banach spaces, evidently isomorphic to their cubes, but for which it is not clear whether or not they are isomorphic to their square.* For a given integer k larger than 2, the construction can be modified such as to result in Banach spaces F with F^n isomorphic to F^m if $m = n \,(\mathrm{mod}\ k)$ while it is unclear whether or not F^n is isomorphic to F^m in case $m \neq n \,(\mathrm{mod}\ k)$. We refrain from giving the details concerning this refinement.

We now begin with the construction which, as one will realize, is inspired by that of the familiar Cantor set. The starting point is a non-empty topological space U, not necessarily compact or metrizable. Suppose U_-, U_0 and U_+ are mutually disjoint subspaces of U, all three homeomorphic to U, hence non-empty. Let $\phi_j : U \to U_j$ be a homeomorphism from U onto U_j. Here $j \in \{-, 0, +\}$. For $n = 1, 2, 3, \ldots$ and $j_1, j_2, \ldots, j_n \in \{-, 0, +\}$, write

$$U_{j_1, j_2, \ldots, j_n} = \phi_{j_1} \phi_{j_2} \ldots \phi_{j_n}[U]. \tag{12}$$

Note that, as far as the expressions U_-, U_0 and U_+ are concerned, no confusion is possible. Indeed, the sets $U_- = \phi_-[U]$, $U_0 = \phi_0[U]$ and $U_+ = \phi_+[U]$ coming from (12), coincide with the originally given U_-, U_0 and U_+. Clearly, all the subspaces $U_{j_1, j_2, \ldots, j_n}$ are non-empty,

$$U_{j_1, j_2, \ldots, j_n} \subset U_{j_1, j_2, \ldots, j_{n-1}} \subset \cdots \subset U_{j_1, j_2} \subset U_{j_1} \subset \bigcup_{j \in \{-, 0, +\}} U_j,$$

and $U_{j_1, j_2, \ldots, j_n} = \phi_{j_1}[U_{j_2, \ldots, j_n}]$. We also have

$$U_{j_1, j_2, \ldots, j_n} \cap U_j = \begin{cases} U_{j, j_2, \ldots, j_n}, & j = j_1, \\ \emptyset, & j \neq j_1. \end{cases}$$

Note further that

$$U_{j_1, j_2, \ldots, j_n} \cap U_{i_1, i_2, \ldots, i_n} \neq \emptyset \iff j_k = i_k, \ k = 1, \ldots, n,$$

and so $U_{j_1, j_2, \ldots, j_n}$ and $U_{i_1, i_2, \ldots, i_n}$ coincide if and only if they are not disjoint.

For $n = 1, 2, 3 \ldots$, introduce

$$S_n = \bigcup_{j_1, j_2, \ldots, j_n \in \{-, 0, +\}} U_{j_1, j_2, \ldots, j_n}.$$

Then S_n is non-empty and $S_1 = U_- \cup U_0 \cup U_+$. Regardless of whether j_{n+1} is $-, 0$ or $+$, the inclusion $U_{j_1, j_2, \ldots, j_n, j_{n+1}} \subset U_{j_1, j_2, \ldots, j_n}$ holds. Hence

$$S_{n+1} \subset \bigcup_{j_1, j_2, \ldots, j_n \in \{-, 0, +\}} U_{j_1, j_2, \ldots, j_n} = S_n,$$

so $S_1 \supset S_2 \supset S_3 \supset \ldots$. We also note that $S_{n+1} = \phi_-[S_n] \cup \phi_0[S_n] \cup \phi_+[S_n]$, and this true for $n = 0$ too when we interpret S_0 as U. The identity

$$\phi_j[S_n] = S_{n+1} \cap U_j, \quad j \in \{-, 0, +\}, \ n = 1, 2, 3, \ldots,$$

needed later, is now immediate.

Let $S = \bigcap_{n=1}^{\infty} S_n$ be the intersection of the descending chain of sets S_1, S_2, S_3, ... In order to directly relate S to the sets $U_{j_1, j_2, \ldots, j_n}$ we do the following. Write $\mathcal{J}$ for the collection of (infinite) sequences with entries from $\{-, 0, +\}$. With an element $\{j_1, j_2, j_3 \ldots\}$ from $\mathcal{J}$, we associate the intersection of the descending sequence $U_{j_1} \supset U_{j_1, j_2} \supset U_{j_1, j_2, j_3} \supset \ldots$, i.e., the set

$$U_{j_1, j_2, j_3, \ldots} = \bigcap_{n=1}^{\infty} U_{j_1, j_2, \ldots, j_n}.$$

As $U_{j_1} \subset S_1$, $U_{j_1, j_2} \subset S_2$, $U_{j_1, j_2, j_3} \subset S_n$ and so on, we have that $U_{j_1, j_2, j_3, \ldots} \subset S$. So the union of all the sets $U_{j_1, j_2, j_3, \ldots}$ is contained in S. In fact there is equality:

$$S = \bigcup_{\{j_1, j_2, j_3, \ldots\} \in \mathcal{J}} U_{j_1, j_2, j_3, \ldots} = \bigcup_{\{j_1, j_2, j_3, \ldots\} \in \mathcal{J}} \bigcap_{n=1}^{\infty} U_{j_1, j_2, \ldots, j_n}.$$

For completeness we mention that, given $s \in S$, there is precisely one sequence $\{j_1, j_2, j_3, \ldots\} \in \mathcal{J}$ such that $s \in U_{j_1, j_2, j_3, \ldots}$. As we shall see below in an example, it may happen that the set $U_{j_1, j_2, j_3, \ldots}$ contains more than one point. It can be empty too.

Next introduce $S_- = S \cap U_-$, $S_0 = S \cap U_0$ and $S_+ = S \cap U_+$. Then S, being a subset of the disjoint union $U_- \cup U_0 \cup U_+$, is the disjoint union of S_-, S_0 and S_+. Also, for $j \in \{-, 0, +\}$ we have (using the injectivity of ϕ_j in the second equality below)

$$\phi_j[S] = \phi_j \left[\bigcap_{n=1}^{\infty} S_n \right] = \bigcap_{n=1}^{\infty} \phi_j[S_n] = \bigcap_{n=1}^{\infty} [S_{n+1} \cap U_j]$$

$$= \left[\bigcap_{n=1}^{\infty} S_{n+1} \right] \cap U_j = S \cap U_j = S_j.$$

Hence the restriction ψ_j of ϕ_j to S, viewed as a mapping $\psi_j : S \to S_j$, is a homeomorphism from S onto S_j. Consequently, for $i, j \in \{-, 0, +\}$, the mapping $\psi_j \psi_i^{-1} : S_i \to S_j$ is a homeomorphism from S_i onto S_j. For completeness we mention that $\psi_i^{-1} : S_i \to S$ is the restriction of ϕ_i^{-1} to S_i, considered as a mapping onto S.

The statements in the previous paragraph are only of interest when the set S is non-empty. A relevant special case in which this necessarily holds is when the underlying topological space U is compact and the sets U_-, U_0 and U_- are closed in U. In that case we can even deduce the non-emptiness of all sets $U_{j_1, j_2, j_3, \ldots}$ with the sequence $\{j_1, j_2, j_3, \ldots\}$ taken from $\mathcal{J}$. Closedness of U_-, U_0 and U_- is guaranteed when U, in addition to being compact, is also Hausdorff.

In order to conclude that S is non-empty, sometimes fixed point theorems can be employed too. Here is such a case, which applies, for instance, to non-empty closed subset U of real or complex Banach spaces. Suppose U is a complete metric space (possibly non-compact) and let ϱ denote the metric on U. Further assume that the homeomorphisms $\phi_- : U \to U_-$, $\phi_0 : U \to U_0$ and $\phi_+ : U \to U+$ are contractions and there exists a constant $c \in (0, 1)$ such that

$$\varrho\big(\phi_j(x), \phi_j(y)\big) \le c\varrho(x, y), \qquad x, y \in U.$$

Take $j \in \{-, 0, +\}$. Then by the Banach fixed point theorem, ϕ_j has a (unique) fixed point. Now let $u_{-,\infty}$, $u_{0,\infty}$ and $u_{+,\infty}$ be the fixed points of ϕ_-, ϕ_0 and ϕ_+, respectively. Clearly these belong to $U_- = \phi_-[U]$, $U_0 = \phi_0[U]$ and $U_+ = \phi_+[U]$, respectively. Note now that $u_{-,\infty} \in U_{-,-,-,\ldots} \subset S$, $u_{0,\infty} \in U_{0,0,0,\ldots} \subset S$ and $u_{+,\infty} \in U_{+,+,+,\ldots} \subset S$. In particular S is non-empty.

In what follows we shall assume that S is non-empty. Along with S, the homeomorphic images S_-, S_0 and S_+ of S under, respectively, the homeomorphisms $\psi_- = \phi_-|_S : S \to S_-$, $\psi_0 = \phi_0|_S : S \to S_0$ and $\psi_+ = \phi_+|_S : S \to S_+$, are then non-empty as well. Recall now that S is the disjoint union of S_-, S_0 and S_+. Thus, if the latter three sets happen to be clopen in S, we have that S is homeomorphic to the topological direct sum of three copies of itself. But then, given a Banach space Z, we can conclude from Proposition 7.7 that the Banach space $\mathcal{C}_\infty(S; Z)$ is isomorphic to its cube $\mathcal{C}_\infty(S; Z)^3$, hence the operator algebra $\mathcal{L}\big(\mathcal{C}_\infty(S; Z)\big)$ has the Cuntz 3-property and is (therefore) spectrally irregular. It is not clear whether or not $\mathcal{C}_\infty(S; Z)$ is isomorphic to its square $\mathcal{C}_\infty(S; Z)^2$. This might among other things, depend on the choice of the Banach space Z.

The requirement, featuring in the above paragraph, that S_-, S_0 and S_+ are clopen in S, is met when the three sets U_-, U_0 and U_+ are open in the underlying space U. Here is the argument. Clearly under this assumption $S_- = S \cap U_-$, $S_0 = S \cap U_0$ and $S_+ = S \cap U_+$ are open in S. But then $S_- = S \setminus [S_0 \cup S_+]$ is closed in S. Similarly S_0 and S_+ are closed in S too. The same type of reasoning shows that S_-, S_0 and S_+ are clopen in S whenever U_-, U_0 and U_+ are closed in U.

By way of illustration, we now present an example in which the underlying topological space U is not compact.

Example B. Let U be the open interval $(0, 1)$ and take for U_-, U_0 and U_+ the open intervals

$$U_- = \left(0, \frac{1}{3}\right), \qquad U_0 = \left(\frac{1}{3}, \frac{2}{3}\right), \qquad U_+ = \left(\frac{2}{3}, 1\right).$$

Further, define $\phi_- : U \to U_-$, $\phi_0 : U \to U_0$ and $\phi_+ : U \to U_+$ by

$$\phi_-(t) = \frac{1}{3}t, \qquad \phi_+(t) = \frac{2}{3} + \frac{1}{3}t, \qquad 0 < t < 1,$$

$$\phi_0(t) = \begin{cases} \dfrac{1}{3} + \dfrac{1}{5}t, & 0 < t < \dfrac{5}{12}, \\[2mm] t, & \dfrac{5}{12} \le t \le \dfrac{7}{12}, \\[2mm] \dfrac{7}{15} + \dfrac{1}{5}t, & \dfrac{7}{12} < t < 1. \end{cases}$$

Obviously these mappings are homeomorphisms. Observe that ϕ_0 acts as the identity mapping on the closed interval $\left[\frac{5}{12}, \frac{7}{12}\right]$. Hence this interval is contained in (and actually equal) to the set $U_{0,0,0,\dots}$ (notation as above). But then the closed interval $\left[\frac{5}{12}, \frac{7}{12}\right]$ is a subset of every set S_n and so $\left[\frac{5}{12}, \frac{7}{12}\right] \subset S$. In particular S is non-empty, in fact even uncountable. As the sets U_-, U_0 and U_+ are open in U, we may conclude that, given a Banach space Z, the Banach space $\mathcal{C}_\infty(S; Z)$ is isomorphic to its cube $\mathcal{C}_\infty(S; Z)^3$, hence the operator algebra $\mathcal{L}\big(\mathcal{C}_\infty(S; Z)\big)$ has the Cuntz 3-property and is (therefore) spectrally irregular. It is not clear whether or not $\mathcal{C}_\infty(S; Z)$ is isomorphic to its square $\mathcal{C}_\infty(S; Z)^2$. This is an open question, even for the case $Z = \mathbb{C}$, so for the Banach space $\mathcal{C}_\infty(S; \mathbb{C})$. Note here that Theorem 7.13 does not apply because S is uncountable. Further Theorem 7.12 cannot be used for the space S, containing the closed interval $\left[\frac{5}{12}, \frac{7}{12}\right]$, is not nowhere locally compact. For that matter, it is not zero-dimensional neither. Finally, Theorem 7.11 cannot be employed to show that $\mathcal{L}\big(\mathcal{C}_\infty(S; \mathbb{C})\big)$ has the Cunz 2-property. The reason is that S is not compact. Indeed, S is not a closed subset of the real line; zero is an accumulation point of S which does not belong to S.

As was said before, the construction presented above is inspired by that of the familiar Cantor set, earlier denoted by K. It is illuminating to observe that, in fact, a slight modification of K is a subset of S. The set in question, here denoted by $K_\circ$, is obtained as follows. Start with the open interval $(0, 1)$ and leave out the closed middle third interval $\left[\frac{1}{3}, \frac{2}{3}\right]$. What remains is the union of the two open intervals $\left(0, \frac{1}{3}\right)$ and $\left(\frac{2}{3}, 1\right)$. Next in each of them cut out the closed intervals $\left[\frac{1}{9}, \frac{2}{9}\right]$ and $\left[\frac{7}{9}, \frac{8}{9}\right]$, which leaves us with the four open intervals $\left(0, \frac{1}{9}\right)$, $\left(\frac{2}{9}, \frac{1}{3}\right)$, $\left(\frac{2}{3}, \frac{7}{9}\right)$ and $\left(\frac{8}{9}, 1\right)$. And so on (formal definition by induction), resulting in

$$K_\circ = (0, 1) \setminus \bigcup_{m=1}^{\infty} \bigcup_{k=1}^{3^{m-1}} \left[\frac{3k-2}{3^m}, \frac{3k-1}{3^m}\right].$$

Clearly $K_\circ$ is contained in the usual Cantor middle-third set K which admits the representation

$$K = [0, 1] \setminus \bigcup_{m=1}^{\infty} \bigcup_{k=1}^{3^{m-1}} \left(\frac{3k-2}{3^m}, \frac{3k-1}{3^m}\right).$$

Along with K, the set $K_\circ$ is uncountable. Indeed,

$$K \setminus K_\circ \subset \{0,1\} \cup \bigcup_{m=1}^{\infty} \bigcup_{k=1}^{3^{m-1}} \left\{ \frac{3k-2}{3^m}, \frac{3k-1}{3^m} \right\},$$

and the set in the right-hand side of this inclusion is countable. (Actually, $K_\circ$ coincides with the complement in K of the countable subset of K consisting of the end points of the intervals left out.) One sees that $K_\circ \subset S$ by ignoring the presence of U_0 and ϕ_0. More precisely, by looking at the sets $U_{j_1, j_2, \ldots, j_n}$ with $j_1, \ldots, j_n$ taken from $\{-,+\}$, so avoiding the use of the index 0. In sharp contrast to K and $K_\circ$, the Cantor type set S contains (countably many) closed intervals.

The topological space $K_\circ$ is neither countable nor compact. So neither Theorem 7.13 nor Theorem 7.11 applies. However, as $K_\circ$ is homeomorphic to the topological direct sum of two copies of itself, we can conclude from Proposition 7.7 that for every Banach space Z, the Banach space $\mathcal{C}_\infty(K_\circ; Z)$ is isomorphic to its square $\mathcal{C}_\infty(K_\circ; Z)^2$. Hence, when Z is non-trivial, the operator algebra $\mathcal{L}(\mathcal{C}_\infty(K_\circ; Z))$ has the Cuntz 2-property and is (therefore) spectrally irregular. Theorem 7.12 does apply, and its proof indicates that $K_\circ$ is homeomorphic to $\mathbb{P}$, the space of irrational real numbers. □

We give another example, this one along lines suggested by Van Mill.

Example C. Start with the half closed, half open square $U = [0,1] \times (0,1)$. Then for U_-, U_0 and U_+, take

$$U_- = \left[0, \frac{1}{5}\right] \times (0,1), \quad U_0 = \left[\frac{2}{5}, \frac{3}{5}\right] \times (0,1), \quad U_+ = \left[\frac{4}{5}, 1\right] \times \left(\frac{2}{5}, \frac{3}{5}\right).$$

Further, for $0 \leq x \leq 1$, $0 < y < 1$, put

$$\phi_-(x,y) = \left(\frac{1}{5}\sqrt{x}, y\right), \quad \phi_0(x,y) = \left(\frac{2}{5} + \frac{1}{5}x, y\right), \quad \phi_+(x,y) = \left(\frac{4}{5} + \frac{1}{5}x, \frac{2}{5} + \frac{1}{5}y\right).$$

Then $\phi_- : U \to U_-$, $\phi_0 : U \to U_0$ and $\phi_+ : U \to U_+$ are homeomorphisms.

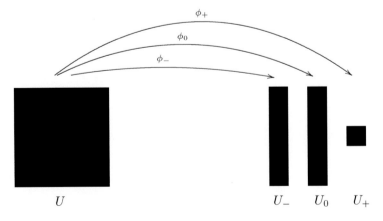

By repeated application of these homomorphisms, the sets U_-, U_0, U_+ are "squeezed" into, respectively,

the half closed, half open rectangle $\left[0, \dfrac{1}{25}\right] \times (0,1)$, $(= U_{-,-,-,\ldots})$,

the open line segment $\left\{\dfrac{1}{2}\right\} \times (0,1)$, $(= U_{0,0,0,\ldots})$,

the singleton set $\left\{\left(1, \dfrac{1}{2}\right)\right\}$, $(= U_{+,+,+,\ldots})$.

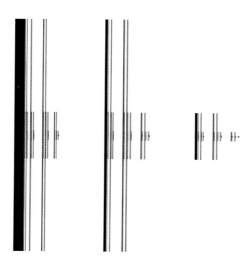

The topological space S resulting from the construction described above is uncountable but not compact, so Theorems 7.13 and 7.11 do not apply. Note that Theorem 7.12 does not apply either. As U_-, U_0 and U_+ are closed in U, the space S is homeomorphic to the topological direct sum of three of its copies. So, if Z is a Banach space, then $\mathcal{C}_\infty(S; Z)$ is isomorphic to $\mathcal{C}_\infty(S; Z)^3$. It is not clear whether or not $\mathcal{C}_\infty(S; Z)$ is isomorphic to $\mathcal{C}_\infty(S; Z)^2$. So on this basis, and assuming Z to be non-trivial, we can conclude that $\mathcal{L}(\mathcal{C}_\infty(S; Z))$ has the Cuntz 3-property, but not (yet) that it has the Cuntz 2-property. □

We close with one more remark. Both Examples B and C feature a topological space S such that (evidently!) S is homeomorphic to the topological direct sum of three copies of S while it is not clear whether or not S is homeomorphic to the topological direct sum of two copies of S. There do exist topological spaces that are homeomorphic to the direct sum of three copies of itself but not to that of two. One such example is given in [26]. However, that example is not of primary interest to us because the conditions of Theorem 7.10 are satisfied. This is different for a space constructed by W. Hanf for whose description (modulo Stone duality, i.e., in the language of Boolean algebras) we refer to [27], Section 6.2 (see also [25], [32], [35] and [36]). That space, here for the moment denoted by T, although uncountable, compact and Hausdorff, is not metrizable; hence Theorem 7.10 does not apply. The question arises whether or not $\mathcal{C}_\infty(T; \mathbb{C})$ or, more generally, $\mathcal{C}_\infty(T; Z)$ with Z a Banach space, is isomorphic to its square. Evidently it is isomorphic to its cube. Recall here (cf. the first paragraph of this section) that a very sophisticated example of a Banach space homeomorphic to its cube but not to its square has been given by W.T. Gowers [21].

Acknowledgement

The authors gratefully acknowledge stimulating contacts with Jan van Mill from the Free University in Amsterdam about the subject matter of Sections 7 and 8 of the present paper.

References

[1] S.A. Argyros, D. Freeman, R.G. Haydon, E. Odell, Th. Raikoftsalis, Th. Schlump-recht, D. Zisimopoulou, Embedding uniformly convex spaces into spaces with very few operators. *J. Funct. Anal.* **262** (2012), 825–849.

[2] S.A. Argyros, R.G. Haydon, A hereditarily indecomposable $\mathcal{L}_\infty$-space that solves the scalar-plus-compact problem, *Acta Math.* **206** (2011), 1–54.

[3] P. Alexandroff and P. Urysohn, Über nulldimensionale Punktmengen, *Math. Ann.* **98** (1928), 89–106.

[4] B.A. Barnes, G.J. Murphy, M.R.F. Smyth, T.T. West, *Riesz and Fredholm Theory in Banach Algebras*, Research Notes in Mathematics, **Vol. 67**, Pitman (Advanced Publishing Program), Boston, London, Melbourne 1982.

[5] H. Bart, T. Ehrhardt, B. Silbermann, Zero sums of idempotents in Banach algebras, *Integral Equations and Operator Theory* **19** (1994), 125–134.

[6] H. Bart, T. Ehrhardt, B. Silbermann, Logarithmic residues in Banach algebras, *Integral Equations and Operator Theory* **19** (1994), 135–152.

[7] H. Bart, T. Ehrhardt, B. Silbermann, Sums of idempotents and logarithmic residues in matrix algebras, In: *Operator Theory: Advances and Applications*, Vol. 122, Birkhäuser, Basel 2001, 139–168.

[8] H. Bart, T. Ehrhardt, B. Silbermann, Logarithmic residues of Fredholm operator valued functions and sums of finite rank projections, In: *Operator Theory: Advances and Applications*, Vol. 130, Birkhäuser, Basel 2001, 83–106.

[9] H. Bart, T. Ehrhardt, B. Silbermann, Logarithmic residues of analytic Banach algebra valued functions possessing a simply meromorphic inverse, *Linear Algebra Appl.* **341** (2002), 327–344.

[10] H. Bart, T. Ehrhardt, B. Silbermann, Sums of idempotents in the Banach algebra generated by the compact operators and the identity, In: *Operator Theory: Advances and Applications*, Vol. 135, Birkhäuser, Basel 2002, 39–60.

[11] H. Bart, T. Ehrhardt, B. Silbermann, Logarithmic residues in the Banach algebra generated by the compact operators and the identity, *Mathematische Nachrichten* **268** (2004), 3–30.

[12] H. Bart, T. Ehrhardt, B. Silbermann, Trace conditions for regular spectral behavior of vector-valued analytic functions, *Linear Algebra Appl.* **430** (2009), 1945–1965.

[13] H. Bart, T. Ehrhardt, B. Silbermann, Spectral regularity of Banach algebras and non-commutative Gelfand theory, In: H. Dym et al. (eds.): *Operator Theory: Advances and Applications. The Israel Gohberg Memorial Volume*, Vol. 218, Birkhäuser, Basel 2012, 123–153.

[14] H. Bart, T. Ehrhardt, B. Silbermann, Families of homomorphisms in non-commutative Gelfand theory: comparisons and counterexamples, In: W. Arendt et al. (eds.), *Spectral Theory, Mathematical System Theory, Evolution Equations, Differential and Difference Equations, Operator Theory: Advances and Applications*, **OT 221**, Birkhäuser, Springer Basel AG, 2012, 131–160.

[15] H. Bart, T. Ehrhardt, B. Silbermann, Logarithmic residues, Rouché's Theorem, spectral regularity, and zero sums of idempotents: the C^*-algebra case, *Indag. Math.* **23** (2012), 816–847.

[16] D.S. Bernstein, *Matrix Mathematics*, Second Edition, Princeton University Press, Princeton and Oxford, 2009.

[17] J. Cuntz, Simple C^*-algebras generated by isometries, *Commun. Math. Physics*, **57** (1977), 173–185.

[18] K.R. Davidson, C^*-algebras by Example, Fields Institute Monographs, 6. American Mathematical Society, Providence, Rhode Island, 1996.

[19] T. Ehrhardt, V. Rabanovich, Yu. Samoĭlenko, B. Silbermann, On the decomposition of the identity into a sum of idempotents, *Methods Funct. Anal. Topology* **7** (2001), 1–6.

[20] I. Gohberg, S. Goldberg, M.A. Kaashoek, *Classes of Linear Operators, Vol. I*, Operator Theory: Advances and Applications, Vol. 49, Birkhäuser, Basel 1990.

[21] W.T. Gowers, A solution to the Schroeder-Bernstein problem for Banach spaces, *Bull. London Math. Soc.* **28**, No. 3 (1996), 297–304.

[22] W.T. Gowers, B. Maurey, The unconditional basic sequence problem, *Journal A.M.S.* **6** (1993), 851–874.

[23] W.T. Gowers, B. Maurey, Banach spaces with small spaces of operators, *Math. Ann.* **307** (1997), 543–568.

[24] R. Hagen, S. Roch, B. Silbermann, C^*-algebras and Numerical Analysis, Marcel Dekker, New York, 2001.

[25] W. Hanf, On some fundamental problems concerning isomorphism of Boolean algebras, *Math. Scand.* **5** (1957),205–217.

[26] J. Ketonen, The structure of countable Boolean algebras, *Ann.of Math.* (2) **108**, No. 1 (1978), 41–89.

[27] S. Koppelberg, *Handbook of Boolean algebras,* Vol. 1 (J.D. Monk and R. Bonnet, eds.), North-Holland Publishing Co., Amsterdam, 1989.

[28] A.A. Miljutin, Isomorphisms of the spaces of continuous functions over compact sets of the cardinality of the continuum, Teor. Funkciĭ Funkcional Anal. i Priložen. Vyp. **2** (1966), 150–156 (Russian).

[29] J. van Mill, *The infinite-dimensional topology of function spaces*, North-Holland Mathematical Library, 64, North-Holland Publishing Co., Amsterdam, 2001.

[30] C. Pearcy, D. Topping, Sums of small numbers of idempotents, *Michigan Math. J.* **14** (1967), 453–465.

[31] S. Roch, P.A. Santos, B. Silbermann, *Non-commutative Gelfand Theories*, Springer Verlag, London Dordrecht, Heidelberg, New York 2011.

[32] B.M. Scott, On an example of Sundaresan, The Proceedings of the 1980 Topology Conference (Univ. Alabama, Birmingham, Ala., 1980), *Topology Proc.* **5** (1980), 185–186.

[33] W. Sierpiński, Sur une propriété topologique des ensembles dénombrables denses en soi, *Fund. Math.* **1** (1920), 11–16.

[34] A.E. Taylor, D.C. Lay, *Introduction to Functional Analysis*, Second Edition, John Wiley and Sons, New York 1980.

[35] V. Trnková, Isomorphisms of sums of countable Boolean algebras, *Proc. Amer. Math. Soc.*, **80**, No. 3 (1980), 389–392.

[36] V. Trnková, V. Koubek, Isomorphisms of sums of Boolean algebras, *Proc. Amer. Math. Soc.*, **66**, No. 2 (1977), 231–236.

H. Bart
Econometric Institute
Erasmus University Rotterdam
P.O. Box 1738
NL-3000 DR Rotterdam, The Netherlands
e-mail: bart@ese.eur.nl

T. Ehrhardt
Mathematics Department
University of California
Santa Cruz, CA-95064, USA
e-mail: tehrhard@ucsc.edu

B. Silbermann
Fakultät für Mathematik
Technische Universität Chemnitz
D-09107 Chemnitz, Germany
e-mail: silbermn.toeplitz@googlemail.com

Operator Theory:
Advances and Applications, Vol. 237, 79–106
© 2013 Springer Basel

Fast Inversion of Polynomial-Vandermonde Matrices for Polynomial Systems Related to Order One Quasiseparable Matrices

T. Bella, Y. Eidelman, I. Gohberg$^{\text{Z"L}}$, V. Olshevsky and E. Tyrtyshnikov

Dedicated to Leonia Lerer on the occasion of his seventieth birthday

Abstract. While Gaussian elimination is well known to require $\mathcal{O}(n^3)$ operations to invert an arbitrary matrix, Vandermonde matrices may be inverted using $\mathcal{O}(n^2)$ operations by a method of Traub [24]. While this original version of the Traub algorithm was noticed to be unstable, it was shown in [12] that with a minor modification, the Traub algorithm can typically yield a very high accuracy. This approach has been extended from classical Vandermonde matrices to polynomial-Vandermonde matrices involving real orthogonal polynomials [3], [10], and Szegő polynomials [19]. In this paper we present an algorithm for inversion of a class of polynomial-Vandermonde matrices with special structure related to order one quasiseparable matrices, generalizing monomials, real orthogonal polynomials, and Szegő polynomials. We derive a fast $\mathcal{O}(n^2)$ inversion algorithm applicable in this general setting, and explore its reduction in the previous special cases. Some very preliminary numerical experiments are presented, demonstrating that, as observed by our colleagues in previous work, good forward accuracy is possible in some circumstances, which is consistent with previous work of this type.

Mathematics Subject Classification (2010). 15A09; 65F05.

Keywords. Inversion of vandermonde matrices; polynomial vandermonde matrices; quasiseparable matrices.

1. Introduction

Let $R = \{r_0(x), r_1(x), \ldots, r_{n-1}(x)\}$ be a sequence of polynomials satisfying $\deg(r_k) = k$, and $x_1, \ldots, x_n$ a set of pairwise distinct values. Then the corresponding *polynomial-Vandermonde* matrix is given by

$$
V_R(x) = \begin{bmatrix}
r_0(x_1) & r_1(x_1) & \cdots & r_{n-1}(x_1) \\
r_0(x_2) & r_1(x_2) & \cdots & r_{n-1}(x_2) \\
\vdots & \vdots & & \vdots \\
r_0(x_n) & r_1(x_n) & \cdots & r_{n-1}(x_n)
\end{bmatrix}.
\tag{1.1}
$$

In this paper the problem of inversion of the matrix $V_R(x)$ for a given system of polynomials R satisfying some special recurrence relations is considered. While the structure-ignoring approach of Gaussian elimination for inversion of $V_R(x)$ requires $\mathcal{O}(n^3)$ operations, the special structure allows algorithms to be derived exploiting that structure, resulting in fast algorithms that can compute the n^2 entries of the inverse in only $\mathcal{O}(n^2)$ operations.

In the simplest case where $R = \{1, x, x^2, \ldots, x^{n-1}\}$, $V_R(x)$ reduces to a classical Vandermonde matrix and the inversion algorithm is due to Traub [24]. In addition to the order of magnitude decrease in complexity, it was observed in [12] that a minor modification of the original Traub algorithm results in very good accuracy. The derivation of this fast and accurate algorithm attracted attention in the community, and several results were published giving fast algorithms for inversion of $V_R(x)$ for various special cases of the polynomial system R. This previous work is listed in Table 1.

Matrix $V_R(x)$	Polynomials R	Fast inversion algorithm
Classical Vandermonde	monomials	Traub [24]
Chebyshev–Vandermonde	Chebyshev	Gohberg–Olshevsky [10]
Three-Term Vandermonde	Real orthogonal	Calvetti–Reichel [3]
Szegő–Vandermonde	Szegő	Olshevsky [19]

TABLE 1. Fast $\mathcal{O}(n^2)$ inversion algorithms for polynomial-Vandermonde matrices.

In this paper, we consider a more general class of polynomials that contains all of those listed in Table 1 as special cases. This more general class of polynomials is related to a class of rank structured matrices called *quasiseparable matrices*, and hence we refer to them as quasiseparable polynomials. As quasiseparable polynomials generalize monomials, real orthogonal polynomials, and Szegő polynomials, the resulting inversion algorithm for polynomial-Vandermonde matrices $V_R(x)$ whose defining polynomials R are quasiseparable polynomials generalizes the previous work in Table 1 in addition to providing new results.

In addition to generalizing these results, it is also applicable to some interesting new classes of polynomials for which no fast inversion algorithm of this type is currently available. The algorithm that is derived in this paper relies on the use of perturbed recurrence relations for associated polynomials for the computational speedup, and thus the classes of polynomials for which it may be used are best referred to in terms of the recurrence relations that they satisfy. One such class of polynomials are those satisfying the three recurrence relations

$$r_k(x) = (\alpha_k x - \delta_k) \cdot r_{k-1}(x) - (\beta_k x + \gamma_k) \cdot r_{k-2}(x). \tag{1.2}$$

As collected in Table 2, special cases of these recurrence relations are satisfied by monomials, real orthogonal polynomials (including Chebychev polynomials), and Szegő polynomials. It is of interest that although three-term recurrence relations for Szegő polynomials (shown in Table 2) do exist in most cases [9], far more often two-term recurrence relations are used for computations with Szegő polynomials. Two-term recurrence relations for the Szegő polynomials $\left\{\phi_k^\#\right\}$ in terms of the reflection coefficients ρ_k and complimentary parameters μ_k are

$$\begin{bmatrix} \phi_0(x) \\ \phi_0^\#(x) \end{bmatrix} = \frac{1}{\mu_0} \begin{bmatrix} 1 \\ 1 \end{bmatrix}, \quad \begin{bmatrix} \phi_k(x) \\ \phi_k^\#(x) \end{bmatrix} = \frac{1}{\mu_k} \begin{bmatrix} 1 & -\rho_k^* \\ -\rho_k & 1 \end{bmatrix} \begin{bmatrix} \phi_{k-1}(x) \\ x\phi_{k-1}^\#(x) \end{bmatrix},$$
$$\tag{1.3}$$

which involve an auxiliary sequence of polynomials $\{\phi_k\}$. In this paper, generalizations of these two-term recurrence relations of the form

$$\begin{bmatrix} G_k(x) \\ r_k(x) \end{bmatrix} = \begin{bmatrix} \alpha_k & \beta_k \\ \gamma_k & 1 \end{bmatrix} \begin{bmatrix} G_{k-1}(x) \\ (\delta_k x + \theta_k)r_{k-1}(x) \end{bmatrix}, \tag{1.4}$$

which we will refer to as *Szegő-type recurrence relations*, are also considered. Finally, motivated by the most generally applicable recurrence relations available for the class of quasiseparable polynomials that we will consider [7], the [EGO05]-*type recurrence relations*

$$\begin{bmatrix} G_k(x) \\ r_k(x) \end{bmatrix} = \begin{bmatrix} \alpha_k & \beta_k \\ \gamma_k & \delta_k x + \theta_k \end{bmatrix} \begin{bmatrix} G_{k-1}(x) \\ r_{k-1}(x) \end{bmatrix}, \tag{1.5}$$

are considered as well. Details about these classes and their corresponding recurrence relations will be given later.

Polynomial System R	Recurrence relations
monomials	$r_k(x) = x \cdot r_{k-1}(x)$
Chebyshev polynomials	$r_k(x) = 2x \cdot r_{k-1}(x) - r_{k-2}(x)$
Real orthogonal polynomials	$r_k(x) = (\alpha_k x - \delta_k)r_{k-1}(x) - \gamma_k \cdot r_{k-2}(x)$
Szegő polynomials	$r_k(x) = \left(\frac{1}{\mu_k}x + \frac{\rho_k}{\rho_{k-1}}\frac{1}{\mu_k}\right)r_{k-1}(x)$ $\qquad - \left(\frac{\rho_k}{\rho_{k-1}}\frac{\mu_{k-1}}{\mu_k} \cdot x\right)r_{k-2}(x)$

TABLE 2. Systems of polynomials and corresponding recurrence relations.

1.1. Structure of the paper

In Section 2 an inversion formula valid for a general system of polynomials (although expensive in general) is presented. The formula presented there reduces the problem of inversion of $V_R(x)$ to that of evaluating the so-called associated polynomials $\widehat{R}$ corresponding to the polynomial system R. A relation between the polynomial systems R and $\widehat{R}$ is presented in terms of their confederate matrices. This relation suggests a procedure for evaluating the associated polynomials $\widehat{R}$. In Section 3 quasiseparable matrices and polynomials are defined and shown to generalize the confederate matrices of the motivating special cases. Conversions are given between the polynomial language (i.e., polynomials satisfying recurrence relations) and the matrix language (i.e., generators of a quasiseparable matrix), and the motivating recurrence relations are identified in terms of the generators of their quasiseparable confederate matrices. In Section 4, perturbed recurrence relations are presented for the associated polynomials $\widehat{R}$. Three different sets of recurrence relations are given, two generalizing known formulas for real orthogonal polynomials and Szegő polynomials, and a third that produces new formulas for these cases. We briefly describe in Section 5 a fast algorithm for computing the coefficients of the master polynomial, which are required for computing the perturbations of the recurrence relations. In Section 6 the reduction of the described algorithms in the special cases of monomials, real orthogonal polynomials, and Szegő polynomials are examined in detail as well. Section 7 consists of some results of preliminary numerical experiments with the proposed algorithm, and conclusions are offered in the final section.

2. Confederate matrices and associated polynomials

In this section we present the formula that will be used to invert a polynomial-Vandermonde matrix. Such a matrix is completely determined by n polynomials $R = \{r_0(x), \ldots, r_{n-1}(x)\}$ and n nodes $x = (x_1, \ldots, x_n)$. The desired inverse $V_R(x)^{-1}$ is given by the formula

$$V_R(x)^{-1} = \tilde{I} \cdot V_{\widehat{R}}^T(x) \cdot \mathrm{diag}(c_1, \ldots, c_n), \tag{2.1}$$

with

$$c_i = \prod_{\substack{k=1 \\ k \neq i}}^{n} (x_k - x_i)^{-1}$$

(see [18], [19]) where $\tilde{I}$ is the antidiagonal matrix (with ones on the antidiagonal and zeros elsewhere), and $\widehat{R}$ is the system of *associated* (*generalized Horner*) *polynomials*, defined as follows: if we define the *master polynomial* $P(x)$ by $P(x) = (x - x_1) \cdots (x - x_n)$, then for the polynomial sequence $R = \{r_0(x), \ldots, r_{n-1}(x), P(x)\}$, the associated polynomials $\widehat{R} = \{\widehat{r}_0(x), \ldots, \widehat{r}_{n-1}(x), P(x)\}$ are those satisfying the relations

$$\frac{P(x) - P(y)}{x - y} = \sum_{k=0}^{n-1} r_k(x) \cdot \widehat{r}_{n-k-1}(y), \qquad (2.2)$$

see [16]. It can be shown that for any polynomials R, a corresponding sequence of polynomials $\widehat{R}$ satisfying (2.2) exist, and can be understood as a generalization of the Horner polynomials associated with the monomials; see, for instance, [2].

This discussion gives a relation between the inverse $V_R(x)^{-1}$ and the polynomial-Vandermonde matrix $V_{\widehat{R}}(x)$, where $\widehat{R}$ is the system of polynomials associated with R. The next definition from [17] provides a connection between recurrence relations for R with those for $\widehat{R}$.

Definition 2.1. Let the sequence of polynomials $R = \{r_0(x), r_1(x), \ldots, r_n(x)\}$ with $\deg(r_k) = k$ satisfy the n-term recurrence relations

$$r_k(x) = (\alpha_k x - a_{k-1,k}) \cdot r_{k-1}(x) - a_{k-2,k} \cdot r_{k-2}(x) - \cdots - a_{0,k} \cdot r_0(x) \qquad (2.3)$$

for $k = 1, \ldots, n$, and let

$$P(x) = P_0 \cdot r_0(x) + P_1 \cdot r_1(x) + \cdots + P_{n-1} \cdot r_{n-1}(x) + P_n \cdot r_n(x) \qquad (2.4)$$

for $P_n \neq 0$. Then the *confederate matrix* of $P(x)$ with respect to R is given by

$$C_R(P) = \begin{bmatrix} \frac{a_{01}}{\alpha_1} & \frac{a_{02}}{\alpha_2} & \frac{a_{03}}{\alpha_3} & \cdots & \frac{a_{0,k}}{\alpha_k} & \cdots & \cdots & \frac{a_{0,n}}{\alpha_n} - \frac{P_0}{\alpha_n P_n} \\ \frac{1}{\alpha_1} & \frac{a_{12}}{\alpha_2} & \frac{a_{13}}{\alpha_3} & \cdots & \frac{a_{1,k}}{\alpha_k} & \cdots & \cdots & \frac{a_{1,n}}{\alpha_n} - \frac{P_1}{\alpha_n P_n} \\ 0 & \frac{1}{\alpha_2} & \frac{a_{23}}{\alpha_3} & \cdots & \vdots & \cdots & \cdots & \frac{a_{2,n}}{\alpha_n} - \frac{P_2}{\alpha_n P_n} \\ 0 & 0 & \frac{1}{\alpha_3} & \ddots & \frac{a_{k-2,k}}{\alpha_k} & \ddots & & \vdots \\ \vdots & \vdots & \ddots & \ddots & \frac{a_{k-1,k}}{\alpha_k} & \ddots & \ddots & \vdots \\ \vdots & \vdots & \ddots & \ddots & \frac{1}{\alpha_k} & \ddots & \ddots & \vdots \\ \vdots & \vdots & \ddots & \ddots & \ddots & \ddots & \ddots & \vdots \\ 0 & 0 & \cdots & \cdots & \cdots & 0 & \frac{1}{\alpha_{n-1}} & \frac{a_{n-1,n}}{\alpha_n} - \frac{P_{n-1}}{\alpha_n P_n} \end{bmatrix}. \qquad (2.5)$$

We refer to [17] for many useful properties of the confederate matrix and only recall here that $\det(xI - C_R(P)) = P(x)/(\alpha_0 \cdot \alpha_1 \cdot \ldots \cdot \alpha_n)$, and that similarly, the characteristic polynomial of the $k \times k$ leading submatrix of $C_R(P)$ is equal to $r_k(x)/\alpha_0 \cdot \alpha_1 \cdot \ldots \cdot \alpha_k$.

The motivation for considering confederate matrices is that they will allow the computation of the polynomials associated with the given system of polynomials. The confederate matrices of R and $\widehat{R}$ are related by

$$C_{\widehat{R}}(P) = \tilde{I} \cdot C_R(P)^T \cdot \tilde{I}. \qquad (2.6)$$

Recurrence Relations	Confederate matrix $C_R(r_n)$
$r_k(x) = xr_{k-1}(x)$ Monomials	$$\begin{bmatrix} 0 & 0 & \cdots & \cdots & 0 \\ 1 & \ddots & \ddots & & 0 \\ 0 & \ddots & \ddots & \ddots & \vdots \\ \vdots & \ddots & \ddots & \ddots & \vdots \\ 0 & \cdots & 0 & 1 & 0 \end{bmatrix}$$ Companion matrix
$r_k(x) = (\alpha_k x - \delta_k)r_{k-1}(x)$ $- \gamma_k r_{k-2}(x)$ Real orthogonal polynomials	$$\begin{bmatrix} \frac{\delta_1}{\alpha_1} & \frac{\gamma_2}{\alpha_2} & 0 & \cdots & 0 \\ \frac{1}{\alpha_1} & \frac{\delta_2}{\alpha_2} & \ddots & \ddots & \vdots \\ 0 & \frac{1}{\alpha_2} & \ddots & \frac{\gamma_{n-1}}{\alpha_{n-1}} & 0 \\ \vdots & & \ddots & \frac{\delta_{n-1}}{\alpha_{n-1}} & \frac{\gamma_n}{\alpha_n} \\ 0 & \cdots & 0 & \frac{1}{\alpha_{n-1}} & \frac{\delta_n}{\alpha_n} \end{bmatrix}$$ Tridiagonal matrix
$r_k(x)$ $= \left[\frac{1}{\mu_k} \cdot x + \frac{\rho_k}{\rho_{k-1}}\frac{1}{\mu_k}\right]r_{k-1}(x)$ $- \frac{\rho_k}{\rho_{k-1}}\frac{\mu_{k-1}}{\mu_k} \cdot xr_{k-2}(x)$ Szegő polynomials	$$\begin{bmatrix} -\rho_0^*\rho_1 & -\rho_0^*\mu_1\rho_2 & \cdots & -\rho_0^*\mu_1\cdots\mu_{n-1}\rho_n \\ \mu_1 & -\rho_1^*\rho_2 & \cdots & -\rho_1^*\mu_2\cdots\mu_{n-1}\rho_n \\ & \mu_2 & \cdots & -\rho_2^*\mu_3\cdots\mu_{n-1}\rho_n \\ & & \ddots & \vdots \\ & & \mu_{n-1} & -\rho_{n-1}^*\rho_n \end{bmatrix}$$ Unitary Hessenberg matrix

TABLE 3. Polynomial systems and corresponding confederate matrices.

(see [18], [19]). The passage from $C_R(P)$ to $C_{\widehat{R}}(P)$ in (2.6) can be seen as a transposition across the antidiagonal, or a pertransposition.

In Table 3, the confederate matrices corresponding to the polynomials considered in previous work are given, including monomials, real orthogonal polynomials, and Szegő polynomials. These equivalences are all well known. We will show later that all of these confederate matrices in Table 3 are special cases of order one quasiseparable matrices, and use properties of these confederate matrices to derive the fast algorithm.

In accordance with (2.1), the main computational burden in inversion is to compute $V_{\widehat{R}}$, which requires evaluating n associated polynomials $\{\hat{r}_k(x)\}_{k=0}^{n-1}$ at n points $\{x_k\}_{k=1}^n$. Using (2.6) directly to accomplish this is expensive for arbitrary R, since it leads to the full n-term recurrence relations. However, in special cases where sparse recurrence relations for R may be found, this leads to a fast algorithm. For instance, in the monomial case, $R = \{1, x, x^2, \ldots, x^{n-1}\}$ satisfy the obvious recurrence relations $x^k = x \cdot x^{k-1}$ and hence the confederate matrix (2.5) becomes

$$C_R(P) = \begin{bmatrix} 0 & 0 & \cdots & 0 & -P_0 \\ 1 & 0 & \cdots & 0 & -P_1 \\ 0 & 1 & \ddots & \vdots & \vdots \\ \vdots & \ddots & \ddots & 0 & \vdots \\ 0 & \cdots & 0 & 1 & -P_{n-1} \end{bmatrix} \tag{2.7}$$

which is the well-known companion matrix. Using the pertransposition rule (2.6), we obtain the confederate matrix

$$C_{\widehat{R}}(\widehat{r}_n) = \begin{bmatrix} -P_{n-1} & -P_{n-2} & \cdots & -P_1 & -P_0 \\ 1 & 0 & \cdots & 0 & 0 \\ 0 & 1 & \ddots & \vdots & 0 \\ \vdots & \ddots & \ddots & 0 & \vdots \\ 0 & \cdots & 0 & 1 & 0 \end{bmatrix}. \tag{2.8}$$

for the associated polynomials $\widehat{r}_k(x)$. Using the formula (2.3), we read from the matrix (2.8) the familiar Horner recurrence relations

$$\widehat{r}_0(x) = 1, \quad \widehat{r}_k(x) = x \cdot \widehat{r}_{k-1}(x) + P_{n-k}. \tag{2.9}$$

Thus the use of the Horner recurrence relations provides the computational speedup in the original Traub algorithm. In the next sections we use the quasiseparability of the confederate matrices to derive corresponding recurrence relations that accomplish the same speedup in a more general setting.

3. Quasiseparable matrices and polynomials

Definition 3.1 (Quasiseparable matrices and polynomials).

- A matrix A is called (H,m)-*quasiseparable* if

  (i) it is strongly upper Hessenberg (upper Hessenberg with a nonzero first subdiagonal),

  and

  (ii) $\max(\operatorname{rank} A_{12}) = m$ where the maximum is taken over all symmetric partitions of the form

  $$A = \left[\begin{array}{c|c} * & A_{12} \\ \hline * & * \end{array} \right]; \tag{3.1}$$

for instance, the low-rank blocks of a 5×5 (H, m)-quasiseparable matrix would be those shaded below:

$$
\begin{bmatrix}
\star & \star & \star & \star & \star \\
\star & \star & \star & \star & \star \\
0 & \star & \star & \star & \star \\
0 & 0 & \star & \star & \star \\
0 & 0 & 0 & \star & \star
\end{bmatrix}
\qquad
\begin{bmatrix}
\star & \star & \star & \star & \star \\
\star & \star & \star & \star & \star \\
0 & \star & \star & \star & \star \\
0 & 0 & \star & \star & \star \\
0 & 0 & 0 & \star & \star
\end{bmatrix}
$$

$$
\begin{bmatrix}
\star & \star & \star & \star & \star \\
\star & \star & \star & \star & \star \\
0 & \star & \star & \star & \star \\
0 & 0 & \star & \star & \star \\
0 & 0 & 0 & \star & \star
\end{bmatrix}
\qquad
\begin{bmatrix}
\star & \star & \star & \star & \star \\
\star & \star & \star & \star & \star \\
0 & \star & \star & \star & \star \\
0 & 0 & \star & \star & \star \\
0 & 0 & 0 & \star & \star
\end{bmatrix}
$$

- Let $A = [a_{ij}]$ be an (H, m)-quasiseparable matrix. Then the system of polynomials $\{r_k(x)\}$ related to A via

$$
r_k(x) = \alpha_1 \cdots \alpha_k \det (xI - A)_{(k \times k)} \qquad \text{(where} \qquad \alpha_i = 1/a_{i+1,i} \text{)}
$$

is called a system of (H, m)-*quasiseparable polynomials*. That is, (H, m)-quasiseparable polynomials are those polynomials with an (H, m)-quasiseparable confederate matrix.

The low-rank property described in this definition means there is redundancy in the definition of the n^2 entries of an (H, m)-quasiseparable matrix, and these entries may be described by a smaller number $\mathcal{O}(mn)$ of parameters. If m is sufficiently small and independent of n, then this provides a significant reduction. The following well-known result may be found, for instance, in [4], and it provides this smaller set of parameters, called the *generators* of the (H, m)-quasiseparable matrix. An $n \times n$ matrix A is (H, m)-quasiseparable if and only if it may be written in the form

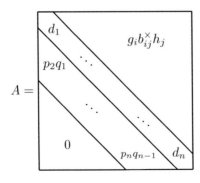

with

$$
b_{ij}^{\times} = (b_{i+1}) \cdots (b_{j-1}), \qquad b_{i,i+1} = I. \tag{3.2}
$$

Here p_k, q_k, d_k are scalars, the elements g_k are row vectors of maximal size m, h_k are column vectors of maximal size m, and b_k are matrices of maximal size $m \times m$ such that all products are defined. The elements $\{p_k, q_k, d_k, g_k, b_k, h_k\}$ are called the *generators* of the matrix A.

The elements in the upper part of the matrix $g_i b_{ij}^\times h_j$ are products of a row vector, a (possibly empty) sequence of matrices possibly of different sizes, and finally a column vector, as depicted here:

$$g_i b_{ij}^\times h_j =$$

(3.3)

with $u_k \leqslant m$ for each $k = 1, \ldots, n-1$.

The generator definition and formula (2.3) together give the following n-term recurrence relations for $(H, 1)$-quasiseparable polynomials. These n-term recurrence relations are not useful for computations, as they would not provide fast algorithms, but rather are used theoretically in the derivations.

Lemma 3.2. *Let R be a sequence of $(H, 1)$-quasiseparable polynomials specified by the generators of the corresponding $(H, 1)$-quasiseparable confederate matrix. Then R satisfies the n-term recurrence relations*

$$r_k(x) = \frac{1}{p_{k+1}q_k}\left[(x - d_k)r_{k-1}(x) - \sum_{j=0}^{k-2}\left(g_{j+1}b_{j+1,k}^\times h_k r_j(x)\right)\right], \qquad (3.4)$$

for $k = 1, \ldots, n-1$.

It is easily verified that each of the confederate matrices in Table 3 is $(H, 1)$-quasiseparable. Therefore, the class of $(H, 1)$-quasiseparable polynomials includes as special cases the important classical polynomial classes of real orthogonal polynomials and Szegő polynomials. The next theorems show that, like these motivating examples, the confederate matrices corresponding to the polynomials satisfying the recurrence relations (1.2), (1.4), and (1.5) are also $(H, 1)$-quasiseparable. Furthermore, explicit expressions for the generators of the confederate matrices are given in terms of the recurrence relations coefficients. This is useful in case the input to the algorithms is to be these recurrence relations coefficients, as the algorithms are given in terms of the quasiseparable generators. We omit the proofs, which follow from repeated use of the appropriate recurrence relations and given generators to produce the n-term recurrence relations of Lemma 3.2.

Theorem 3.3. *Let $R = \{r_0(x), \ldots, r_n(x)\}$ be a sequence of polynomials satisfying $\deg(r_k) = k$ and the recurrence relations (1.2). Then the confederate matrix of*

$r_n(x)$ *with respect to* R *is given by*

$$
C_R(r_n) =
\begin{bmatrix}
\frac{\delta_1}{\alpha_1} & \frac{\frac{\delta_1}{\alpha_1}\beta_2+\gamma_2}{\alpha_2} & \cdots & \frac{\frac{\delta_1}{\alpha_1}\beta_2+\gamma_2}{\alpha_2}\left(\frac{\beta_3}{\alpha_3}\right)\left(\frac{\beta_4}{\alpha_4}\right)\cdots\left(\frac{\beta_n}{\alpha_n}\right) \\[2mm]
\frac{1}{\alpha_1} & \frac{\delta_2}{\alpha_2}+\frac{\beta_2}{\alpha_1\alpha_2} & \cdots & \frac{\left(\frac{\delta_2}{\alpha_2}+\frac{\beta_2}{\alpha_1\alpha_2}\right)\beta_3+\gamma_3}{\alpha_3}\left(\frac{\beta_4}{\alpha_4}\right)\cdots\left(\frac{\beta_n}{\alpha_n}\right) \\[2mm]
 & & \ddots & \\[2mm]
 & & \ddots & \frac{\left(\frac{\delta_{n-1}}{\alpha_{n-1}}+\frac{\beta_{n-1}}{\alpha_{n-2}\alpha_{n-1}}\right)\beta_n+\gamma_n}{\alpha_n} \\[2mm]
 & & \frac{1}{\alpha_{n-1}} & \frac{\delta_n}{\alpha_n}+\frac{\beta_n}{\alpha_{n-1}\alpha_n}
\end{bmatrix}.
$$

Furthermore, $C_R(r_n)$ *is an* $(H,1)$-*quasiseparable matrix with generators*

$$
d_1 = \frac{\delta_1}{\alpha_1}, \qquad d_k = \frac{\delta_k}{\alpha_k} + \frac{\beta_k}{\alpha_{k-1}\alpha_k}, \qquad k = 2,\ldots,n
$$

$$
p_{k+1}q_k = \frac{1}{\alpha_k}, \qquad g_k = \frac{d_k\beta_{k+1}+\gamma_{k+1}}{\alpha_{k+1}}, \qquad k = 1,\ldots,n-1 \tag{3.5}
$$

$$
b_k = \frac{\beta_{k+1}}{\alpha_{k+1}}, \qquad k = 2,\ldots,n-1 \qquad h_k = 1, \quad k = 2,\ldots,n.
$$

Theorem 3.4. *Let* $R = \{r_0(x),\ldots,r_n(x)\}$ *be a sequence of polynomials satisfying* $\deg(r_k) = k$ *and the recurrence relations* (1.4). *Then the confederate matrix of* $r_n(x)$ *with respect to* R *is given by*

$$
\begin{bmatrix}
-\frac{\theta_1+\gamma_1}{\delta_1} & -(\alpha_1-\beta_1\gamma_1)\frac{\gamma_2}{\delta_2} & \cdots & -(\alpha_1-\beta_1\gamma_1)\cdots(\alpha_{n-1}-\beta_{n-1}\gamma_{n-1})\frac{\gamma_n}{\delta_n} \\[2mm]
\frac{1}{\delta_1} & -\frac{\theta_2+\gamma_2\beta_1}{\delta_2} & \cdots & -\beta_1(\alpha_2-\beta_2\gamma_2)\cdots(\alpha_{n-1}-\beta_{n-1}\gamma_{n-1})\frac{\gamma_n}{\delta_n} \\[2mm]
 & \frac{1}{\delta_2} & \ddots & \vdots \\[2mm]
 & & \ddots & -\beta_{n-1}(\alpha_{n-1}-\beta_{n-1}\gamma_{n-1})\frac{\gamma_n}{\delta_n} \\[2mm]
 & & \frac{1}{\delta_{n-1}} & -\frac{\theta_n+\gamma_n\beta_{n-1}}{\delta_n}
\end{bmatrix}.
$$
$$\tag{3.6}$$

Furthermore, $C_R(r_n)$ *is an* $(H,1)$-*quasiseparable matrix with generators*

$$
d_1 = -\frac{\theta_1+\gamma_1}{\delta_1}, \qquad d_k = -\frac{\theta_k+\gamma_k\beta_{k-1}}{\delta_k}, \qquad k = 2,\ldots,n
$$

$$
p_k = 1, \qquad k = 2,\ldots,n
$$

$$
q_k = \frac{1}{\delta_k}, \qquad k = 1,\ldots,n-1
$$

$$
g_1 = 1, \qquad g_k = \beta_{k-1}, \qquad k = 2,\ldots,n-1
$$

$$
b_k = \alpha_{k-1} - \beta_{k-1}\gamma_{k-1}, \qquad k = 2,\ldots,n-1
$$

$$
h_k = -\frac{\gamma_k}{\delta_k}\left(\alpha_{k-1}-\beta_{k-1}\gamma_{k-1}\right), \qquad k = 2,\ldots,n
$$

We next give a detailed example of the specification of this result to the classical Szegő case.

Example 3.5 (Classical Szegő polynomials). With the choices

$$\alpha_k = \frac{1}{\mu_k}, \quad \beta_k = -\rho_k^*, \quad \gamma_k = -\frac{\rho_k}{\mu_k}, \quad \delta_k = \frac{1}{\mu_k}, \quad \theta_k = 0$$

from which it follows that

$$\alpha_k - \beta_k \gamma_k = \frac{1 - |\rho_k|^2}{\mu_k} = \mu_k, \quad \frac{\gamma_k}{\delta_k} = -\rho_k$$

the two-term recurrence relations (1.4) become

$$\begin{bmatrix} \phi_k(x) \\ \phi_k^\#(x) \end{bmatrix} = \frac{1}{\mu_k} \begin{bmatrix} 1 & -\rho_k^* \\ -\rho_k & 1 \end{bmatrix} \begin{bmatrix} \phi_{k-1}(x) \\ x\phi_{k-1}^\#(x) \end{bmatrix} \qquad (3.7)$$

and the matrix (3.6) reduces to the matrix

$$\begin{bmatrix} \rho_1 & \mu_1\rho_2 & \mu_1\mu_2\rho_3 & \cdots & \mu_1\cdots\mu_{n-1}\rho_n \\ \mu_1 & -\rho_1^*\rho_2 & -\rho_1^*\mu_2\rho_3 & \cdots & -\rho_1^*\mu_2\cdots\mu_{n-1}\rho_n \\ 0 & \mu_2 & -\rho_2^*\rho_3 & \cdots & -\rho_2^*\mu_3\cdots\mu_{n-1}\rho_n \\ \vdots & \ddots & \ddots & \ddots & \vdots \\ 0 & \cdots & 0 & \mu_{n-1} & -\rho_{n-1}^*\rho_n \end{bmatrix}.$$

Using the convention that $\rho_0 := -1$ to insert $1 = -\rho_0^*$ throughout the first row, this matrix becomes exactly the unitary Hessenberg matrix displayed in Table 3. This demonstrates that the Szegő polynomials are a special case of polynomials satisfying (1.4), and likewise the unitary Hessenberg matrix is a special case of those of the form (3.6).

We also note that the condition $b_k \neq 0$ is *not* satisfied by the real orthogonal polynomials, and hence the form (3.6) cannot be used for them.

Theorem 3.6. *Let $R = \{r_0(x), \ldots, r_n(x)\}$ be a sequence of polynomials satisfying* $\deg(r_k) = k$ *and the recurrence relations (1.5). Then the confederate matrix of* $r_n(x)$ *with respect to R is given by*

$$\begin{bmatrix} -\frac{\theta_1}{\delta_1} & -\beta_1\left(\frac{\gamma_2}{\delta_2}\right) & -\beta_1\alpha_2\left(\frac{\gamma_3}{\delta_3}\right) & -\beta_1\alpha_2\alpha_3\left(\frac{\gamma_4}{\delta_4}\right) & \cdots & -\beta_1\alpha_2\alpha_3\alpha_4\cdots\alpha_{n-1}\left(\frac{\gamma_n}{\delta_n}\right) \\ \frac{1}{\delta_1} & -\frac{\theta_2}{\delta_2} & -\beta_2\left(\frac{\gamma_3}{\delta_3}\right) & -\beta_2\alpha_3\left(\frac{\gamma_4}{\delta_4}\right) & \cdots & -\beta_2\alpha_3\alpha_4\cdots\alpha_{n-1}\left(\frac{\gamma_n}{\delta_n}\right) \\ 0 & \frac{1}{\delta_2} & -\frac{\theta_3}{\delta_3} & -\beta_3\left(\frac{\gamma_4}{\delta_4}\right) & \ddots & -\beta_3\alpha_4\cdots\alpha_{n-1}\left(\frac{\gamma_n}{\delta_n}\right) \\ 0 & 0 & \frac{1}{\delta_3} & -\frac{\theta_4}{\delta_4} & \ddots & \vdots \\ \vdots & \ddots & \ddots & \ddots & \ddots & -\beta_{n-1}\left(\frac{\gamma_n}{\delta_n}\right) \\ 0 & \cdots & 0 & 0 & \frac{1}{\delta_{n-1}} & -\frac{\theta_n}{\delta_n} \end{bmatrix}$$

$$(3.8)$$

Furthermore, $C_R(r_n)$ *is an* $(H, 1)$-*quasiseparable matrix with generators*

$$d_k = -\frac{\theta_k}{\delta_k}, \quad k = 1, \ldots, n, \qquad p_k = 1, \quad k = 2, \ldots, n$$

$$q_k = \frac{1}{\delta_k}, \quad k = 1, \ldots, n-1, \qquad g_k = \beta_k, \quad k = 1, \ldots, n-1$$

$$b_k = \alpha_k, \quad k = 2, \ldots, n-1, \qquad h_k = -\frac{\gamma_k}{\delta_k}, \quad k = 2, \ldots, n$$

Example 3.7 (Szegő polynomials). If we choose

$$\alpha_k = \mu_k, \quad \beta_k = \rho_{k-1}^* \mu_k, \quad \gamma_k = \frac{\rho_k}{\mu_k}, \quad \delta_k = \frac{1}{\mu_k}, \quad \theta_k = \frac{\rho_{k-1}^* \rho_k}{\mu_k}$$

the two-term recurrence relations (1.5) do *not* reduce to the known two-term recurrence relations for the Szegő polynomials (1.3), but become instead the new relations

$$\begin{bmatrix} G_0(x) \\ \phi_0^\#(x) \end{bmatrix} = \begin{bmatrix} 0 \\ 1 \end{bmatrix}$$

$$\begin{bmatrix} G_k(x) \\ \phi_k^\#(x) \end{bmatrix} = \begin{bmatrix} \mu_k & \rho_{k-1}^* \mu_k \\ \frac{\rho_k}{\mu_k} & \frac{x + \rho_{k-1}^* \rho_k}{\mu_k} \end{bmatrix} \begin{bmatrix} G_{k-1}(x) \\ \phi_{k-1}^\#(x) \end{bmatrix}. \tag{3.9}$$

The matrix (3.8) does in fact reduce to the classical unitary Hessenberg matrix displayed in (6.8).

Both the classical Szegő formula (3.7) and the new formula (3.9) describe, of course, the same Szegő polynomials $\{\phi_k^\#(x)\}$. However, the auxilary polynomials $\{G_k(x)\}$ differ from $\{\phi_k(x)\}$ used in (3.7). Indeed, it is well known that the auxiliary polynomials involved in (3.7) satisfy

$$\phi_k(x) = x^n \cdot \left[\phi_k^\#\left(\frac{1}{x^*}\right) \right]^*,$$

and in particular, $\deg \phi_k(x) = \deg \phi_k^\#(x)$. At the same time, it is easy to see that the auxiliary polynomials $\{\phi_k(x)\}$ of the new formula (3.9) are different; in particular $\deg G_k(x) = \deg \phi_k^\#(x) - 1$.

Example 3.8 (Real orthogonal polynomials). For systems with $\alpha_k = 0$, the matrix (3.8) becomes tridiagonal, and the corresponding system of polynomials are orthogonal on a real interval. Indeed, $\alpha_k = 0$ implies $G_{k-1} = \beta_{k-1} r_{k-2}(x)$ and hence the relations (1.5) become just the familiar three-term recurrence relations

$$r_k(x) = (\delta_k x + \theta_k) r_{k-1}(x) + \gamma_k \beta_{k-1} r_{k-2}(x).$$

4. Sparse recurrence relations for associated polynomials

At this point we can give an overview of the procedure for inversion of a polynomial-Vandermonde matrix whose polynomials are quasiseparable polynomials. Given the generators of an $(H, 1)$-quasiseparable matrix (or using the theorems of the previous section to obtain these generators given the recurrence relations coefficients) which is the confederate matrix with respect to the master polynomial $P(x) = \prod(x - x_k)$ defined by the nodes $x_k, k = 1, \ldots, n$, with (2.4), we have

$$
C_R(P) = \left[\begin{array}{ccccc} d_1 & & g_i b_{ij}^\times h_j & & \\ p_2 q_1 & \ddots & & & \\ & \ddots & & \ddots & \\ & & & \ddots & \\ 0 & & & p_n q_{n-1} & d_n \end{array}\right] \left[\begin{array}{cccc} & & & P_0 \\ -\frac{1}{P_n} & & 0 & \vdots \\ & & & P_{n-1} \end{array}\right] \tag{4.1}
$$

Applying (2.6) gives us the confederate matrix for the associated polynomials as

$$
C_{\widehat{R}}(P) = \left[\begin{array}{ccccc} d_n & & g_{n-j} b_{n-j,n-i}^\times h_{n-i} & & \\ p_n q_{n-1} & \ddots & & & \\ & \ddots & & \ddots & \\ & & & \ddots & \\ 0 & & & p_2 q_1 & d_1 \end{array}\right] \left[\begin{array}{cccc} P_{n-1} & \cdots & & P_0 \\ -\frac{1}{P_n} & & 0 & \\ & & & \end{array}\right] \tag{4.2}
$$

From this last equation we can see that the n-term recurrence relations satisfied by the associated polynomials $\widehat{R}$ are given by

$$
\widehat{r}_k(x) = \frac{1}{\widehat{p}_{k+1}\widehat{q}_k} \Bigg[\underbrace{(x - \widehat{d}_k)\widehat{r}_{k-1}(x) - \sum_{j=0}^{k-2}\left(\widehat{g}_{j+1}\widehat{b}_{j+1,k}^\times \widehat{h}_k \widehat{r}_j(x)\right)}_{\text{typical term as in (3.4)}} - \underbrace{\frac{P_{n-k}}{P_n}\widehat{r}_0(x)}_{\text{perturbation}} \Bigg]
$$

$$\tag{4.3}$$

where, in order to simplify the formulas, we have introduced the notation

$$
\begin{array}{lll}
\widehat{p}_k = q_{n-k+1}, & \widehat{q}_k = p_{n-k+1}, & \widehat{d}_k = d_{n-k+1}, \\
\widehat{g}_k = h_{n-k+1}, & \widehat{b}_k = b_{n-k+1}, & \widehat{h}_k = g_{n-k+1}.
\end{array} \tag{4.4}
$$

Notice that the *nonzero top row* of the second matrix in (4.2) introduces perturbation terms into the recurrence relations for all of the associated polynomials.

Having found explicit n-term recurrence relations for the sequence of polynomials associated with the given polynomials, the next goal is to find *sparse* recurrence relations. The motivation is that the n-term recurrence relations are slow; they lead to $\mathcal{O}(n^3)$ algorithms, while two- and three-term recurrence relations lead to $\mathcal{O}(n^2)$ algorithms.

Sparse recurrence relations are not, of course, available for all polynomial sequences R; this is a special property. In this section we consider the case where R is a system of $(H, 1)$-quasiseparable polynomials, and we derive sparse recurrence relations for the associated system of polynomials.

For certain polynomial systems whose confederate matrix is not Hessenberg, such recurrence relations are derived in [7]. Obtaining similar formulas for the leading minors of $C_{\widehat{R}}(P)$ of the form shown in (4.2) is not simple, as the second term in (4.2) now affects each column in such a way that the resulting leading submatrices become only $(H, 2)$-quasiseparable, as opposed to the submatrices of (4.1), which are $(H, 1)$-quasiseparable.

A summary of the results obtained in this section is presented in the next Table 4. It is worth noting that in this paper we generalize all of the previous algorithms, and not just the most widely applicable one. As a result, some of the results that are derived have some restrictions on their use. This leads to generalizations of the classical algorithms as well as some new ones below.

Generators of $R \Rightarrow$ Perturbed recurrence relations for $\widehat{R}$	Type of recurrence relations derived	Restrictions on applicability
Theorem 4.1	Perturbed 3-term	$g_k \neq 0$
Theorem 4.2	Perturbed Szegő-type	$b_k \neq 0$
Theorem 4.3	Perturbed [EGO05]-type	none

TABLE 4. Perturbed recurrence relations for the system of associated polynomials $\widehat{R}$.

Theorem 4.1 (Perturbed three-term recurrence relations). *Let $R = \{r_0(x), \ldots, r_{n-1}(x), P(x)\}$ be a system of $(H, 1)$-quasiseparable polynomials corresponding to an $(H, 1)$-quasiseparable matrix of size $n \times n$ with generators $\{p_k, q_k, d_k, g_k, b_k, h_k\}$, with the convention that $g_n = 1, b_n = 0$. Suppose further that $g_k \neq 0$ for $k = 1, \ldots, n - 1$. Then the system of polynomials $\widehat{R}$ associated with R satisfies the recurrence relations*

$$\widehat{r}_0(x) = P_n, \qquad \widehat{r}_1(x) = \frac{1}{\widehat{p}_2 \widehat{q}_1} \left(x - \widehat{d}_1 \right) \widehat{r}_0(x) + \frac{1}{\widehat{p}_2 \widehat{q}_1} P_{n-1}$$

$$\widehat{r}_k(x) = \underbrace{(\widehat{\alpha}_k x - \widehat{\delta}_k) \cdot \widehat{r}_{k-1}(x) - (\widehat{\beta}_k x + \widehat{\gamma}_k) \cdot \widehat{r}_{k-2}(x)}_{\text{typical three-term recurrence relation terms}} + \underbrace{\widehat{\alpha}_k P_{n-k} - \widehat{\beta}_k P_{n-k+1}}_{\text{perturbation term}},$$

(4.5)

for $k = 2, \ldots, n-1$, where

$$\widehat{\alpha}_k = \frac{1}{\widehat{p}_{k+1}\widehat{q}_k}, \qquad \widehat{\delta}_k = \frac{1}{\widehat{p}_{k+1}\widehat{q}_k} \left(\widehat{d}_k - \frac{\widehat{p}_k \widehat{q}_{k-1} \widehat{h}_k \widehat{b}_{k-1}}{\widehat{h}_{k-1}} \right)$$

(4.6)

$$\widehat{\beta}_k = \frac{1}{\widehat{p}_{k+1}\widehat{q}_k} \frac{\widehat{h}_k \widehat{b}_{k-1}}{\widehat{h}_{k-1}}, \qquad \widehat{\gamma}_k = \frac{1}{\widehat{p}_{k+1}\widehat{q}_k} \frac{\widehat{h}_k}{\widehat{h}_{k-1}} \left(\widehat{h}_{k-1}\widehat{g}_{k-1} - \widehat{d}_{k-1}\widehat{b}_{k-1} \right),$$

(4.7)

and the coefficients $P_k, k = 0, \ldots, n$ are as defined in (2.4).

Proof. Let $S = \{s_0(x), s_1(x), \ldots, s_{n-1}(x)\}$ be the system of polynomials corresponding to the $(H, 2)$-quasiseparable matrix $C_{\widehat{R}}(P)$ of the form in (4.2). Then from (2.3) and (4.2), we have for $k = 1, 2, \ldots, n-1$

$$s_k(x) = \frac{1}{\widehat{p}_{k+1}\widehat{q}_k} \Big[(x - \widehat{d}_k)s_{k-1}(x) - \widehat{g}_{k-1}\widehat{h}_k s_{k-2}(x) - \widehat{g}_{k-2}\widehat{b}_{k-1}\widehat{h}_k s_{k-3}(x)$$

$$- \cdots - \widehat{g}_2\widehat{b}_3 \cdots \widehat{b}_{k-1}\widehat{h}_k s_1(x) - \widehat{g}_1\widehat{b}_2 \cdots \widehat{b}_{k-1}\widehat{h}_k s_0(x) + P_{n-k} \Big]. \quad (4.8)$$

It suffices to show that the system of polynomials $\{\widehat{r}_0(x), \widehat{r}_1(x), \ldots, \widehat{r}_{n-1}(x)\}$ defined by the recurrence relations in (4.5)–(4.7) coincide with the system S; that is, that $\widehat{r}_k(x) = s_k(x)$ for each k. We present this proof by induction. By direct confirmation, it is seen that $\widehat{r}_0(x) = s_0(x)$ and $\widehat{r}_1(x) = s_1(x)$.

Next suppose that the conclusion is true for each index less than or equal to $k-1$ for some $2 \leqslant k \leqslant n-1$. Then (4.8) for $k-1$ yields

$$x\widehat{r}_{k-2}(x) = \widehat{p}_k\widehat{q}_{k-1}\widehat{r}_{k-1}(x) + \widehat{d}_{k-1}\widehat{r}_{k-2}(x) \qquad (4.9)$$

$$+ \widehat{g}_{k-2}\widehat{h}_{k-1}\widehat{r}_{k-3}(x) + \widehat{g}_{k-3}\widehat{b}_{k-2}\widehat{h}_{k-1}\widehat{r}_{k-4}(x)$$

$$+ \cdots + \widehat{g}_2\widehat{b}_3 \cdots \widehat{b}_{k-2}\widehat{h}_{k-1}\widehat{r}_1(x)$$

$$+ \widehat{g}_1\widehat{b}_2 \cdots \widehat{b}_{k-2}\widehat{h}_{k-1}\widehat{r}_0(x) - P_{n-k+1}.$$

Next, the polynomial $\widehat{r}_k(x)$ satisfies the recurrence relations (4.5), noting that by hypothesis, $\widehat{h}_k = g_{n-k+1} \neq 0$ for each k. Inserting (4.9) into (4.5) and using the inductive hypothesis, we arrive at exactly (4.8) for $r_k(x)$, which completes the proof. $\qquad \square$

Theorem 4.2 (Perturbed Szegő-type recurrence relations). *Let $R = \{r_0(x), \ldots, r_{n-1}(x), P(x)\}$ be a system of $(H, 1)$-quasiseparable polynomials corresponding to an $(H, 1)$-quasiseparable matrix of size $n \times n$ with generators $\{p_k, q_k, d_k, g_k, b_k, h_k\}$, with the convention that $g_n = 0, b_n = 1$. Suppose further that $b_k \neq 0$ for $k = 2, \ldots, n-1$. Then the system of polynomials $\widehat{R}$ associated with R satisfy the recurrence relations*

$$\begin{bmatrix} G_0(x) \\ \widehat{r}_0(x) \end{bmatrix} = \begin{bmatrix} -\widehat{g}_1 P_n \\ P_n \end{bmatrix},$$

(4.10)

$$\begin{bmatrix} G_k(x) \\ \widehat{r}_k(x) \end{bmatrix} = \frac{1}{\widehat{p}_{k+1}\widehat{q}_k} \begin{bmatrix} v_k & -\widehat{g}_{k+1} \\ \widehat{h}_k/\widehat{b}_k & 1 \end{bmatrix} \begin{bmatrix} G_{k-1}(x) \\ u_k(x)\widehat{r}_{k-1}(x) & \underbrace{+P_{n-k}} \end{bmatrix}$$

$$\text{perturbation term}$$

$$(4.11)$$

for $k = 1, \dots, n-1$, *with auxiliary polynomials* $G_k(x)$, *and the coefficients* $P_k, k = 0, \dots, n$ *are as defined in* (2.4), *with the notations*

$$u_k(x) = (x - \widehat{d}_k) + \frac{\widehat{g}_k\widehat{h}_k}{\widehat{b}_k}, \qquad v_k = \widehat{p}_{k+1}\widehat{b}_{k+1}\widehat{q}_k - \frac{\widehat{g}_{k+1}\widehat{h}_k}{\widehat{b}_k}. \qquad (4.12)$$

Proof. Suppose first that the generators are such that $g_k \neq 0$ for each k. The proof in this case will be given by showing that the system of polynomials generated by the perturbed two-term recurrence relations (4.10)–(4.11) coincide with those given by Theorem 4.1. From (4.11),

$$\left(v_k + \frac{\widehat{g}_{k+1}\widehat{h}_k}{\widehat{b}_k}\right) \begin{bmatrix} G_{k-1}(x) \\ u_k(x)\widehat{r}_{k-1}(x) + P_{n-k} \end{bmatrix} = \widehat{p}_{k+1}\widehat{q}_k \begin{bmatrix} 1 & \widehat{g}_{k+1} \\ -\frac{\widehat{h}_k}{\widehat{b}_k} & v_k \end{bmatrix} \begin{bmatrix} G_k(x) \\ \widehat{r}_k(x) \end{bmatrix},$$

$$(4.13)$$

and using (4.11) for $k+1$, we have

$$G_k(x) = \left(\frac{\widehat{b}_{k+1}}{\widehat{h}_{k+1}}\right)(\widehat{p}_{k+2}\widehat{q}_{k+1}\widehat{r}_{k+1}(x) - u_{k+1}\widehat{r}_k(x) - P_{n-k-1})$$

Together with this, (4.13) produces (4.5) as desired. As by assumption $g_k \neq 0$, for each k, Theorem 4.1 implies the result.

For the case of a polynomial system R where $g_j = 0$ for some j, note that the coefficients of the polynomials generated by the two-term recurrence relations depend continuously on the entries of the 2×2 transfer matrix. Let $\{\epsilon_k\}$ be a sequence tending to zero with $\epsilon_k \neq 0$ for each k, and consider a sequence R_k with $g_j = \epsilon_k$ for each j such that $g_j = 0$ in the original polynomial system R, and all other generators the same as in R. Then the result of the theorem holds for the system R_k for every k by above, and $R_k \to R$, so by continuity, the result must hold for R as well. This completes the proof. $\qquad\square$

The formulas of the previous two theorems generalize the classical formulas for monomials, real-orthogonal polynomials, and Szegő polynomials (demonstrated below). We emphasize at this point that these formulas have limitations in the general case: Theorem 4.1 requires nonzero g_k for each k, and Theorem 4.2 requires nonzero b_k for each k. The next theorem is more general, and does not have any such limitations.

Theorem 4.3 (Perturbed [EGO05]-type recurrence relations). *Let* $R = \{r_0(x), \dots, r_{n-1}(x), P(x)\}$ *be a system of* $(H, 1)$-*quasiseparable polynomials corresponding to an* $(H, 1)$-*quasiseparable matrix of size* $n \times n$ *with generators* $\{p_k, q_k, d_k, g_k, b_k, h_k\}$, with the convention that $q_n = 0, b_n = 0$. Then the system of polynomials $\widehat{R}$ associated with R satisfy the recurrence relations

$$\begin{bmatrix} F_0(x) \\ \widehat{r}_0(x) \end{bmatrix} = \begin{bmatrix} 0 \\ P_n \end{bmatrix}, \tag{4.14}$$

$$\begin{bmatrix} \widehat{F}_k(x) \\ \widehat{r}_k(x) \end{bmatrix} = \underbrace{\frac{1}{\widehat{p}_{k+1}\widehat{q}_k} \begin{bmatrix} \widehat{q}_k\widehat{p}_k\widehat{b}_k & -\widehat{q}_k\widehat{g}_k \\ \widehat{p}_k\widehat{h}_k & x - \widehat{d}_k \end{bmatrix} \begin{bmatrix} \widehat{F}_{k-1}(x) \\ \widehat{r}_{k-1}(x) \end{bmatrix}}_{\text{typical terms}} + \underbrace{\frac{1}{\widehat{p}_{k+1}\widehat{q}_k} \begin{bmatrix} 0 \\ P_{n-k} \end{bmatrix}}_{\text{perturbation term}}$$

$$\tag{4.15}$$

with auxiliary polynomials $\widehat{F}_k(x)$, and the coefficients $P_k, k = 0, \ldots, n$ are as defined in (2.4).

Proof. The recurrence relations (4.15) define a system of polynomials which satisfy the n-term recurrence relations

$$\widehat{r}_k(x) = (\alpha_k x - a_{k-1,k}) \cdot \widehat{r}_{k-1}(x) - a_{k-2,k} \cdot \widehat{r}_{k-2}(x) - \cdots - a_{0,k} \cdot \widehat{r}_0(x) \tag{4.16}$$

for some coefficients $\alpha_k, a_{k-1,k}, \ldots, a_{0,k}$. The proof is presented by showing that these n-term recurrence relations coincide exactly with (4.3). Using relations for $\widehat{r}_k(x)$ and $\widehat{F}_{k-1}(x)$ from (4.15), we have

$$\widehat{r}_k(x) = \frac{1}{\widehat{p}_{k+1}\widehat{q}_k} \Big[(x - \widehat{d}_k)\widehat{r}_{k-1}(x) - \widehat{g}_{k-1}\widehat{h}_k\widehat{r}_{k-2}(x)$$
$$+ \widehat{h}_k\widehat{p}_{k-1}\widehat{b}_{k-1}\widehat{F}_{k-2}(x) + \frac{P_{n-k}}{P_n}\widehat{r}_0(x) \Big]. \tag{4.17}$$

Notice that again using (4.15) to eliminate $\widehat{F}_{k-2}(x)$ from the equation (4.17) will result in an expression for $\widehat{r}_k(x)$ in terms of $\widehat{r}_{k-1}(x)$, $\widehat{r}_{k-2}(x)$, $\widehat{r}_{k-3}(x)$, $\widehat{F}_{k-3}(x)$, and $\widehat{r}_0(x)$ without modifying the coefficients of $\widehat{r}_{k-1}(x), \widehat{r}_{k-2}(x)$, or $\widehat{r}_0(x)$. Again applying (4.15) to eliminate $\widehat{F}_{k-3}(x)$ results in an expression in terms of $\widehat{r}_{k-1}(x)$, $\widehat{r}_{k-2}(x)$, $\widehat{r}_{k-3}(x)$, $\widehat{r}_{k-4}(x)$, $\widehat{F}_{k-4}(x)$, and $\widehat{r}_0(x)$ without modifying the coefficients of $\widehat{r}_{k-1}(x)$, $\widehat{r}_{k-2}(x)$, $\widehat{r}_{k-3}(x)$, or $\widehat{r}_0(x)$. Continuing in this way, the n-term recurrence relations of the form (4.16) are obtained without modifying the coefficients of the previous ones. Suppose that for some $0 < j < k - 1$ the expression for $\widehat{r}_k(x)$ is of the form

$$\widehat{r}_k(x) = \frac{1}{\widehat{p}_{k+1}\widehat{q}_k} \Big[(x - \widehat{d}_k)\widehat{r}_{k-1}(x) - \widehat{g}_{k-1}\widehat{h}_k\widehat{r}_{k-2}(x) - \cdots$$
$$\cdots - \widehat{g}_{j+1}\widehat{b}^\times_{j+1,k}\widehat{h}_k\widehat{r}_j(x) + \widehat{p}_{j+1}\widehat{b}^\times_{j,k}\widehat{h}_k\widehat{F}_j(x) + \frac{P_{n-k}}{P_n}\widehat{r}_0(x) \Big]. \tag{4.18}$$

Using (4.15) for $\widehat{F}_j(x)$ gives the relation

$$\widehat{r}_k(x) = \frac{1}{\widehat{p}_{k+1}\widehat{q}_k} \Big[(x - \widehat{d}_k)\widehat{r}_{k-1}(x) - \widehat{g}_{k-1}\widehat{h}_k\widehat{r}_{k-2}(x) - \cdots$$
$$\cdots - \widehat{g}_j\widehat{b}^\times_{j,k}\widehat{h}_k\widehat{r}_{j-1}(x) + \widehat{p}_j\widehat{b}^\times_{j-1,k}\widehat{h}_k\widehat{F}_{j-1}(x) + \frac{P_{n-k}}{P_n}\widehat{r}_0(x) \Big]. \tag{4.19}$$

Therefore since (4.17) is the case of (4.18) for $j = k - 2$, (4.18) is true for each $j = k - 2, k - 3, \ldots, 0$, and for $j = 0$, using the fact that $\widehat{F}_0 = 0$ we have exactly (4.3) as desired. □

5. Computing the coefficients P_k of the master polynomial $P(x)$

Note that in order to use the recurrence relations of the previous section it is necessary to decompose the master polynomial $P(x)$ into the R basis; that is, the coefficients P_k as in (2.4) must be computed. To this end, an efficient method of calculating these coefficients follows.

It is easily seen that the last polynomial $r_n(x)$ in the system R does not affect the resulting confederate matrix $C_R(P)$. Thus, if

$$\bar{R} = \{r_0(x), \ldots, r_{n-1}(x), xr_{n-1}(x)\},$$

we have $C_R(P) = C_{\bar{R}}(P)$. Decomposing the polynomial $P(x)$ into the $\bar{R}$ basis can be done recursively by setting $r_n^{(0)}(x) = 1$ and then for $k = 0, \ldots, n - 1$ updating $r_n^{(k+1)}(x) = (x - x_{k+1}) \cdot r_n^{(k)}(x)$.

Lemma 5.1. Let $R = \{r_0(x), \ldots, r_n(x)\}$ be given by (2.3), and $f(x) = \sum_{i=1}^{k} a_i \cdot r_i(x)$, where $k < n - 1$. Then the coefficients of $x \cdot f(x) = \sum_{i=1}^{k+1} b_i \cdot r_i(x)$ can be computed by

$$
\begin{bmatrix}
b_0 \\
\vdots \\
b_k \\
b_{k+1} \\
0 \\
\vdots \\
0
\end{bmatrix}
=
\left[
\begin{array}{ccc|c}
 & & & 0 \\
 & C_R(r_n) & & \vdots \\
 & & & 0 \\
\hline
0 & \cdots & 0 & \frac{1}{\alpha_n} & 0 \\
\end{array}
\right]
\begin{bmatrix}
a_0 \\
\vdots \\
a_k \\
0 \\
0 \\
\vdots \\
0
\end{bmatrix}.
\tag{5.1}
$$

Proof. It can be easily checked that

$$
x \cdot \begin{bmatrix} r_0(x) & \cdots & r_n(x) \end{bmatrix} - \begin{bmatrix} r_0(x) & \cdots & r_n(x) \end{bmatrix} \cdot \left[\begin{array}{ccc|c} & C_R(r_n) & & 0 \\ \hline 0 & \cdots & 0 & \frac{1}{\alpha_n} & 0 \end{array} \right]
$$

$$= \begin{bmatrix} 0 & \cdots & 0 & x \cdot r_n(x) \end{bmatrix}.$$

Multiplying the latter equation by the column of the coefficients, we obtain (5.1). □

This lemma suggests the following algorithm for computing coefficients $\{P_0, P_1, \ldots, P_{n-1}, P_n\}$ in (2.4) of the master polynomial.

Algorithm 5.2 (Coefficients of the master polynomial in the R basis). Cost: $\mathcal{O}(n \times m(n))$, where $m(n)$ is the cost of multiplication of an $n \times n$ quasiseparable matrix by a vector.

Input: *A quasiseparable confederate matrix* $C_R(r_n)$ *and* n *nodes* $x = (x_1, x_2, \dots, x_n)$.

1. *Set* $\begin{bmatrix} P_0^{(0)} & \cdots & P_{n-1}^{(0)} & P_n^{(0)} \end{bmatrix} = \begin{bmatrix} 1 & 0 & \cdots & 0 \end{bmatrix}$

2. *For* $k = 1 : n$,

$$\begin{bmatrix} P_0^{(k)} \\ \vdots \\ P_{n-1}^{(k)} \\ P_n^{(k)} \end{bmatrix} = \left(\left[\begin{array}{c|c} C_{\bar{R}}(xr_{n-1}(x)) & 0 \\ \hline 0 \ \cdots \ 0 \ 1 & 0 \end{array} \right] - x_k I \right) \begin{bmatrix} P_0^{(k-1)} \\ \vdots \\ P_{n-1}^{(k-1)} \\ P_n^{(k-1)} \end{bmatrix}$$

where $\bar{R} = \{r_0(x), \dots, r_{n-1}(x), xr_{n-1}(x)\}$.

3. *Take* $\begin{bmatrix} P_0 & \cdots & P_{n-1} & P_n \end{bmatrix} = \begin{bmatrix} P_0^{(n)} & \cdots & P_{n-1}^{(n)} & P_n^{(n)} \end{bmatrix}$

Output: *Coefficients* $\{P_0, P_1, \dots, P_{n-1}, P_n\}$ *such that* (2.4) *is satisfied.*

It is clear that the computational burden in implementing this algorithm is in multiplication of the matrix $C_{\bar{R}}(r_n)$ by the vector of coefficients. The cost of each such step is $\mathcal{O}(m(n))$, where $m(n)$ is the cost of multiplication of an $n \times n$ quasiseparable matrix by a vector, thus the cost of computing the n coefficients is $\mathcal{O}(n \times m(n))$. Using a fast $\mathcal{O}(n)$ algorithm for multiplication of a quasiseparable matrix by a vector from [5], the cost of this algorithm is $\mathcal{O}(n^2)$.

6. Special cases of these new inversion algorithms

In what follows we show how these algorithms (as the previous section contains a choice of three perturbed recurrence relations, each leading to an inversion algorithm for the corresponding polynomial-Vandermonde matrix) generalizes the previous work in the important special cases of monomials, real orthogonal polynomials, and Szegő polynomials. The reductions in all three special cases are summarized in Table 5.

6.1. First special case. Monomials and the classical Traub algorithm

As shown earlier, the well-known companion matrix (2.7) results when the polynomial system R is simply a system of monomials. By choosing the generators $p_k = 1, q_k = 1, d_k = 0, g_k = 1, b_k = 1$, and $h_k = 0$, the matrix (4.1) reduces to (2.7), and also (4.2) reduces to the confederate matrix for the Horner polynomials (2.8). In this special case, the perturbed three-term recurrence relations of Theorem 4.1 become

$$\hat{r}_0(x) = P_n, \quad \hat{r}_k(x) = x\hat{r}_{k-1}(x) + P_{n-k}, \tag{6.1}$$

coinciding with the known recurrence relations for the Horner polynomials, used in the evaluation of the polynomial

$$P(x) = P_0 + P_1 x + \cdots + P_{n-1} x^{n-1} + P_n x^n. \tag{6.2}$$

Special Case	R.R. Type	Resulting R.R.
Monomials	Theorem 4.1 – 3-term r.r.	(6.1)
	Theorem 4.2 – Szegő-type r.r.	(6.1)
	Theorem 4.3 – [EGO05]-type r.r.	(6.1)
Real orthogonal	Theorem 4.1 – 3-term r.r.	(6.6)
	Theorem 4.2 – Szegő-type r.r.	N/A, $b_k = 0$.
	Theorem 4.3 – [EGO05]-type r.r.	(6.6)
Szegő polynomials	Theorem 4.1 – 3-term r.r.	(6.13)
	Theorem 4.2 – Szegő-type r.r.	(6.11)
	Theorem 4.3 – [EGO05]-type r.r.	(6.12)

TABLE 5. Reduction of derived recurrence relations in special cases.

In fact, after eliminating the auxiliary polynomials present in Theorems 4.2 and 4.3, these recurrence relations also reduce to (6.1). Thus all of the presented recurrence relations generalize those used in the classical Traub algorithm.

6.2. Second special case. Real orthogonal polynomials and the Calvetti–Reichel algorithm

Consider the almost tridiagonal confederate matrix

$$
C_R(P) = \begin{bmatrix}
d_1 & h_2 & 0 & \cdots & 0 & -P_0/P_n \\
q_1 & d_2 & h_3 & \ddots & \vdots & -P_1/P_n \\
0 & q_2 & d_3 & h_4 & 0 & \vdots \\
0 & 0 & q_3 & d_4 & \ddots & -P_{n-3}/P_n \\
\vdots & \ddots & \ddots & \ddots & \ddots & h_n - P_{n-2}/P_n \\
0 & \cdots & 0 & 0 & q_{n-1} & d_n - P_{n-1}/P_n
\end{bmatrix}.
\tag{6.3}
$$

The corresponding system of polynomials R satisfy three-term recurrence relations. Such confederate matrices can be seen as special cases of our general class by choosing the generators $p_k = 1, b_k = 0$, and $g_k = 1$, and in this case the matrix (4.1) reduces to (6.3).

To invert the corresponding polynomial-Vandermonde matrix by our algorithm, we first find the confederate matrix $C_{\widehat{R}}(P)$ of the polynomial system $\widehat{R}$ associated with R. That is, we must evaluate the polynomials corresponding to the confederate matrix $C_R(P)$ given by

$$
\begin{bmatrix}
d_n - P_{n-1}/P_n & h_n - P_{n-2}/P_n & -P_{n-3}/P_n & \cdots & & -P_1/P_n & -P_0/P_n \\
q_{n-1} & d_{n-1} & h_{n-1} & \ddots & & & \vdots \\
0 & q_{n-2} & d_{n-2} & h_{n-2} & 0 & & \vdots \\
0 & 0 & q_{n-3} & d_{n-3} & \ddots & & \vdots \\
\vdots & \ddots & \ddots & \ddots & \ddots & & h_2 \\
0 & \cdots & & 0 & 0 & q_1 & d_1
\end{bmatrix}.
$$

(6.4)

Note that the highlighted column corresponds to the full recurrence relation

$$
\widehat{r}_3(x) = \frac{1}{q_{n-3}}(x - d_{n-2})\widehat{r}_2(x) - \frac{h_{n-1}}{q_{n-3}}\widehat{r}_1(x) + \frac{1}{q_{n-3}}\frac{P_{n-3}}{P_n}\widehat{r}_0(x) \tag{6.5}
$$

In this case the perturbed three-term recurrence relations from Theorem 4.1 as well as the two-term recurrence relations from Theorem 4.3 both become

$$
\widehat{r}_k(x) = \frac{1}{q_{n-k}}(x - d_{n-k})\widehat{r}_{k-1}(x) - \frac{q_{n-k+1}}{q_{n-k}}h_{n-k+1}\widehat{r}_{k-2}(x) + \frac{1}{q_{n-k}}P_{n-k} \tag{6.6}
$$

which coincides with the Clenshaw rule for evaluating

$$
P(x) = P_0 r_0(x) + P_1 r_1(x) + \cdots + P_{n-1}r_{n-1}(x) + P_n r_n(x). \tag{6.7}
$$

Thus our formula generalizes both the Clenshaw rule and the algorithms designed for inversion of three-term-Vandermonde matrices in [3] and [10].

Notice that the Szegő-like two-term recurrence relations of Theorem 4.2 are inapplicable as $b_k = 0$ is a necessary choice of generators.

6.3. Third special case. Szegő polynomials and the algorithm of [18]

Next consider the important special case of the almost unitary Hessenberg matrix of Table 2,

$$
C_R(P) =
\begin{bmatrix}
-\rho_0^* \rho_1 & -\rho_0^* \mu_1 \rho_2 & -\rho_0^* \mu_1 \mu_2 \rho_3 & \cdots & -\rho_0^* \mu_1 \cdots \mu_{n-1}\rho_n \\
\mu_1 & -\rho_1^* \rho_2 & -\rho_1^* \mu_2 \rho_3 & \cdots & -\rho_1^* \mu_2 \cdots \mu_{n-1}\rho_n \\
0 & \mu_2 & -\rho_2^* \rho_3 & \cdots & -\rho_2^* \mu_3 \cdots \mu_{n-1}\rho_n \\
\vdots & \ddots & \ddots & \ddots & \vdots \\
0 & \cdots & 0 & \mu_{n-1} & -\rho_{n-1}^* \rho_n
\end{bmatrix} \tag{6.8}
$$

that corresponds to the Szegő polynomials (represented by the reflection coefficients ρ_k and complimentary parameters μ_k), and polynomial $P(x)$. The Szegő polynomials are known to satisfy the two-term recurrence relations (1.3) as well as the three-term recurrence relations

$$
\phi_0^\#(x) = 1, \quad \phi_1^\#(x) = \frac{1}{\mu_1} \cdot x \phi_0^\#(x) - \frac{\rho_1}{\mu_1}\phi_0^\#(x)
$$

$$
\phi_k^\#(x) = \left[\frac{1}{\mu_k} \cdot x + \frac{\rho_k}{\rho_{k-1}}\frac{1}{\mu_k}\right]\phi_{k-1}^\#(x) - \frac{\rho_k}{\rho_{k-1}}\frac{\mu_{k-1}}{\mu_k} \cdot x \cdot \phi_{k-2}^\#(x)
$$

(6.9)

(see [13], [9]). As above, the polynomials associated with the system of Szegő polynomials are determined by the confederate matrix $C_{\widehat{R}}(P)$ given by

$$
\begin{bmatrix}
-\rho_n\rho_{n-1}^* - \dfrac{P_{n-1}}{P_n} & -\rho_n\mu_{n-1}\rho_{n-2}^* - \dfrac{P_{n-2}}{P_n} & \cdots & -\rho_n\mu_{n-1}\cdots\mu_1\rho_0^* - \dfrac{P_0}{P_n} \\
\mu_{n-1} & -\rho_{n-1}\rho_{n-2}^* & \cdots & -\rho_{n-1}\mu_{n-2}\cdots\mu_1\rho_0^* \\
0 & \ddots & \ddots & \vdots \\
\vdots & \ddots & \ddots & \vdots \\
0 & \cdots & \mu_1 & -\rho_1\rho_0^*
\end{bmatrix}.
$$

$$(6.10)$$

For this special case, let $p_k = 1$, $q_k = \mu_k$, $d_k = -\rho_k\rho_{k-1}^*$, $g_k = \rho_{k-1}^*$, $b_k = \mu_{k-1}$, and $h_k = -\mu_{k-1}\rho_k$ (alternatively $g_k = \rho_{k-1}^*\mu_k$, $b_k = \mu_k$, $h_k = -\rho_k$). This choice of generators reduces (4.1) to the matrix (6.8) as well as (4.2) to (6.10), and in this case the perturbed two-term recurrence relations of Theorem 4.2 become

$$
\begin{bmatrix} \widehat{\phi}_0(x) \\ \widehat{\phi}_0^{\#}(x) \end{bmatrix} = \frac{1}{\mu_n} \begin{bmatrix} -\rho_n \\ 1 \end{bmatrix},
$$

$$
\begin{bmatrix} \widehat{\phi}_k(x) \\ \widehat{\phi}_k^{\#}(x) \end{bmatrix} = \frac{1}{\mu_{n-k}} \begin{bmatrix} 1 & -\rho_{n-k}^* \\ -\rho_{n-k} & 1 \end{bmatrix} \begin{bmatrix} \widehat{\phi}_{k-1}(x) \\ x\widehat{\phi}_{k-1}^{\#}(x) + P_{n-k} \end{bmatrix},
$$

$$(6.11)$$

coinciding with those recurrence relations derived in [18]. The recurrence relations from Theorem 4.3 reduce to new two-term recurrence relations; that is, relations that do not generalize those derived in [18]. They become

$$
\begin{bmatrix} \widehat{F}_k(x) \\ \widehat{\phi}_k^{\#}(x) \end{bmatrix} = \frac{1}{\mu_{n-k}} \begin{bmatrix} \mu_{n-k}\mu_{n-k+1} & -\mu_{n-k}\rho_{n-k+1}^* \\ -\mu_{n-k+1}\rho_{n-k} & x + \rho_{n-k}\rho_{n-k+1}^* \end{bmatrix} \begin{bmatrix} \widehat{F}_{k-1}(x) \\ \widehat{\phi}_{k-1}^{\#}(x) \end{bmatrix}
$$
$$
+ \begin{bmatrix} 0 \\ P_{n-k} \end{bmatrix}.
$$

$$(6.12)$$

Also, the perturbed three-term recurrence relations of Theorem 4.1 reduce to

$$
\widehat{\phi}_0(x) = \frac{1}{\mu_n}, \quad \widehat{\phi}_1(x) = \left\{ \frac{1}{\mu_{n-1}} \cdot x\widehat{\phi}_0(x) - \frac{\rho_{n-1}\rho_n^*}{\mu_{n-1}}\widehat{\phi}_0(x) \right\} + \frac{P_{n-1}}{\mu_{n-1}}.
$$

$$
\widehat{\phi}_k(x) = \left[\frac{1}{\mu_{n-k}} \cdot x + \frac{\rho_{n-k}}{\rho_{n-k+1}} \frac{1}{\mu_{n-k}} \right] \widehat{\phi}_{k-1}(x) - \frac{\rho_{n-k}}{\rho_{n-k+1}} \frac{\mu_{n-k+1}}{\mu_{n-k}} \cdot x \cdot \widehat{\phi}_{k-2}(x)
$$
$$
+ \frac{P_{n-k} - P_{n-k+1}\mu_{n-k+1}\frac{\rho_{n-k}}{\rho_{n-k+1}}}{\mu_{n-k}}
$$

$$(6.13)$$

in this case, also coinciding with the perturbed three-term recurrence relations in [18]. Thus both of these theorems generalize the recurrence relations derived in [18] as well.

7. Numerical experiments

The numerical properties of the Traub algorithm and its generalizations (that are the special cases of our general algorithm) were studied by many different authors. It was noticed in [12] that a version of the Traub algorithm can yield high accuracy in certain cases if the algorithm is preceded with the *Leja ordering* of the nodes; that is, ordering such that

$$|x_1| = \max_{1 \leqslant i \leqslant n} |x_i|, \qquad \prod_{j=1}^{k-1} |x_k - x_j| = \max_{k \leqslant i \leqslant n} \prod_{j=1}^{k-1} |x_i - x_j|, \quad k = 2, \ldots, n-1$$

(see [22], [15], [20]). It was noticed in [12] that the same is true for Chebyshev–Vandermonde matrices.

No error analysis was done, but the conclusions of the above authors was that in many cases the Traub algorithm and its extensions can yield much better accuracy than Gaussian elimination, even for very ill-conditioned matrices.

We made our preliminary experiments with the general algorithm, and our conclusions are consistent with the experience of our colleagues. In all cases we studied the proposed algorithm yields better accuracy than Gaussian elimination, e.g., in the new special cases of Szegő–Vandermonde and $(H,1)$-quasiseparable-Vandermonde matrices. However, our experiments need to be done for different special cases and also the numerical properties of different recurrence relations are worth analyzing. This is a topic for future study.

The algorithm has been implemented in MATLAB version 7. The results of the algorithm using standard MATLAB code, and hence double precision arithmetic, were compared with exact solutions calculated using the MATLAB Symbolic Toolbox command vpa(), which allows software-implemented precision of arbitrary numbers of digits. The number of digits was set to 64, however in cases where the condition number of the coefficient matrix exceeded 10^{30}, this was raised to 100 digits to maintain accuracy.

We compare the forward accuracy of the inverse computed by the algorithm with respect to the inverse computed in high precision, defined by

$$e = \frac{\|V_R(x)^{-1} - \widehat{V_R}(x)^{-1}\|_2}{\|V_R(x)^{-1}\|_2} \tag{7.1}$$

where $\widehat{V_R}(x)^{-1}$ is the solution computed by each algorithm in MATLAB in double precision, and $V_R(x)^{-1}$ is the exact solution. In the tables, TraubQS denotes the proposed Traub-like algorithm, and inv() indicates MATLAB's inversion command. Finally, cond(V) denotes the condition number of the matrix V computed via the MATLAB command cond().

Experiment 1. In this experiment, the problem was chosen by choosing the generators that define the recurrence relations of the polynomial system randomly in $(-1, 1)$, and the nodes x_k were selected equidistant on $(-1, 1)$ via the formula

$$x_k = -1 + 2\left(\frac{k}{n-1}\right), \quad k = 0, 1, \ldots, n-1.$$

We test the accuracy of the inversion algorithm for various sizes n of matrices generated in this way. Some results are tabulated in Table 6, and shown graphically in Figure 1.

n	cond(V)	inv()	TraubQS
10	4.2e04	4.1e-14	3.4e-15
	2.2e05	2.5e-14	6.3e-15
	3.7e08	1.0e-13	8.9e-14
15	1.1e10	3.5e-11	3.5e-11
	1.1e11	1.5e-12	4.8e-13
	4.7e11	1.3e-13	7.7e-14
20	7.6e14	1.1e-10	3.4e-12
	1.2e15	4.2e-11	1.1e-11
	7.8e17	1.2e-09	1.7e-15
25	4.8e19	1.2e-09	1.7e-13
	1.1e24	5.9e-07	1.3e-11
	1.5e27	8.4e-08	2.4e-09
30	3.3e24	7.2e-07	1.1e-13
	5.0e27	2.8e-06	1.7e-11
	1.8e30	1.3e-03	9.5e-10
35	7.3e23	2.4e-04	6.9e-10
	8.3e26	2.6e-03	1.2e-06
	1.4e27	2.9e-05	1.4e-08
40	1.1e31	8.2e-02	2.4e-13
	2.4e32	3.4e+00	9.9e-12
	1.7e33	1.2e-01	1.0e-08
45	4.3e30	1.7e-01	1.7e-05
	1.7e31	5.9e-01	1.0e-08
	3.9e35	1.0e-02	2.4e-08
50	2.1e42	1.0e+00	4.7e-06
	3.9e44	1.0e+00	7.0e-06
	6.6e45	1.0e+00	6.3e-06

Table 6. Equidistant nodes on $(-1, 1)$.

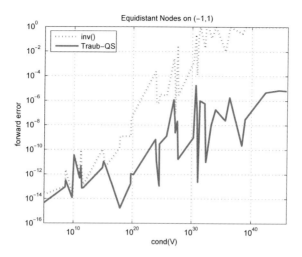

FIGURE 1. Equidistant nodes on $(-1, 1)$.

Notice that the performance of the proposed inversion algorithm is an improvement over that of MATLAB's standard inversion command `inv()` in this specific case.

Experiment 2. Next, the values for the generators and the nodes were chosen randomly on the unit disc. We test the accuracy for various 30×30 matrices generated in this way, and present some results in Table 7 and Figure 2.

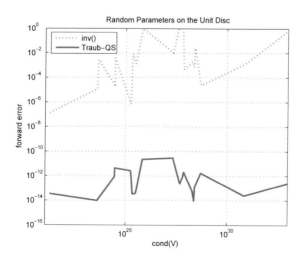

FIGURE 2. Random parameters on the unit disc.

T. Bella et al.

cond(V)	inv()	TraubQS
1.7e21	1.3e-07	3.5e-14
3.9e23	1.2e-05	9.6e-15
4.3e23	2.7e-03	1.2e-14
2.8e24	2.4e-05	8.4e-13
2.9e24	3.9e-03	4.3e-12
1.8e25	6.8e-07	2.6e-12
2.2e25	8.9e-03	3.4e-14
3.1e25	1.3e-03	3.6e-14
3.5e25	2.9e-03	7.9e-14
6.8e25	1.0e+00	2.2e-11
2.2e27	1.0e-02	2.9e-11
4.9e27	3.6e+00	2.3e-13
6.6e27	9.9e+00	7.6e-13
7.6e27	4.6e-04	2.0e-12
2.0e28	1.9e-03	5.7e-14
2.4e28	6.9e-04	9.6e-15
2.6e28	2.5e-02	1.2e-13
5.2e28	2.4e-05	1.7e-12
6.9e30	1.2e-03	2.5e-14
1.4e33	1.0e+00	2.9e-13

Table 7. Random parameters on the unit disc.

8. Conclusions

In this paper we used properties of confederate matrices to extend the classical Traub algorithm for inversion of Vandermonde matrices to the general polynomial-Vandermonde case. The relation between polynomial systems satisfying some recurrence relations and quasiseparable matrices allowed an order of magnitude computational savings in this case, resulting in an $\mathcal{O}(n^2)$ algorithm as opposed to Gaussian elimination, which requires $\mathcal{O}(n^3)$ operations. Finally, some numerical experiments were presented that indicate that, under some circumstances, the resulting algorithm can give better performance than Gaussian elimination.

References

[1] M. Bakonyi and T. Constantinescu, *Schur's algorithm and several applications*, in Pitman Research Notes in Mathematics Series, vol. 61, Longman Scientific and Technical, Harlow, 1992.

[2] T. Bella, Y. Eidelman, I. Gohberg, I. Koltracht and V. Olshevsky, *A Björck–Pereyra-type algorithm for Szegő–Vandermonde matrices based on properties of unitary Hessenberg matrices*, Linear Algebra and Applications, Volume 420, Issues 2-3 pp. 634–647, 2007.

[3] D. Calvetti and L. Reichel, *Fast inversion of Vandermonde-like matrices involving orthogonal polynomials*, BIT, 1993.

[4] Y. Eidelman and I. Gohberg, *On a new class of structured matrices*, Integral Equations and Operator Theory, **34** (1999), 293–324.

[5] Y. Eidelman and I. Gohberg, *Linear complexity inversion algorithms for a class of structured matrices*, Integral Equations and Operator Theory, **35** (1999), 28–52.

[6] Y. Eidelman and I. Gohberg, *A modification of the Dewilde–van der Veen method for inversion of finitestructured matrices*, Linear Algebra Appl., **343-344** (2002), 419–450.

[7] Y. Eidelman, I. Gohberg and V. Olshevsky, *Eigenstructure of Order-One-Quasiseparable Matrices. Three-term and Two-term Recurrence Relations*, Linear Algebra and its Applications, Volume 405, 1 August 2005, pages 1–40.

[8] G. Forney, *Concatenated codes*, The M.I.T. Press, 1966, Cambridge.

[9] L.Y. Geronimus, *Polynomials orthogonal on a circle and their applications*, Amer. Math. Translations, **3** pp. 1–78, 1954 (Russian original 1948).

[10] I. Gohberg and V. Olshevsky, *Fast inversion of Chebyshev–Vandermonde matrices*, Numerische Mathematik, **67, No. 1** (1994), 71–92.

[11] I. Gohberg and V. Olshevsky, *A fast generalized Parker–Traub algorithm for inversion of Vandermonde and related matrices*, Journal of Complexity, **13(2)** (1997), 208–234.
A short version in pp. in *Communications, Computation, Control and Signal Processing: A tribute to Thomas Kailath*, eds. A. Paulraj, V. Roychowdhury and C. Shaper, Kluwer Academic Publishing, 1996, 205–221.

[12] I. Gohberg and V. Olshevsky, *The fast generalized Parker–Traub algorithm for inversion of Vandermonde and related matrices*, J. of Complexity, 13(2) (1997), 208–234.

[13] U. Grenader and G. Szegő, *Toeplitz forms and Applications*, University of California Press, 1958.

[14] W.G. Horner, *A new method of solving numerical equations of all orders by continuous approximation*, Philos. Trans. Roy. Soc. London, (1819), 308–335.

[15] N. Higham, *Stability analysis of algorithms for solving confluent Vandermonde-like systems*, SIAM J. Matrix Anal. Appl., **11**(1) (1990), 23–41.

[16] T. Kailath and V. Olshevsky, *Displacement structure approach to polynomial Vandermonde and related matrices*, Linear Algebra and Its Applications, **261** (1997), 49–90.

[17] J. Maroulas and S. Barnett, Polynomials with respect to a general basis. I. Theory, J. of Math. Analysis and Appl., **72** (1979), 177–194.

[18] V. Olshevsky, *Eigenvector computation for almost unitary Hessenberg matrices and inversion of Szegő–Vandermonde matrices via Discrete Transmission lines.* Linear Algebra and Its Applications, 285 (1998), 37–67.

[19] V. Olshevsky, *Associated polynomials, unitary Hessenberg matrices and fast generalized Parker–Traub and Bjorck–Pereyra algorithms for Szegő–Vandermonde matrices* invited chapter in the book "Structured Matrices: Recent Developments in Theory and Computation," 67–78, (D. Bini, E. Tyrtyshnikov, P. Yalamov, eds.), 2001, NOVA Science Publ., USA.

[20] V. Olshevsky, *Pivoting for structured matrices and rational tangential interpolation,* in Fast Algorithms for Structured Matrices: Theory and Applications, CONM/323, pp. 1–75, AMS publications, May 2003.

[21] F. Parker, *Inverses of Vandermonde matrices,* Amer. Math. Monthly, **71** (1964), 410–411.

[22] L. Reichel and G. Opfer, *Chebyshev–Vandermonde systems,* Math. of Comp., **57** (1991), 703–721.

[23] J. Stoer, R. Bulirsch, *Introduction to Numerical Analysis,* Springer-Verlag, 1992, 277–301.

[24] J. Traub, *Associated polynomials and uniform methods for the solution of linear problems,* SIAM Review, **8, No. 3** (1966), 277–301.

T. Bella
Department of Mathematics
University of Rhode Island
Kingston, RI 02881, USA
e-mail: `tombella@math.uri.edu`

Y. Eidelman and I. Gohberg[Z"L]
School of Mathematical Sciences
Raymond and Beverly Sackler Faculty of Exact Sciences
Tel Aviv University
Ramat-Aviv 69978, Israel
e-mail: `eideyu@post.tau.ac.il`

V. Olshevsky
Department of Mathematics
University of Connecticut
Storrs, CT 06269, USA
e-mail: `olshevsky@math.uconn.edu`

E. Tyrtyshnikov
Institute of Numerical Mathematics
Russian Academy of Sciences
Gubkina Street, 8
Moscow, 119991, Russia
e-mail: `tee@inm.ras.ru`

Operator Theory:
Advances and Applications, Vol. 237, 107–125

Long Proofs of Two Carlson–Schneider Type Inertia Theorems

Harry Dym and Motke Porat

To Leonid Lerer on the occasion of his retirement from the faculty of the Department of Mathematics at the Technion, with affection and respect.

Abstract. This expository note is devoted to a discussion of the equivalence of inertia theorems of the Carlson–Schneider type with the existence of finite-dimensional reproducing kernel Krein spaces of the de Branges type. The first five sections focus on an inertia theorem connected with a Lyapunov equation. A sixth supplementary section sketches an analogous treatment of the Stein equation. The topic was motivated by a question raised by Leonid Lerer.

Mathematics Subject Classification (2010). 46C20, 46E22, 47B32, 47B50, 93B20.

Keywords. Inertia theorems, realization theory, reproducing kernel spaces, finite-dimensional de Branges–Krein spaces, factorization of rational matrix-valued functions, Lyapunov–Stein equations.

1. Introduction

The papers [9] by Lerer and Tismenetsky and [5] by Dym and Volok both study the zero distribution of matrix polynomials, but by very different methods. In particular, [9] rests heavily on the spectral theory of matrix polynomials that is conveniently summarized in the monograph [6] and on the Carlson–Schneider inertia theorem, whereas [5] uses reproducing kernel space methods. Some years ago, in an exchange of E-mails with the first listed author of this note, Leonid wondered how the analysis in [5] managed to avoid invoking the Carlson–Schneider inertia theorem. The purpose of the first five sections is to suggest an answer to this question by showing how to deduce Carlson–Schneider type theorems from reproducing kernel formulas. To be more precise, the theorem we consider is weaker than the Carlson–Schneider theorem because of the observability assumption; it corresponds to Corollary 1 on p. 449 of [8], which is credited to C.T. Chen [2] and H.K. Wimmer [10]. The proof is a bit on the long side, but is completely self-contained and uses only elementary ideas from linear system theory. A sixth section deals with an analogous inertia theorem for the disc.

The notation Π_+ (resp., Π_-) for the open right (resp., left) half-plane and $\mathcal{E}_-(A), \mathcal{E}_0(A)$ and $\mathcal{E}_+(A)$ for the sum of the algebraic multiplicities of the eigenvalues of a matrix $A \in \mathbb{C}^{n\times n}$ in $\Pi_-, i\mathbb{R}$ and Π_+, respectively, will be needed to formulate the version of the Carlson–Schneider inertia theorem under consideration:

Theorem 1.1. *If $A \in \mathbb{C}^{n\times n}, C \in \mathbb{C}^{m\times n}, (C, A)$ is an observable pair and $P = P^* \in \mathbb{C}^{n\times n}$ is a solution of the Lyapunov equation*

$$A^*P + PA + C^*C = 0, \tag{1.1}$$

then

(1) $\sigma(A) \cap i\mathbb{R} = \emptyset$ *and P is invertible, i.e., $\mathcal{E}_0(A) = \mathcal{E}_0(P) = 0$,*
(2) $\mathcal{E}_+(A) = \mathcal{E}_-(P)$ *and* $\mathcal{E}_-(A) = \mathcal{E}_+(P)$.

This note is organized as follows: In Section 2 we present some preliminary facts and notation for subsequent use. A short well-known proof of Theorem 1.1 based on elementary tools of linear system theory is presented in Section 3. In Section 4 Theorem 1.1 is used to establish a finite-dimensional reproducing kernel Krein space (RKKS) with a reproducing kernel (RK) of the special form

$$K_\omega(\lambda) = \frac{I_m - \Theta(\lambda)\Theta(\omega)^*}{\rho_\omega(\lambda)} \tag{1.2}$$

where $\rho_\omega(\lambda) = \lambda + \omega^*$ and Θ admits a factorization $\Theta = \Theta_1^{-1}\Theta_2$, with Θ_1 and Θ_2 both inner rmvf's (rational matrix-valued functions) with respect to Π_+. A proof of Theorem 1.1 that is based on methods and formulas of reproducing kernel spaces is presented in Section 5. An analogous treatment of an inertia theorem for the Stein equation is sketched in the sixth section.

2. Preliminaries

We begin with a lemma that verifies assertion (1) of Theorem 1.1 and reduces the problem of establishing (2) to the case where the matrix A is of particular form. Subsequently, we present some notation and formulas that will be used frequently in the later sections.

Lemma 2.1. *If the assumptions in Theorem 1.1 are in force, then $\sigma(A) \cap i\mathbb{R} = \emptyset$ and P is invertible and without loss of generality A may be assumed to be of the form*

$$A = A_+, \text{ or } A = A_-, \text{ or } A = \begin{bmatrix} A_+ & 0 \\ 0 & A_- \end{bmatrix}, \quad \text{with} \quad \sigma(A_\pm) \subset \Pi_\pm. \tag{2.1}$$

Proof. The proof is separated into parts:

1. If $Pu = 0$ for some $u \in \mathbb{C}^n$, then, by (1.1),

$$0 = u^* A^* Pu + u^* PAu + u^* C^* Cu = u^* C^* Cu \Longrightarrow Cu = 0.$$

Therefore, (1.1) implies that $PAu = 0$ and hence, by repeating the previous argument, that $CAu = 0$ and $PA^2u = 0$. Thus, $CA^i u = 0$ for every nonnegative integer i and hence, as (C, A) is observable, $u = 0$, i.e., P is invertible.

2. If $Au = \lambda u$ for some $u \in \mathbb{C}^n$ and $\lambda \in \mathbb{C}$, then $u^* A^* = \lambda^* u^*$ and hence (1.1) implies that

$$0 = u^* A^* Pu + u^* PAu + u^* C^* Cu = (\lambda + \lambda^*) u^* Pu + u^* C^* Cu$$
$$= u^* C^* Cu \quad \text{if } \lambda \in i\mathbb{R}$$

Thus, $CA^k u = \lambda^k Cu = 0$ in this case, which implies that $u = 0$, since the pair (C, A) is observable. Therefore, $\sigma(A) \cap i\mathbb{R} = \emptyset$.

3. As $\sigma(A) \cap i\mathbb{R} = \emptyset$, $A = UJU^{-1}$, where J is a Jordan matrix with $\sigma(J) \subset \Pi_+$ or $\sigma(A) \subset \Pi_+$ or is of the form

$$J = \begin{bmatrix} A_+ & 0 \\ 0 & A_- \end{bmatrix} \quad \text{where} \quad \sigma(A_\pm) \subset \Pi_\pm.$$

Then P is an invertible Hermitian solution of equation (1.1) if and only if $U^* PU$ a Hermitian invertible solution of the equation

$$J^* (U^* PU) + (U^* PU)J + (CU)^* CU = 0.$$

Moreover, $\sigma(J) \cap i\mathbb{R} = \emptyset$, $\mathcal{E}_\pm(J) = \mathcal{E}_\pm(A)$, $\mathcal{E}_\pm(U^* PU) = \mathcal{E}_\pm(P)$ and the pair (CU, J) is observable, since

$$\text{rank} \begin{bmatrix} \lambda I_n - A \\ C \end{bmatrix} = \text{rank} \begin{bmatrix} U & 0 \\ 0 & I_m \end{bmatrix} \begin{bmatrix} \lambda I_n - J \\ CU \end{bmatrix} U^{-1} = \text{rank} \begin{bmatrix} \lambda I_n - J \\ CU \end{bmatrix}. \qquad \square$$

In view of the preceding lemma, we can without loss of generality assume that A is of the form (2.1) and $k := \mathcal{E}_+(A)$. In this note we shall **always assume that** $1 \le k \le n - 1$ and shall leave the cases $k = 0$ and $k = n$ to the reader. Correspondingly, let $C = \begin{bmatrix} C_1 & C_2 \end{bmatrix}$ with $C_1 \in \mathbb{C}^{m \times k}, C_2 \in \mathbb{C}^{m \times (n-k)}$ and $P = \begin{bmatrix} P_{11} & P_{12} \\ P_{21} & P_{22} \end{bmatrix}$ with $P_{11} \in \mathbb{C}^{k \times k}, P_{22} \in \mathbb{C}^{(n-k) \times (n-k)}$ be the decompositions of the matrices C and P that are conformal with that of A. Then the Lyapunov equation (1.1) is equivalent to the first, second and fourth of the following four equations:

$$A_+^* P_{11} + P_{11} A_+ + C_1^* C_1 = 0, \tag{2.2}$$
$$A_-^* P_{21} + P_{21} A_+ + C_2^* C_1 = 0, \tag{2.3}$$
$$A_+^* P_{12} + P_{12} A_- + C_1^* C_2 = 0 \tag{2.4}$$

and

$$A_-^* P_{22} + P_{22} A_- + C_2^* C_2 = 0. \tag{2.5}$$

Finally, let $\mathcal{S}_k^{m \times m}(\Pi_+)$ denote the generalized Schur class of mvf's (matrix-valued functions) $s(\lambda)$, that are meromorphic in Π_+ and for which the kernel

$$\Lambda_\omega^s(\lambda) = \frac{I_m - s(\lambda)s(\omega)^*}{\rho_\omega(\lambda)}$$

has k negative squares on $\mathfrak{h}_s^+ \times \mathfrak{h}_s^+$, and $\mathfrak{h}_s^+$ denotes the domain of analyticity of s in Π_+.

3. A quick proof of Theorem 1.1

In this section we present a variant of the well-known short proof of Theorem 1.1 (see, e.g., [8] or [3]).

Proof. Assertion (1) of the theorem is verified in Lemma 2.1. Next, as $\sigma(A_\pm) \subset \Pi_\pm$, (2.2) and (2.5) both have unique solutions:

$$P_{11} = -\int_0^\infty e^{-tA_+^*} C_1^* C_1 e^{-tA_+} dt \le 0 \quad \text{and} \quad P_{22} = \int_0^\infty e^{tA_-^*} C_2^* C_2 e^{tA_-} dt \ge 0,$$

respectively. The observability of the pair (C, A) implies that the pairs (C_1, A_+) and (C_2, A_-) are observable, as is perhaps verified most easily by the Popov–Belevitch–Hautus test. Then Lemma 2.1 implies that P_{11} and P_{22} are invertible matrices and hence, by Schur complements, that

$$P = \begin{bmatrix} I_\kappa & P_{11}^{-1}P_{12} \\ 0 & I_{n-k} \end{bmatrix}^* \begin{bmatrix} P_{11} & 0 \\ 0 & P_{22} - P_{21}P_{11}^{-1}P_{12} \end{bmatrix} \begin{bmatrix} I_\kappa & P_{11}^{-1}P_{12} \\ 0 & I_{n-k} \end{bmatrix}.$$

Thus, by the Sylvester law of inertia,

$$\begin{aligned} \mathcal{E}_+(P) &= \mathcal{E}_+(P_{11}) + \mathcal{E}_+(P_{22} - P_{21}P_{11}^{-1}P_{12}) \\ &= \mathcal{E}_+(P_{22} - P_{21}P_{11}^{-1}P_{12}) \\ &= n - k = \mathcal{E}_-(A) \end{aligned}$$

as $P_{22} > 0, P_{11} < 0$ and $P_{12} = P_{21}^*$, and

$$\begin{aligned} \mathcal{E}_-(P) &= \mathcal{E}_-(P_{11}) + \mathcal{E}_-(P_{22} - P_{21}P_{11}^{-1}P_{12}) \\ &= \mathcal{E}_-(P_{11}) \\ &= k = \mathcal{E}_+(A). \end{aligned} \qquad \square$$

4. From the inertia theorem to a RKKS

In this section we will establish a finite-dimensional reproducing kernel Krein space $\mathcal{M}$ with a RK of the form (1.2) and then use the inertia theorem to obtain a coprime factorization formula for the rmvf $\Theta \in \mathcal{S}_{\mathcal{E}_-(P)}^{m \times m}(\Pi_+)$ of the form

$$\Theta = \Theta_1^{-1}\Theta_2,$$

where Θ_1 and Θ_2 are both inner rational matrix-valued functions (rmvf's) with respect to Π_+, and hence Blaschke–Potapov products. Moreover, we will give explicit realization formulas for Θ_1 and Θ_2 that are minimal.

Theorem 4.1. If $A \in \mathbb{C}^{n \times n}$, $C \in \mathbb{C}^{m \times n}$, the pair (C, A) is observable and $P = P^* \in \mathbb{C}^{n \times n}$ is a solution of the Lyapunov equation (1.1), then:

(1) $\sigma(A) \cap i\mathbb{R} = \emptyset$ and P is invertible, i.e., $\mathcal{E}_0(A) = \mathcal{E}_0(P) = 0$,

(2) The space

$$\mathcal{M} = \{ F(\lambda)u : u \in \mathbb{C}^n \} \tag{4.1}$$

with

$$F(\lambda) = C(\lambda I_n - A)^{-1}, \quad \lambda \in \mathbb{C} \setminus \sigma(A) \tag{4.2}$$

and indefinite inner product

$$\langle F(\lambda)u, F(\lambda)v \rangle_{\mathcal{M}} = v^* P u, \tag{4.3}$$

is an n-dimensional RKKS with RK

$$K_\omega(\lambda) = F(\lambda) P^{-1} F(\omega)^*, \quad \lambda, \omega \in \mathbb{C} \setminus \sigma(A). \tag{4.4}$$

(3) The RK may be expressed in the form

$$K_\omega(\lambda) = \frac{I_m - \Theta(\lambda)\Theta(\omega)^*}{\rho_\omega(\lambda)}, \quad \lambda, \omega \in \mathbb{C} \setminus \sigma(A), \tag{4.5}$$

where the $m \times m$ rmvf

$$\Theta(\lambda) = I_m - C(\lambda I_n - A)^{-1} P^{-1} C^* \tag{4.6}$$

admits the factorization

$$\Theta(\lambda) = \Theta_1(\lambda)^{-1} \Theta_2(\lambda), \tag{4.7}$$

with

$$\Theta_1(\lambda) = I_m + C_1(\lambda I_k - A_1)^{-1} P_{11}^{-1} C_1^*, \tag{4.8}$$

$$\Theta_2(\lambda) = I_m + \widetilde{C}(\lambda I_{n-k} - A_-)^{-1} Y \widetilde{C}^*, \tag{4.9}$$

$$A_1 = -P_{11}^{-1} A_+^* P_{11}, \widetilde{C} = C_2 - C_1 P_{11}^{-1} P_{12}$$

and

$$Y = -(P_{22} - P_{21} P_{11}^{-1} P_{12})^{-1}.$$

(4) The rmvf's Θ_1 and Θ_2 are finite Blaschke–Potapov products that are both inner with respect to Π_+ and are left coprime.

(5) The realizations (4.8) and (4.9) are minimal.

Proof. Assertion (1) is verified in Lemma 2.1. Next, since $\mathcal{M}$ is a RKKS with RK that is given by formula (4.4) and (C, A) is an observable pair, the columns of the rmvf F are linearly independent. Therefore, (2) holds.

The formula (4.5) in (3) may be verified by a straightforward calculation that uses the Lyapunov equation (1.1). Another direct calculation serves to verify the factorization formula and it is easily seen that P_{11} is invertible (just as in the proof in Section 3) and hence that $P_{22} - P_{12}P_{11}^{-1}P_{21}$ is also invertible (by the Schur complements formula for P) and Y is well defined. Thus, (3) holds.

The observability of the pairs (C_1, A_+) and (C_2, A_-) is inherited from the observability of the pair (C, A) (as is most easily seen by the Popov–Belevitch–Hautus test). Therefore, successive applications of Theorem 1.1 to equation (2.2) for the $k \times k$ matrix P_{11} and to equation (2.5) for the $(n - k) \times (n - k)$ matrix P_{22}, yields the implications

$$\mathcal{E}_\pm(A_+) = \mathcal{E}_\mp(P_{11}) \Longrightarrow \mathcal{E}_-(P_{11}) = k \Longrightarrow P_{11} < 0.$$

and

$$\mathcal{E}_\pm(A_-) = \mathcal{E}_\mp(P_{22}) \Longrightarrow \mathcal{E}_+(P_{22}) = n - k \Longrightarrow P_{22} > 0.$$

Thus, as $P_{11} < 0$ and $P_{12} = P_{21}^*$,

$$Y = -(P_{22} - P_{21}P_{11}^{-1}P_{12})^{-1} < 0.$$

A direct calculation shows that

$$\Theta_1(\lambda)^*\Theta_1(\lambda) - I_m = (\lambda + \lambda^*)C_1(\lambda I_k + A_+)^{-1}P_{11}(\lambda^* I_k + A_+^*)^{-1}C_1^*$$

and hence

$$\Theta_1(\lambda)^*\Theta_1(\lambda) = I_m \text{ on } i\mathbb{R} \quad \text{and} \quad \Theta_1(\lambda)^*\Theta_1(\lambda) \leq I_m \text{ in } \Pi_+.$$

A similar calculation shows that, if $Z = -Y^{-1}$, then

$$\Theta_2(\lambda)^*\Theta_2(\lambda) - I_m = -(\lambda + \lambda^*)\widetilde{C}Z^{-1}(\lambda^* I_{n-k} - A_-^*)^{-1}Z(\lambda I_{n-k} - A_-)^{-1}Z^{-1}\widetilde{C}^*$$
$$+ \widetilde{C}Z^{-1}(\lambda^* I_{n-k} - A_-^*)^{-1}X(\lambda I_{n-k} - A_-)^{-1}Z^{-1}\widetilde{C}^*$$

with $X = A_-^*Z + ZA_- + \widetilde{C}^*\widetilde{C}$. Now, multiplying equation (2.3) on the right by $P_{11}^{-1}P_{12}$ and subtracting it from equation (2.5), yields the equation

$$0 = A_-^*(P_{22} - P_{21}P_{11}^{-1}P_{12}) + P_{22}A_- - P_{21}A_+P_{11}^{-1}P_{12} + C_2^*C_2 - C_2^*C_1P_{11}^{-1}P_{12}$$
$$= A_-^*(-Y^{-1}) + (P_{22} - P_{21}P_{11}^{-1}P_{12} + P_{21}P_{11}^{-1}P_{12}A_- - P_{21}A_+P_{11}^{-)A-1}P_{12} + C_2^*\widetilde{C}$$
$$= A_-^*Z + ZA_- + P_{21}P_{11}^{-1}P_{12}A_- - P_{21}A_+P_{11}^{-1}P_{12} + C_2^*\widetilde{C} = \mathbf{M}.$$

Next, multiplying equation (2.4) on the right by $P_{21}P_{11}^{-1}$ and subtracting it from $\mathbf{M}$,

$$0 = A_-^*Z + ZA_- - P_{21}A_+P_{11}^{-1}P_{12} + C_2^*\widetilde{C} - P_{21}P_{11}^{-1}A_+^*P_{12} - P_{21}P_{11}^{-1}C_1^*C_2$$
$$= A_-^*Z + ZA_- - P_{21}P_{11}^{-1}(P_{11}A_+ + A_+^*P_{11})P_{11}^{-1}P_{12} + C_2^*\widetilde{C} + (\widetilde{C}^* - C_2^*)C_2$$
$$= A_-^*Z + ZA_- + P_{21}P_{11}^{-1}C_1^*C_1P_{11}^{-1}P_{12} - (\widetilde{C} - C_2)^*(\widetilde{C} - C_2) + \widetilde{C}^*\widetilde{C}$$
$$= A_-^*Z + ZA_- + \widetilde{C}^*\widetilde{C} + P_{21}P_{11}^{-1}C_1^*C_1P_{11}^{-1}P_{12} - (C_1P_{11}^{-1}P_{12})^*(C_1P_{11}^{-1}P_{12})$$
$$= A_-^*Z + ZA_- + \widetilde{C}^*\widetilde{C}.$$

Therefore,

$$\Theta_2(\lambda)^*\Theta_2(\lambda) - I_m = -(\lambda + \lambda^*)\widetilde{C}Z^{-1}(\lambda^* I_{n-k} - A_-^*)^{-1}Z(\lambda I_{n-k} - A_-)^{-1}Z^{-1}\widetilde{C}^*$$

and hence

$$\Theta_2(\lambda)^*\Theta_2(\lambda) = I_m \text{ on } i\mathbb{R} \quad \text{and} \quad \Theta_2(\lambda)^*\Theta_2(\lambda) \leq I_m \text{ in } \Pi_+.$$

As $\sigma(A_1), \sigma(A_-) \subset \Pi_-$, the rmvf's Θ_1 and Θ_2 are holomorphic in Π_+ and hence Θ_1 and Θ_2 are inner with respect to Π_+. Therefore, as all inner rmvf's are finite Blaschke–Potapov products, it follows that Θ_1 and Θ_2 are finite Blaschke–Potapov products.

Next, (5) is verified by establishing the observability and controllability of four pairs of matrices. This will be done in four steps.

5.1 *The pair (C_1, A_1) is observable.*

If there exist $\lambda \in \mathbb{C}$ and $u \in \mathbb{C}^k$, such that $C_1 u = 0$ and $A_1 u = \lambda u$, then $-P_{11}^{-1}A_+^* P_{11}u = \lambda u$ and equation (2.2) implies that

$$\begin{aligned}
0 &= A_+^* P_{11}u + P_{11}A_+u + C_1^* C_1 u \\
&= -\lambda P_{11}u + P_{11}A_+u = P_{11}(A_+u - \lambda u).
\end{aligned}$$

Thus, $A_+u = \lambda u$, as P_{11} is invertible, and hence the observability of the pair (C_1, A_+) implies $u = 0$.

5.2 *The pair $(A_1, P_{11}^{-1}C_1^*)$ is controllable.*

If there exist $\lambda \in \mathbb{C}$ and $0 \neq u \in \mathbb{C}^k$, such that $C_1 P_{11}^{-1}u = 0$ and $A_1^* u = \lambda u$, i.e., $-P_{11}A_+P_{11}^{-1}u = \lambda u$, then equation (2.2) implies that

$$0 = A_+^* u + P_{11}A_+P_{11}^{-1}u + C_1^* C_1 P_{11}^{-1}u = A_+^* u - \lambda u.$$

Thus, $\lambda \in \sigma(A_+^*) \cap \sigma(-P_{11}A_+P_{11}^{-1}) = \sigma(A_+^*) \cap \sigma(-A_+) = \emptyset$, as $\sigma(A_+) \subset \Pi_+$, a contradiction. Thus $u = 0$, i.e., the pair $(C_1 P_{11}^{-1}, A_1^*)$ is observable.

5.3 *The pair $(\widetilde{C}, A_-)$ is observable.*

If there exist $\lambda \in \mathbb{C}$ and a vector $u \in \mathbb{C}^{n-k}$, such that $\widetilde{C}u = 0$ and $A_-u = \lambda u$, then $C_2 u = C_1 P_{11}^{-1}P_{12}u$ and equations (2.2) and (2.4) imply that

$$\begin{aligned}
0 &= \lambda P_{12}u + A_+^* P_{12}u + C_1^* C_1 P_{11}^{-1}P_{12}u \\
&= (\lambda I_{n-k} + A_+^* - (A_+^* P_{11} + P_{11}A_+)P_{11}^{-1})P_{12}u \\
&= (\lambda I_{n-k} - P_{11}A_+P_{11}^{-1})P_{12}u = 0.
\end{aligned}$$

If $P_{12}u = 0$, then $C_2 u = 0$ and thus the observability of the pair (C_2, A_-) implies $u = 0$. If $P_{12}u \neq 0$, then $\lambda \in \sigma(P_{11}A_+P_{11}^{-1}) = \sigma(A_+)$. Then, $\lambda \in \sigma(A_+) \cap \sigma(A_-) \subset \Pi_+ \cap \Pi_- = \emptyset$, a contradiction and so $u = 0$.

5.4 *The pair $(A_-, Y\widetilde{C}^*)$ is controllable.*

The equation $A_-^* Z + ZA_- + \widetilde{C}^*\widetilde{C} = 0$, with $Z = -Y^{-1}$, implies

$$-YA_-^* - A_-Y + Y\widetilde{C}^*\widetilde{C}Y = 0. \tag{4.10}$$

If there exist $\lambda \in \mathbb{C}$ and $u \in \mathbb{C}^{n-k}$, such that $\widetilde{C}Yu = 0$ and $A_-^* u = \lambda u$, then, as follows from (4.10),

$$0 = -YA_-^* u - A_- Yu + Y\widetilde{C}^* \widetilde{C}Yu$$
$$= -\lambda Yu - A_- Yu = -(A_- + \lambda I_{n-k})Yu.$$

Thus, if $Yu \neq 0$, then $\lambda \in \sigma(-A_-)$. Therefore $\lambda \in \sigma(-A_-) \cap \sigma(A_-^*) = \emptyset$, as $\sigma(A_-) \subset \Pi_-$, and so $u = 0$ and the pair $(\widetilde{C}Y, A_-^*)$ is observable.

Finally, since the realization of Θ is also minimal and

$$\deg \Theta = \deg \Theta_1^{-1} + \deg \Theta_2,$$

the factorization (4.7) is left coprime.   □

Remark 4.2. *The factorization* (4.7) *is a special case of a general theorem by Krein and Langer (see, e.g.,* [7]) *which states that every* $\Theta \in \mathcal{S}_k^{m \times m}(\Pi_+)$, *admits a factorization*

$$\Theta = \Theta_1^{-1}\Theta_2,$$

where Θ_1 *is a Blaschke–Potapov product of degree* k, Θ_2 *is in the Schur class* $\mathcal{S}^{m \times m}(\Pi_+)$ *and* $\ker \Theta_2(\lambda)^* \cap \ker \Theta_1(\lambda)^* = \{0\}$.

5. From the RKKS to the inertia theorem

Theorem 5.1. *Let* $A \in \mathbb{C}^{n \times n}$ *and* $C \in \mathbb{C}^{m \times n}$. *If* $\sigma(A) \cap i\mathbb{R} = \emptyset$ *(i.e.,* $\mathcal{E}_0(A) = 0$), *the pair* (C, A) *is observable,* $F(\lambda)$ *is the* $m \times n$ *rmvf defined by the realization formula*

$$F(\lambda) = C(\lambda I_n - A)^{-1}, \tag{5.1}$$

$P = P^* \in \mathbb{C}^{n \times n}$ *is invertible (i.e.,* $\mathcal{E}_0(P) = 0$) *and the space*

$$\mathcal{M} = \{F(\lambda)u : u \in \mathbb{C}^n\}$$

endowed with the indefinite inner product

$$\langle F(\lambda)u, F(\lambda)v \rangle_{\mathcal{M}} = v^* Pu \tag{5.2}$$

is an n-*dimensional RKKS, with RK*

$$K_\omega(\lambda) = \frac{I_m - \Theta(\lambda)\Theta(\omega)^*}{\rho_\omega(\lambda)}, \quad \lambda, \omega \in \mathbb{C} \setminus \sigma(A)$$

where

$$\Theta(\lambda) = I_m - C(\lambda I_n - A)^{-1} P^{-1} C^*, \tag{5.3}$$

then P *is a solution of the Lyapunov equation* (1.1) *and* $\mathcal{E}_\pm(A) = \mathcal{E}_\mp(P)$.

Proof. It is readily checked that

$$K_\omega(\lambda) = F(\lambda)P^{-1}F(\omega)^*, \quad \lambda, \omega \in \mathbb{C} \setminus \sigma(A)$$

is a RK for the space $\mathcal{M}$. Therefore, by the uniqueness of the RK,

$$F(\lambda)P^{-1}F(\omega)^* = \frac{I_m - \Theta(\lambda)\Theta(\omega)^*}{\rho_\omega(\lambda)}, \quad \lambda, \omega \in \mathbb{C} \setminus \sigma(A), \tag{5.4}$$

which implies, by a straightforward calculation, that the matrix P is a solution of the Lyapunov equation (1.1).

Let $C = \begin{bmatrix} C_1 & C_2 \end{bmatrix}$ with $C_1 \in \mathbb{C}^{m \times k}$ and $P = \begin{bmatrix} P_{11} & P_{12} \\ P_{21} & P_{22} \end{bmatrix}$ with $P_{11} \in \mathbb{C}^{k \times k}$,

be conformal with the decomposition (2.1) of A with $A_+ \in \mathbb{C}^{k \times k}$ and $\sigma(A_\pm) \subset \Pi_\pm$.

The rest of the proof is divided into steps:

1. P_{11} and $P_{22} - P_{21}P_{11}^{-1}P_{12}$ *are invertible matrices.*

The verification is the same as the proof in Section 3.   ▼

2. *If Θ_1 and Θ_2 are given by formulas (4.8) and (4.9), then*

$$\Theta(\lambda) = \Theta_1(\lambda)^{-1}\Theta_2(\lambda).$$

Moreover, the realization formulas (4.8) and (4.9) are minimal.

The verification is the same as in steps 3 and 5 in the proof of Theorem 4.1.   ▼

3. *There exist $\omega_1, \ldots, \omega_n \in \mathbb{C} \setminus \sigma(A)$ and $v_1, \ldots, v_n \in \mathbb{C}^m$ such that*

$$W = \begin{bmatrix} F(\omega_1)^*\Theta_1(\omega_1)v_1 & \cdots & F(\omega_n)^*\Theta_1(\omega_n)^*v_n \end{bmatrix}$$

is invertible.

This follows from the fact that

$$\mathcal{M}_1 = \{F(\lambda)^*u : \lambda \in \Pi_+ \setminus \sigma(A), u \in \mathbb{C}^m\} = \mathbb{C}^m. \tag{5.5}$$

To verify the equality in (5.5), note that if there is a vector $v \in \mathbb{C}^n$ such that $v \perp \mathcal{M}_1$, then $\langle F(\lambda)^*u, v \rangle = \langle u, F(\lambda)v \rangle = 0$ for every $u \in \mathbb{C}^m$ and $\lambda \in \mathbb{C}_+ \setminus \sigma(A)$, and hence $F(\lambda)v = 0$ for every $\lambda \in \mathbb{C}_+ \setminus \sigma(A)$. But this implies that $v = 0$, since the pair (C, A) is observable. Therefore, $v \perp \mathcal{M}_1$ if and only if $v = 0$ and thus equality holds in (5.5).

Therefore, as the matrices $\Theta_1(\omega)$ and $\Theta_2(\omega)$ are all invertible matrices, except for a finite number of points ω in $\mathbb{C}$, there exist $\omega_1, \ldots, \omega_n \in \mathbb{C}_+ \setminus \sigma(A)$ and $v_1, \ldots, v_n \in \mathbb{C}^m$ such that the vectors

$$w_1 = F(\omega_1)^*\Theta_1(\omega_1)v_1, \ldots, w_n = F(\omega_n)^*\Theta_1(\omega_n)^*v_n$$

are linearly independent and so W is invertible.   ▼

4. *Let*

$$K_\omega^{\Theta_s}(\lambda) = \frac{I_m - \Theta_s(\lambda)\Theta_s(\omega)^*}{\rho_\omega(\lambda)}$$

be the RK of the finite-dimensional RKKS $\mathcal{H}(\Theta_s) = H_2^m \ominus \Theta_s H_2^m$, for $s = 1, 2$. If W_1 and W_2 are the $n \times n$ matrices defined by the formulas

$$(W_1)_{ij} = v_i^* K_{\omega_j}^{\Theta_1}(\omega_i) v_j \quad and \quad (W_2)_{ij} = v_i^* K_{\omega_j}^{\Theta_2}(\omega_i) v_j, \tag{5.6}$$

then $W_1 \geq 0$, $W_2 \geq 0$ and

$$W^* P^{-1} W = W_2 - W_1. \tag{5.7}$$

For every $x \in \mathbb{C}^n$ and $s = 1, 2$,

$$\langle W_s x, x \rangle = \sum_{i,j=1}^n x_j x_i^* (W_s)_{ij} = \sum_{i,j=1}^n x_i^* v_i^* K_{\omega_j}^{\Theta_s}(\omega_i) v_j x_j$$

$$= \left\langle \sum_{i=1}^n K_{\omega_i}^{\Theta_s} v_i x_i, \sum_{j=1}^n K_{\omega_j}^{\Theta_s} v_j x_j \right\rangle_{\mathcal{H}(\Theta_s)} \geq 0,$$

i.e., $W_s \geq 0$ for $s = 1, 2$. Next, equation (5.4) implies, for every $\lambda, \omega \in \mathbb{C} \setminus \sigma(A)$,

$$F(\lambda) P^{-1} F(\omega)^* = \frac{I_m - \Theta_1(\lambda)^{-1}\Theta_2(\lambda)\Theta_2(\omega)^*\Theta_1(\omega)^{-*}}{\rho_\omega(\lambda)},$$

i.e.,

$$\Theta_1(\lambda) F(\lambda) P^{-1} F(\omega)^* \Theta_1(\omega)^* = \frac{\Theta_1(\lambda)\Theta_1(\omega)^* - \Theta_2(\lambda)\Theta_2(\omega)^*}{\rho_\omega(\lambda)} \tag{5.8}$$

for every $\lambda, \omega \in \mathbb{C} \setminus \sigma(A)$. Therefore,

$$(W^* P^{-1} W)_{ij} = v_i^* \Theta_1(\omega_i) F(\omega_i) P^{-1} F(\omega_j)^* \Theta_1(\omega_j)^* v_j$$

$$= v_i^* \left[\frac{\Theta_1(\omega_i)\Theta_1(\omega_j)^* - \Theta_2(\omega_i)\Theta_2(\omega_j)^*}{\rho_{\omega_j}(\omega_i)} \right] v_j$$

$$= v_i^* \left[\frac{I_m - \Theta_2(\omega_i)\Theta_2(\omega_j)^*}{\rho_{\omega_j}(\omega_i)} \right] v_j - v_i^* \left[\frac{I_m - \Theta_1(\omega_i)\Theta_1(\omega_j)^*}{\rho_{\omega_j}(\omega_i)} \right] v_j$$

$$= v_i^* K_{\omega_j}^{\Theta_2}(\omega_i) v_j - v_i^* K_{\omega_j}^{\Theta_1}(\omega_i) v_j$$

$$= (W_2)_{ij} - (W_1)_{ij}$$

and thus (5.7) holds. ▼

5. $\mathcal{E}_+(P) \leq \operatorname{rank} W_2$ and $\mathcal{E}_-(P) \leq \operatorname{rank} W_1$.

Formula (5.7) can be reexpressed as

$$W^* P^{-1} W = \begin{bmatrix} I_n & I_n \end{bmatrix} \begin{bmatrix} -W_1 & 0 \\ 0 & W_2 \end{bmatrix} \begin{bmatrix} I_n \\ I_n \end{bmatrix}$$

and since W is invertible, the Sylvester law of inertia implies

$$\mathcal{E}_\pm(P) = \mathcal{E}_\pm(P^{-1}) = \mathcal{E}_\pm(W^* P^{-1} W) \leq \mathcal{E}_\pm \left(\begin{bmatrix} -W_1 & 0 \\ 0 & W_2 \end{bmatrix} \right).$$

Therefore, since $W_1 \geq 0$ and $W_2 \geq 0$,

$$\mathcal{E}_+\left(\begin{bmatrix} -W_1 & 0 \\ 0 & W_2 \end{bmatrix}\right) = \operatorname{rank} W_2 \quad \text{and} \quad \mathcal{E}_-\left(\begin{bmatrix} -W_1 & 0 \\ 0 & W_2 \end{bmatrix}\right) = \operatorname{rank} W_1,$$

as claimed.   ▼

6. $\operatorname{rank} W_1 \leq \deg \Theta_1$ *and* $\operatorname{rank} W_2 \leq \deg \Theta_2$.

Since $K_\omega^{\Theta_s}(\lambda) = F_s(\lambda) R_s F_s(\omega)^*$ for $s = 1, 2$, with

$$F_1(\lambda) = C_1(\lambda I_k - A_1)^{-1}, \quad F_2(\lambda) = \widetilde{C}(\lambda I_{n-k} - A_-)^{-1} \quad \text{and} \quad R_s > 0$$

and the exhibited realizations are minimal, the matrices

$$(W_s)_{ij} = v_i^* K_{\omega_j}^{\Theta_s}(\omega_i) v_j = v_i^* F_s(\omega_i) R_s F_s(\omega_j)^* v_j$$

are of the form $W_s = X_s^* R_s X_s$, where $X_s = \begin{bmatrix} F_s(\omega_1)^* v_1 & \cdots & F_s(\omega_n)^* v_n. \end{bmatrix}$ Therefore,

$$\operatorname{rank} W_s = \operatorname{rank}(X_s^* R_s X_s) \leq \operatorname{rank} R_s = \deg \Theta_s.$$ ▼

7. $\mathcal{E}_\pm(P) = \mathcal{E}_\mp(A)$.

It is well known (see, e.g., [3]) that if the realization

$$\widehat{\Theta}(\lambda) = \widetilde{D} + \widetilde{C}(\lambda I_n - \widetilde{A})^{-1} \widetilde{B}$$

is minimal, then $\deg \widehat{\Theta}$ is equal to the size of the matrix $\widetilde{A}$. Thus, Step 2 guarantees that $\deg \Theta_1 = \mathcal{E}_+(A)$ and $\deg \Theta_2 = \mathcal{E}_-(A)$, and hence, Steps 5 and 6 imply that $\mathcal{E}_+(P) \leq \mathcal{E}_-(A)$ and $\mathcal{E}_-(P) \leq \mathcal{E}_+(A)$. Therefore, as $\mathcal{E}_0(P) = \mathcal{E}_0(A) = 0$,

$$n = \mathcal{E}_+(P) + \mathcal{E}_-(P) \leq \mathcal{E}_-(A) + \mathcal{E}_+(A) = n,$$

which means that $\mathcal{E}_+(P) = \mathcal{E}_-(A)$ and $\mathcal{E}_-(P) = \mathcal{E}_+(A)$.   □

6. A disc analog

In this section we shall sketch an analog of the preceding analysis for the Stein equation

$$P - A^* P A = C^* C, \tag{6.1}$$

The notations $\mathbb{D}$ and $\mathbb{E}$ for the open unit disc and the exterior of the closed unit disc and $\pi_-(A), \pi_0(A)$ and $\pi_+(A)$ for the sum of the algebraic multiplicities of the eigenvalues of a matrix $A \in \mathbb{C}^{n \times n}$ in $\mathbb{E}, \mathbb{T}$ and $\mathbb{D}$, respectively, will be needed to formulate the version of the Carlson–Schneider inertia theorem under consideration. It is equivalent to Theorem 4 on p. 453 of [8].

Theorem 6.1. *If* $A \in \mathbb{C}^{n \times n}, C \in \mathbb{C}^{m \times n}, (C, A)$ *is an observable pair and* $P = P^* \in \mathbb{C}^{n \times n}$ *is a solution of the Stein equation* (6.1), *then*

(1) $\sigma(A) \cap \mathbb{T} = \emptyset$ *and* P *is invertible, i.e.,* $\pi_0(A) = \mathcal{E}_0(P) = 0$,
(2) $\pi_+(A) = \mathcal{E}_+(P)$ *and* $\pi_-(A) = \mathcal{E}_-(P)$.

This section is organized as follows: In subsection 6.1 we present some preliminary facts and notation for subsequent use. A short well-known proof of Theorem 6.1 based on elementary tools of linear system theory is presented in Subsection 6.2. In Subsection 6.3, Theorem 6.1 is used to establish a finite-dimensional reproducing kernel Krein space with a reproducing kernel of the special form

$$K_\omega(\lambda) = \frac{I_m - \Theta(\lambda)\Theta(\omega)^*}{\rho_\omega(\lambda)} \tag{6.2}$$

where $\rho_\omega(\lambda) = 1 - \lambda\omega^*$ and Θ admits a factorization $\Theta = \Theta_1^{-1}\Theta_2$, with Θ_1 and Θ_2 both inner rmvf's with respect to $\mathbb{D}$. Finally, a proof of Theorem 6.1 that is based on methods and formulas of reproducing kernel spaces is presented in Subsection 6.4.

6.1. Preliminaries

Lemma 6.2. *If the assumptions in Theorem 6.1 are in force, then $\sigma(A) \cap \mathbb{T} = \emptyset$ and without loss of generality A may be assumed to be of the form $J = A_\pm$ or*

$$A = \begin{bmatrix} A_- & 0 \\ 0 & A_+ \end{bmatrix}, \quad with \quad \sigma(A_-) \subset \mathbb{E} \quad and \quad \sigma(A_+) \subset \mathbb{D}. \tag{6.3}$$

Proof. The proof is separated into parts:

1. If $Au = \lambda u$ for some $u \in \mathbb{C}^n$ and $\lambda \in \mathbb{T}$, then $u^*A^* = \lambda^*u^*$ and hence (6.1) implies that

    $$0 = u^*Pu - u^*A^*PAu - u^*C^*Cu = -u^*C^*Cu.$$

    Thus, $CA^ku = \lambda^kCu = 0$, which implies that $u = 0$, since the pair (C, A) is observable. Therefore, $\sigma(A) \cap \mathbb{T} = \emptyset$.

2. As $\sigma(A) \cap \mathbb{T} = \emptyset$, $A = UJU^{-1}$, where J is a Jordan matrix of the form $J = A_\pm$ or

    $$J = \begin{bmatrix} A_- & 0 \\ 0 & A_+ \end{bmatrix} \quad with \quad \sigma(A_-) \subset \mathbb{E} \quad and \quad \sigma(A_+) \subset \mathbb{D}.$$

    Then P is an invertible Hermitian solution of equation (6.1) if and only if U^*PU is an invertible Hermitian solution of the equation

    $$U^*PU - J^*(U^*PU)J = (CU)^*CU.$$

    Moreover, $\sigma(J) \cap \mathbb{T} = \emptyset$, $\pi_\pm(J) = \pi_\pm(A)$, $\mathcal{E}_\pm(U^*PU) = \mathcal{E}_\pm(P)$ and the pair (CU, J) is observable, since

    $$\text{rank} \begin{bmatrix} \lambda I_n - A \\ C \end{bmatrix} = \text{rank} \begin{bmatrix} U & 0 \\ 0 & I_m \end{bmatrix} \begin{bmatrix} (\lambda I_n - J) \\ CU \end{bmatrix} U^{-1} = \text{rank} \begin{bmatrix} (\lambda I_n - J) \\ CU \end{bmatrix}. \quad \square$$

In view of the preceding lemma, we can without loss of generality assume that A is of the form (6.3) and let $k := \pi_-(A)$. Just as in the half-plane case, we shall **assume that** $1 \le k \le n - 1$. Correspondingly, let $C = \begin{bmatrix} C_1 & C_2 \end{bmatrix}$ with $C_1 \in \mathbb{C}^{m \times k}$, $C_2 \in \mathbb{C}^{m \times (n-k)}$ and $P = \begin{bmatrix} P_{11} & P_{12} \\ P_{21} & P_{22} \end{bmatrix}$ with $P_{11} \in \mathbb{C}^{k \times k}$, $P_{22} \in \mathbb{C}^{(n-k) \times (n-k)}$ be the decompositions of the matrices C and P that are conformal with that of A. Then the Stein equation (6.1) is equivalent to the first, second and fourth of the following four equations:

$$P_{11} - A_-^* P_{11} A_- = C_1^* C_1, \tag{6.4}$$
$$P_{21} - A_+^* P_{21} A_- = C_2^* C_1, \tag{6.5}$$
$$P_{12} - A_-^* P_{12} A_+ = C_1^* C_2 \tag{6.6}$$

and

$$P_{22} - A_+^* P_{22} A_+ = C_2^* C_2. \tag{6.7}$$

Finally, let $\mathcal{S}_\kappa^{m \times m}(\mathbb{D})$ denote the generalized Schur class of $m \times m$ mvf's $s(\lambda)$, that are meromorphic in $\mathbb{D}$ and for which the kernel

$$\Lambda_\omega^s(\lambda) = \frac{I_m - s(\lambda)s(\omega)^*}{\rho_\omega(\lambda)}$$

has κ negative squares on $\mathfrak{h}_s^+ \times \mathfrak{h}_s^+$, and $\mathfrak{h}_s^+$ denotes the domain of analyticity of s in $\mathbb{D}$.

6.2. A quick proof of Theorem 6.1

In this section we present a variant of the well-known short proof of Theorem 6.1

Proof. It is verified in Lemma 6.2 that $\sigma(A) \cap \mathbb{T} = \emptyset$. Next, as $\sigma(A_-) \subset \mathbb{E}$ and $\sigma(A_+) \subset \mathbb{D}$, (6.4) and (6.7) both have unique solutions:

$$P_{11} = -\sum_{j=0}^{\infty} (A_-^*)^{-j-1} C_1^* C_1 A_-^{-j-1} \leq 0, \text{ and } P_{22} = \sum_{j=0}^{\infty} (A_+^*)^j C_2^* C_2 A_+^j \geq 0$$

respectively. The observability of the pair (C, A) implies that the pairs (C_1, A_-) and (C_2, A_+) are observable, as is perhaps verified most easily by the Popov–Belevitch–Hautus test. Thus,

$$P_{11} u = 0 \Longrightarrow 0 \geq -\sum_{j=0}^{\infty} (C_1 A_-^{-j-1} u)^* (C_1 A_-^{-j-1} u) = 0$$
$$\Longrightarrow C_1 A_-^{-j-1} u = 0, \, j = 0, 1, \ldots \Longrightarrow u = 0,$$

i.e., P_{11} is invertible and by similar calculations it can be proved that P_{22} is also invertible. Therefore, $P_{11} < 0$, $P_{21} = P_{12}^*$ and $P_{22} - P_{12}^* P_{11}^{-1} P_{12} \geq P_{22} > 0$. By Schur complements,

$$P = \begin{bmatrix} I_k & P_{11}^{-1} P_{12} \\ 0 & I_{n-k} \end{bmatrix}^* \begin{bmatrix} P_{11} & 0 \\ 0 & P_{22} - P_{12}^* P_{11}^{-1} P_{12} \end{bmatrix} \begin{bmatrix} I_k & P_{11}^{-1} P_{12} \\ 0 & I_{n-k} \end{bmatrix}$$

is invertible and thus, by the Sylvester law of inertia,

$$\mathcal{E}_-(P) = \mathcal{E}_-(P_{11}) + \mathcal{E}_-(P_{22} - P_{21} P_{11}^{-1} P_{12})$$
$$= \mathcal{E}_-(P_{11}) = k = \pi_-(A)$$

and

$$\mathcal{E}_+(P) = \mathcal{E}_+(P_{11}) + \mathcal{E}_+(P_{22} - P_{21} P_{11}^{-1} P_{12})$$
$$= \mathcal{E}_+(P_{22} - P_{21} P_{11}^{-1} P_{12}) = n - k = \pi_+(A). \qquad \square$$

6.3. From the inertia theorem to a RKKS

In this section we will establish a finite-dimensional reproducing kernel Krein space $\mathcal{M}$ with a RK of the form (6.2) and then use the inertia theorem to obtain a coprime factorization formula for the rmvf $\Theta \in \mathcal{S}_{\mathcal{E}_-(P)}^{n \times n}(\mathbb{D})$ of the form

$$\Theta = \Theta_1^{-1} \Theta_2,$$

with Θ_1 and Θ_2 both inner rmvf's with respect to $\mathbb{D}$ and hence Blaschke–Potapov products. Moreover, we will give explicit realization formulas for Θ_1 and Θ_2 and prove that they are minimal.

Theorem 6.3. *If* $A \in \mathbb{C}^{n \times n}$, $C \in \mathbb{C}^{m \times n}$, *the pair* (C, A) *is observable and* $P = P^* \in \mathbb{C}^{n \times n}$ *is a solution of the Stein equation* (6.1), *then:*

(1) $\sigma(A) \cap \mathbb{T} = \emptyset$ *and* P *is invertible, i.e.,* $\pi_0(A) = \mathcal{E}_0(P) = 0$,

(2) *The space*

$$\mathcal{M} = \{F(\lambda)u : u \in \mathbb{C}^n\} \tag{6.8}$$

with

$$F(\lambda) = C(I_n - \lambda A)^{-1}, \quad \lambda^{-1} \in \mathbb{C} \setminus \sigma(A) \tag{6.9}$$

and indefinite inner product

$$\langle F(\lambda)u, F(\lambda)v \rangle_{\mathcal{M}} = v^* P u, \tag{6.10}$$

is an n-*dimensional RKKS with RK*

$$K_\omega(\lambda) = F(\lambda)P^{-1}F(\omega)^*, \quad \lambda^{-1}, \omega^{-1} \in \mathbb{C} \setminus \sigma(A). \tag{6.11}$$

(3) *The RK may be expressed in the form*

$$K_\omega(\lambda) = \frac{I_m - \Theta(\lambda)\Theta(\omega)^*}{\rho_\omega(\lambda)}, \quad \lambda^{-1}, \omega^{-1} \in \mathbb{C} \setminus \sigma(A), \tag{6.12}$$

where $\mu \in \mathbb{T}$ *and*

$$\Theta(\lambda) = I_m - \rho_\mu(\lambda)C(I_n - \lambda A)^{-1}P^{-1}(I_n - \mu^* A^*)^{-1}C^*. \tag{6.13}$$

The matrices P_{11} *and* $P_{22} - P_{21}P_{11}^{-1}P_{12}$ *are invertible and the rmvf* $\Theta(\lambda)$ *admits the factorization*

$$\Theta(\lambda) = \Theta_1(\lambda)^{-1}\Theta_2(\lambda), \tag{6.14}$$

with A *of the form* (6.3),

$$\Theta_1(\lambda) = I_m - (\lambda - \mu)C_1(I_k - \mu A_-)^{-1}P_{11}^{-1}(\lambda I_k - A_-^*)^{-1}C_1^* \tag{6.15}$$

$$\Theta_2(\lambda) = I_m - \rho_\mu(\lambda)F(\mu)Q_1(I_{n-k} - \mu A_+)(I_{n-k} - \lambda A_+)^{-1}Q_2 Q_1^* F(\mu)^*, \tag{6.16}$$

$$Q_1 = \begin{bmatrix} -P_{11}^{-1}P_{12} \\ I_{n-k} \end{bmatrix} \quad \text{and} \quad Q_2 = (P_{22} - P_{21}P_{11}^{-1}P_{12})^{-1}.$$

(4) *The rmvf's* Θ_1 *and* Θ_2 *are finite Blaschke–Potapov products that are both inner with respect to* $\mathbb{D}$ *and are left coprime.*

(5) *The realizations* (6.15) *and* (6.16) *are minimal.*

Proof. Assertion (1) is verified in Lemma 6.2. Next, since $\mathcal{M}$ is a RKKS with RK that is given by formula (6.11) and (C, A) is an observable pair, the columns of the rmvf F are linearly independent. Therefore, (2) holds.

Formula (6.12) in (3) may be verified by a straightforward calculation that uses the Stein equation (6.1). Moreover, it is easily seen that P_{11} is invertible (just as in the proof in Section 6.2 and hence that $P_{22} - P_{12}P_{11}^{-1}P_{21}$ is also invertible (by the Schur complements formula for P) and Q_1 and Q_2 are well defined. Next, a lengthy calculation that takes advantage of the formulas (6.13), (6.15) and

$$(\lambda I_m - A^*)^{-1}C^*C(I_m - \lambda A)^{-1} = \frac{1}{\lambda}P(I_m - \lambda A)^{-1} + \frac{1}{\lambda}A^*(\lambda I_m - A^*)^{-1}P,$$

serves to verify that $\Theta_1(\lambda)\Theta(\lambda) = \Theta_2(\lambda)$ and hence that (6.14) holds.

The observability of the pairs (C_1, A_-) and (C_2, A_+) is inherited from the observability of the pair (C, A) (as is most easily seen by the Popov–Belevitch–Hautus test). Therefore, successive applications of Theorem 6.1 to equation (6.4) for the $k \times k$ matrix P_{11} and to equation (6.7) for the $(n - k) \times (n - k)$ matrix P_{22}, yields the implications

$$\pi_\pm(A_-) = \mathcal{E}_\pm(P_{11}) \Longrightarrow \mathcal{E}_-(P_{11}) = k \Longrightarrow P_{11} < 0$$

and

$$\pi_\pm(A_+) = \mathcal{E}_\pm(P_{22}) \Longrightarrow \mathcal{E}_+(P_{22}) = n - k \Longrightarrow P_{22} > 0.$$

Thus, as $P_{22} > 0$ and $P_{12} = P_{21}^*$,

$$Q_2 = (P_{22} - P_{21}P_{11}^{-1}P_{12})^{-1} > 0.$$

A direct calculation shows that

$$I_m - \Theta_1(\lambda)\Theta_1(\lambda)^* = -(1 - |\lambda|^2)\Phi_1 P_{11}\Phi_1^*, \tag{6.17}$$

where $\Phi_1 = C_1(I_k - \mu A_-)^{-1}P_{11}^{-1}(\lambda I_k - A_-^*)^{-1}(I_k - \mu^* A_-^*)$. A similar calculation, based on the fact that

$$K := P - P\begin{bmatrix} P_{11}^{-1} & 0 \\ 0 & 0 \end{bmatrix}P = \begin{bmatrix} 0 & 0 \\ 0 & Q_2^{-1} \end{bmatrix} \tag{6.18}$$

is a solution of the Riccati equation

$$K - A^*KA = (I_n - \mu^* A^*)KP^{-1}F(\mu)^*F(\mu)P^{-1}K(I_n - \mu A), \tag{6.19}$$

shows that

$$I_m - \Theta_2(\lambda)^*\Theta_2(\lambda) = (1 - |\lambda|^2)\Phi_2 K\Phi_2^*, \tag{6.20}$$

where $\Phi_2 = F(\mu)P^{-1}(I_m - \lambda^* A^*)^{-1}(I_m - \mu^* A^*)$. Since $\sigma(A_+) \subset \mathbb{D}$ and $\sigma(A_-^*) \subset \mathbb{E}$, the rmvf's Θ_1 and Θ_2 are holomorphic in $\mathbb{D}$ and as $P_{11} < 0$ and $K \geq 0$, it follows from formulas (6.17) and (6.20) that the rmvf's Θ_1 and Θ_2 are inner with respect to $\mathbb{D}$. Therefore, as all inner rmvf's are finite Blaschke–Potapov products, Θ_1 and Θ_2 are finite Blaschke–Potapov products.

Next, since observability and controllability for realizations of the form (6.13), and in particular for the realizations (6.15) and (6.16), may also be verified by applying the Popov–Belevitch–Hautus test to the pairs (C, A) and (A, B), with $B = P^{-1}(I_m - \mu^* A^*)^{-1} C^*$, it suffices to check that

$$\operatorname{rank} \begin{bmatrix} \lambda I_n - A \\ C \end{bmatrix} = n \quad \text{and} \quad \operatorname{rank} \begin{bmatrix} \lambda I_n - A & B \end{bmatrix} = n$$

for all points $\lambda \in \mathbb{C}$ (see, e.g., Theorem 3.5 in [1], where realizations of this form are discussed). This will be done in four steps.

6.1 *The pair* $(C_1(I_k - \mu A_-)^{-1} P_{11}^{-1}, A_-^*)$ *is observable.*

If there exist $u \neq 0$ in $\mathbb{C}^k$ and $\lambda \in \mathbb{C}$ such that $C_1(I_k - \mu A_-)^{-1} P_{11}^{-1} u = 0$ and $A_-^* u = \lambda u$, then

$$\begin{aligned} 0 &= C_1^* C_1 (I_k - \mu A_-)^{-1} P_{11}^{-1} u \\ &= [P_{11}(I_k - \mu A_-) + (\mu I_k - A^*) P_{11} A_-](I_k - \mu A_-)^{-1} P_{11}^{-1} u \\ &= (\mu I_m - A_-^*)[(\mu - \lambda)^{-1} I_k + P_{11} A_-(I_k - \mu A_-)^{-1} P_{11}^{-1}] u \\ &= (\mu I_m - A_-^*)[P_{11}(I_k - \mu A_-) + (\mu - \lambda) P_{11} A_-](I_k - \mu A_-)^{-1} P_{11}^{-1} u \\ &= (\mu I_m - A_-^*) P_{11}(I_k - \lambda A_-)(I_k - \mu A_-)^{-1} P_{11}^{-1} u \end{aligned}$$

which, as $(\mu I_m - A_-^*) P_{11}$ is invertible and $u \neq 0$, implies that $\lambda \in \mathbb{D}$. On the other hand, $\lambda \in \sigma(A_-^*) \implies \lambda \in \mathbb{E}$, which is a contradiction.

6.2 *The pair* (A_-^*, C_1^*) *is controllable.*

This follows immediately from the observability of the pair (C_1, A_-).

6.3 *The pair* $(F(\mu) Q_1 (I_{n-k} - \mu A_+), A_+)$ *is observable.*

If there exist $\xi \neq 0$ in $\mathbb{C}^{n-k}$ and $\lambda \in \mathbb{C}$ such that $F(\mu) Q_1 (I_{n-k} - \mu A_+) \xi = 0$ and $A_+ \xi = \lambda \xi$, then

$$\begin{aligned} 0 &= F(\mu) \begin{bmatrix} -P_{11}^{-1} P_{12} \\ I_{n-k} \end{bmatrix} (1 - \mu\lambda)\xi \\ &= (1 - \mu\lambda)[-C_1(I_k - \mu A_-)^{-1} P_{11}^{-1} P_{12} + C_2(I_{n-k} - \mu A_+)^{-1}]\xi \\ &= (1 - \mu\lambda)[-C_1^* C_1(I_k - \mu A_-)^{-1} P_{11}^{-1} P_{12} + C_1^* C_2(1 - \mu\lambda)^{-1}]\xi \\ &= (\mu\lambda - 1) C_1^* C_1(I_k - \mu A_-)^{-1} P_{11}^{-1} P_{12}\xi + C_1^* C_2\xi. \end{aligned}$$

The equations (6.4) and (6.6) imply that

$$C_1^* C_1 = (I_k - \mu^* A_-^*) P_{11} + \mu^* A_-^* P_{11}(I_k - \mu A_-)$$

and

$$C_1^* C_2 \xi = (I_k - \lambda A_-^*) P_{12}\xi$$

and hence

$$
\begin{aligned}
0 &= [(\mu\lambda - 1)(I_k - \mu^* A_-^*)P_{11}(I_k - \mu A_-)^{-1}P_{11}^{-1} \\
&\quad + (\mu\lambda - 1)\mu^* A_-^* + (I_k - \lambda A_-^*)]P_{12}\xi \\
&= [(\mu\lambda - 1)(I_k - \mu^* A_-^*)P_{11}(I_k - \mu A_-)^{-1}P_{11}^{-1} + (I_k - \mu^* A_-^*)]P_{12}\xi \\
&= (I_k - \mu^* A_-^*)P_{11}[(\mu\lambda - 1)I_k + I_k - \mu A_-](I_k - \mu A_-)^{-1}P_{11}^{-1}P_{12}\xi \\
&= \mu(I_k - \mu^* A_-^*)P_{11}(\lambda I_k - A_-)(I_k - \mu A_-)^{-1}P_{11}^{-1}P_{12}\xi.
\end{aligned}
$$

Therefore, since $\mu \in \mathbb{T}$, $\lambda \in \mathbb{D}$ and P_{11} is invertible, $P_{12}\xi = 0$ and then by the preceding calculation

$$
C_2(I_{n-k} - \mu A_+)^{-1}\xi = 0 \implies C_2(1 - \mu\lambda)^{-1}\xi = 0 \implies C_2\xi = 0.
$$

Finally, $A_+\xi = \lambda\xi$ and $C_2\xi = 0$, together with the observability of (C_2, A_+) lead to contradiction.

6.4 *The pair $(A_+, Q_2 Q_1^* F(\mu)^*)$ is controllable.*

We shall prove that the pair $(F(\mu)Q_1 Q_2, A_+^*)$ is observable: If there exist nonzero $\xi \in \mathbb{C}^{n-k}$ and $\lambda \in \mathbb{C}$ such that $A_+^*\xi = \lambda\xi$ and $\mu F(\mu)Q_1 Q_2\xi = 0$, then, by (6.19),

$$
\begin{aligned}
Q_1 F(\mu)^* &F(\mu)Q_1 \\
&= \begin{bmatrix} 0 & I_{n-k} \end{bmatrix} K P^{-1} F(\mu)^* F(\mu) P^{-1} K \begin{bmatrix} 0 \\ I_{n-k} \end{bmatrix} \\
&= \begin{bmatrix} 0 & I_{n-k} \end{bmatrix} (I_n - \mu^* A^*)^{-1}(K - A^* K A)(I_n - \mu A)^{-1} \begin{bmatrix} 0 \\ I_{n-k} \end{bmatrix} \\
&= (I_{n-k} - \mu^* A_+^*)^{-1}(Q_2^{-1} - A_+^* Q_2^{-1} A_+)(I_{n-k} - \mu A_+)^{-1} \\
&= Q_2^{-1}(I_{n-k} - \mu A_+)^{-1} + (I_{n-k} - \mu^* A_+^*)^{-1}\mu^* A_+^* Q_2^{-1},
\end{aligned}
$$

and hence

$$
\begin{aligned}
0 &= \xi^* Q_2 Q_1^* F(\mu)^* F(\mu) Q_1 Q_2\xi \\
&= \xi^*(I_{n-k} - \mu A_+)^{-1}Q_2\xi + \xi^* Q_2(I_{n-k} - \mu^* A_+^*)^{-1}\mu^* A_+^*\xi \\
&= \xi^*(1 - \mu\lambda^*)^{-1}Q_2\xi + \xi^* Q_2(1 - \mu^*\lambda)^{-1}\mu^*\lambda\xi \\
&= \left(\frac{(1 - \mu^*\lambda) + \mu^*\lambda(1 - \mu\lambda^*)}{|1 - \mu\lambda^*|^2} \right) \xi^* Q_2\xi = \left(\frac{1 - |\lambda|^2}{|1 - \mu\lambda^*|^2} \right) \xi^* Q_2\xi.
\end{aligned}
$$

Therefore, as $\lambda \notin \mathbb{T}$ and $Q_2 > 0$, we have $\xi = 0$, which is a contradiction.

Finally, since the realization of Θ is also minimal and

$$
\deg \Theta = \deg \Theta_1^{-1} + \deg \Theta_2,
$$

the factorization (6.14) is left coprime. $\qquad\square$

6.4. From the RKKS to the inertia theorem

Theorem 6.4. *Let $A \in \mathbb{C}^{n \times n}$ and $C \in \mathbb{C}^{m \times n}$. If $\sigma(A) \cap \mathbb{T} = \emptyset$ (i.e., $\pi_0(A) = 0$), the pair (C, A) is observable, $F(\lambda)$ is the $m \times n$ rmvf defined by the realization formula*

$$F(\lambda) = C(I_n - \lambda A)^{-1}, \tag{6.21}$$

$P = P^* \in \mathbb{C}^{n \times n}$ is invertible (i.e., $\mathcal{E}_0(P) = 0$) and the space

$$\mathcal{M} = \{F(\lambda)u : u \in \mathbb{C}^n\}$$

endowed with the indefinite inner product

$$\langle F(\lambda)u, F(\lambda)v \rangle_{\mathcal{M}} = v^* P u \tag{6.22}$$

is an n-dimensional RKKS, with RK

$$K_\omega(\lambda) = \frac{I_m - \Theta(\lambda)\Theta(\omega)^*}{\rho_\omega(\lambda)}, \quad \lambda^{-1}, \omega^{-1} \in \mathbb{C} \setminus \sigma(A)$$

where

$$\Theta(\lambda) = I_m - \rho_\mu(\lambda)C(I_n - \lambda A)^{-1}P^{-1}(I_n - \mu^* A^*)^{-1}C^*, \tag{6.23}$$

then P is a solution of the Stein equation (6.1) and $\pi_\pm(A) = \mathcal{E}_\pm(P)$.

Proof. It is readily checked that

$$K_\omega(\lambda) = F(\lambda)P^{-1}F(\omega)^*, \quad \lambda^{-1}, \omega^{-1} \in \mathbb{C} \setminus \sigma(A)$$

is a RK for the space $\mathcal{M}$. Therefore, by uniqueness of the RK,

$$F(\lambda)P^{-1}F(\omega)^* = \frac{I_m - \Theta(\lambda)\Theta(\omega)^*}{\rho_\omega(\lambda)}, \quad \lambda^{-1}, \omega^{-1} \in \mathbb{C} \setminus \sigma(A), \tag{6.24}$$

which implies, by a straightforward calculation, that the matrix P is a solution of the Stein equation (6.1).

Let $C = \begin{bmatrix} C_1 & C_2 \end{bmatrix}$ with $C_1 \in \mathbb{C}^{m \times k}$ and $P = \begin{bmatrix} P_{11} & P_{12} \\ P_{21} & P_{22} \end{bmatrix}$ with $P_{11} \in \mathbb{C}^{k \times k}$, be conformal with the decomposition (6.3) of A with $A_- \in \mathbb{C}^{k \times k}$.

The rest of the proof is divided into steps:

1. *P_{11} and $P_{22} - P_{21}P_{11}^{-1}P_{12}$ are invertible matrices.*
The verification is the same as the proof in Section 6.2.   ▼

2. *If Θ_1 and Θ_2 are given by formulas (6.15) and (6.16), then*

$$\Theta(\lambda) = \Theta_1(\lambda)^{-1}\Theta_2(\lambda).$$

Moreover, the realization formulas (6.15) and (6.16) are minimal.
The verification is the same as in steps 3 and 5 in the proof of Theorem 6.3.   ▼

3. *Repeat steps 3 to 6 of Theorem 5.1. They are applicable to this setting, with only minor changes in the proof.*

4. *$\mathcal{E}_\pm(P) = \pi_\pm(A)$.*

It is well known (see, e.g., [1]) that if the realization

$$\Theta(\lambda) = \widetilde{D} + \rho_\mu(\lambda)\widetilde{C}(\lambda I - \widetilde{A})^{-1}\widetilde{B} \quad (\text{or} \quad \Theta(\lambda) = \widetilde{D} + \rho_\mu(\lambda)\widetilde{C}(I - \lambda\widetilde{A})^{-1}\widetilde{B})$$

is minimal, then $\deg\Theta$ is equal to the size of the matrix $\widetilde{A}$. Thus, Step 2 guarantees that $\deg\Theta_1 = \pi_-(A)$ and $\deg\Theta_2 = \pi_+(A)$, and hence, Steps 5 and 6 of Theorem 5.1 imply that $\mathcal{E}_+(P) \leq \pi_+(A)$ and $\mathcal{E}_-(P) \leq \pi_-(A)$. Therefore, as $\mathcal{E}_0(P) = \pi_0(A) = 0$,

$$n = \mathcal{E}_+(P) + \mathcal{E}_-(P) \leq \pi_+(A) + \pi_-(A) = n,$$

which means that $\mathcal{E}_+(P) = \pi_+(A)$ and $\mathcal{E}_-(P) = \pi_-(A)$. □

References

[1] D. Alpay and H. Dym, *On a New Class of Realization Formulas and Their Application*, Linear Algebra and Its Applications, **241-243** (1996), 3–84.

[2] C.T. Chen, *A generalization of the inertia theorem*, SIAM J. Appl. Math. **25** (1973), 158–161.

[3] H. Dym, *Linear Algebra in Action*, Graduate Studies in Mathematics, **78** American Mathematical Society, Providence, 2007.

[4] V. Derkach and H. Dym, *On Linear Fractional Transformations Associated with Generalized J-Inner Matrix Functions*, Integral Equations and Operator Theory, **65** (2009), 1–50.

[5] H. Dym and D. Volok, *Zero Distribution of matrix polynomials*, Linear Algebra and its Applications, **425** (2007), 714–738.

[6] I. Gohberg, P. Lancaster and L. Rodman, *Matrix Polynomials*, Academic Press, New York-London, 1982.

[7] M.G. Krein and H. Langer, *Über die verallgemeinerten Resolventen und die charakteristische Funktion eines isometrischen Operators im Raume Π_k*, Hilbert space Operators and Operator Algebras (*Proc. Intern. Conf., Tihany*, 1970); Colloq. Math. Soc. Janos Bolyai, **5** (1972), 353–399.

[8] P. Lancaster and M. Tismenetsky, *The Theory of Matrices*, second edition with applications, Computer Science and Applied Mathematics, Academic Press, New York, 1985.

[9] L. Lerer and M. Tismenetsky, *The Bezoutian and The Eigenvalue-Separation Problem for Matrix Polynomials*, Integral Equations and Operator Theory, **5** (1982), 386–445.

[10] H.K. Wimmer, *Inertia theorems for matrices, controllability, and linear vibrations*, Linear Algebra and its Applications **8** (1974), 337–343

Harry Dym and Motke Porat
Department of Mathematics
The Weizmann Institute of Science
Rehovot 76100, Israel
e-mail: harry.dym@weizmann.ac.il
        motke.porat@weizmann.ac.il

Operator Theory:
Advances and Applications, Vol. 237, 127–144
© 2013 Springer Basel

On the Kernel and Cokernel of Some Toeplitz Operators

Torsten Ehrhardt and Ilya M. Spitkovsky

To Professor Leonia Lerer, in celebration of his seventieth birthday.

Abstract. We show that the kernel and/or cokernel of a block Toeplitz operator $T(G)$ are trivial if its matrix-valued symbol G satisfies the condition $G(t^{-1})G(t)^* = I_N$. As a consequence, the Wiener–Hopf factorization of G (provided it exists) must be canonical. Our setting is that of weighted Hardy spaces on the unit circle. We extend our result to Toeplitz operators on weighted Hardy spaces on the real line, and also Toeplitz operators on weighted sequence spaces.

Mathematics Subject Classification (2010). Primary 47B35. Secondary 47A68, 47B30.

Keywords. Toeplitz operators, Wiener–Hopf factorization, partial indices, discrete convolution operators.

1. Introduction

The *Wiener algebra* W by definition consists of functions f defined on the unit circle $\mathbb{T} = \{ z \in \mathbb{C} \colon |z| = 1 \}$ and having absolutely convergent Fourier series:

$$f(t) = \sum_{j=-\infty}^{\infty} f_j t^j, \quad \text{where} \quad \sum_{j=-\infty}^{\infty} |f_j| < \infty. \tag{1.1}$$

The functions $f \in W$ with $f_j = 0$ for all $j \le 0$ (resp., $j \ge 0$) form the subalgebra $W_{\pm}$ of W the elements of which admit analytic continuations to the interior (resp., exterior, including the point of infinity) of $\mathbb{T}$. It is a classical result by Gohberg and Krein [9] (see also the monographs [7, 12] and a survey [8] for more detailed bibliographical information and far reaching generalizations) that any invertible matrix function[1] $G \in W^{N \times N}$ admits a representation

$$G(t) = G_-(t)\Lambda(t)G_+(t), \qquad t \in \mathbb{T}, \tag{1.2}$$

where $G_+^{\pm 1} \in W_+^{N \times N}$, $G_-^{\pm 1} \in W_-^{N \times N}$,

$$\Lambda(t) = \mathrm{diag}[t^{\varkappa_1}, \ldots, t^{\varkappa_N}] \tag{1.3}$$

and $\varkappa_1, \ldots, \varkappa_N \in \mathbb{Z}$. Representation (1.2) plays a central role in a variety of applications, including systems of convolution type equations on the half-line and Toeplitz operators $T(G)$ with matrix symbols G. In particular, the *defect numbers* (i.e., dimensions of the kernel and cokernel) of these operators are expressed in terms of the *partial indices* $\varkappa_j$. Namely,

$$\dim \ker T(G) = -\sum_{\varkappa_j \leq 0} \varkappa_j, \qquad \dim \ker T(G)^* = \sum_{\varkappa_j \geq 0} \varkappa_j. \tag{1.4}$$

Note that the partial indices are defined by G uniquely, up to their order. However, for $N > 1$ the partial indices, and the factorization itself, are generically not stable: a necessary and sufficient stability criterion (also going back to [9]) reads

$$\max\{\varkappa_i - \varkappa_j : i, j = 1, \ldots, N\} \leq 1, \tag{1.5}$$

and thus requires the a priori knowledge of the partial indices. The latter is available in some particular cases, e.g., for rational, triangular, or sectorial matrix function, see, e.g., [8], but in general the problem remains open.

One recent result in this direction, obtained by Voronin[2] [13], claims that all the partial indices are equal to zero (and thus (1.5) holds) for matrix functions G satisfying

$$G(t^{-1})G(t)^* = I_N, \quad t \in \mathbb{T}. \tag{1.6}$$

The proof, published in [14], makes use of the description of all factorizations (1.2) of a given matrix function $G \in W^{N \times N}$. In this paper, we propose a different approach, which provides information about the defect numbers of $T(G)$ with G satisfying (1.6) but not necessarily lying in $W^{N \times N}$ and, moreover, not necessarily factorable. Namely, we will show that (under certain mild additional conditions on the spaces where the operators act) at least one of the defect numbers is zero. Note that according to Coburn's lemma this property holds for general Toeplitz operators with scalar non-zero symbols, but fails starting with $N = 2$. Additional conditions on the matrix symbol G under which the property persists are of great interest. Some such conditions, analytic in nature and thus very different from (1.6), were established in [6].

In Section 3 we consider $T(G)$ with measurable bounded symbols G acting on weighted Hardy spaces on the unit circle or the real line. Section 4 deals with operators acting on weighted discrete ℓ^p. Dealing with weighted spaces presents an

---

[1] Here and below we are using the standard notational convention: given any set X, $X^{M \times N}$ stands for the set of all $M \times N$ matrices with the entries in X, and $X^{M \times 1}$ is abbreviated to X^M.
[2] Voronin works with matrix functions defined on the real line $\mathbb{R}$ but the transition between $\mathbb{R}$ and $\mathbb{T}$ is obvious via an appropriate linear fractional transformation.

additional difficulty, and certain nested properties had to be established in order to overcome it. These properties are tackled in Section 2, along with the related results on the kernels of homogeneous Riemann–Hilbert problems.

We end this introduction with some basic observations about condition (1.6). Firstly, the set of matrix functions satisfying (1.6) forms a group under multiplication. Furthermore, such matrix functions can be defined arbitrarily on a half-circle (without loss of generality, say for $\mathrm{Im}(t) > 0$); the values on the complementary half-circle are then determined by (1.6) uniquely. Note also that (1.6) holds if G is even (that is, $G(t^{-1}) = G(t)$) and unitary valued.

2. Homogeneous Riemann–Hilbert problems on the unit circle

For $1 \leq p < \infty$ and a given positive weight ϱ on the unit circle $\mathbb{T}$, let $L^p(\mathbb{T}; \varrho)$ denote the space of all measurable functions f defined on $\mathbb{T}$ and such that

$$\|f\|_{p,\varrho} := \left(\int_0^{2\pi} |f(e^{ix})\varrho(e^{ix})|^p \, dx \right)^{1/p} < \infty. \tag{2.1}$$

In case $\varrho \equiv 1$ we simply write $L^p(\mathbb{T})$. Throughout the paper we will assume that

$$1 < p < \infty, \quad \frac{1}{p} + \frac{1}{q} = 1, \tag{2.2}$$

and require that the weight ϱ satisfies

$$\varrho \in L^p(\mathbb{T}) \quad \text{and} \quad \varrho^{-1} \in L^q(\mathbb{T}). \tag{2.3}$$

The conditions (2.3) imply that $L^\infty(\mathbb{T}) \subseteq L^p(\mathbb{T}; \varrho) \subseteq L^1(\mathbb{T})$. Therefore we can define the Fourier coefficients of a function $f \in L^p(\mathbb{T}; \varrho)$,

$$f_n = \frac{1}{2\pi} \int_0^{2\pi} f(e^{ix}) e^{-inx} \, dx. \tag{2.4}$$

We also introduce the weighted Hardy spaces

$$L_+^p(\mathbb{T}; \varrho) = \left\{ f \in L^p(\mathbb{T}; \varrho) \; : \; f_n = 0 \text{ for all } n < 0 \right\}, \tag{2.5}$$

$$L_-^p(\mathbb{T}; \varrho) = \left\{ f \in L^p(\mathbb{T}; \varrho) \; : \; f_n = 0 \text{ for all } n > 0 \right\}, \tag{2.6}$$

as well as

$$L_{-,0}^p(\mathbb{T}; \varrho) = \left\{ f \in L^p(\mathbb{T}; \varrho) \; : \; f_n = 0 \text{ for all } n \geq 0 \right\}. \tag{2.7}$$

In order to present our result, the following notation will be handy. For a matrix or vector function ϕ defined on $\mathbb{T}$, we introduce the "tilde" operation,

$$\tilde{\phi}(t) = \phi(t^{-1}), \qquad t \in \mathbb{T},$$

as well as the complex adjoint function ϕ^*,

$$\phi^*(t) = \overline{\phi(t)}^T, \qquad t \in \mathbb{T},$$

which is the function obtained by taking the transpose and the complex conjugate pointwise on $\mathbb{T}$. The complex adjoint and the tilde operation commute with each other. Therefore, the notation $\tilde{\phi}^*$ is unambiguous. We also note that

$$\phi \in L_\pm^p(\mathbb{T}; \varrho) \quad \Longrightarrow \quad \tilde{\phi} \in L_\mp^p(\mathbb{T}; \tilde{\varrho})$$

and

$$\phi \in L_\pm^p(\mathbb{T}; \varrho) \quad \Longrightarrow \quad \phi^* \in L_\mp^p(\mathbb{T}; \varrho).$$

The main results of this section are based on sufficient conditions which state that certain spaces are nested. These conditions will be analyzed first in the following lemma. Notice that we only need the much simpler "if" parts. The proof of the "only if" parts is provided for completeness' sake.

Given (2.2), we also introduce $r \in (1, \infty]$ by

$$\frac{1}{r} + \frac{1}{\max\{p, q\}} = \frac{1}{\min\{p, q\}}.$$

Equivalently,

$$r = \begin{cases} \frac{pq}{|p-q|} & \text{if } p \neq q, \\ +\infty & \text{if } p = q \,(= 2). \end{cases} \tag{2.8}$$

Lemma 2.1. *Let (2.2) and (2.8) hold. Then:*
 (a) $L^p(\mathbb{T}, \tilde{\varrho}) \subseteq L^q(\mathbb{T}; \varrho^{-1})$ *if and only if* $p \geq q$ *and* $\varrho^{-1}\tilde{\varrho}^{-1} \in L^r(\mathbb{T})$.
 (b) $L^q(\mathbb{T}; \tilde{\varrho}^{-1}) \subseteq L^p(\mathbb{T}; \varrho)$ *if and only if* $q \geq p$ *and* $\varrho\tilde{\varrho} \in L^r(\mathbb{T})$.

Proof. The "if" parts follow easily from Hölder's inequality using (2.8). Therefore, we restrict ourselves to the "only if" parts.

(a): The inclusion means that

$$|f\tilde{\varrho}|^p \in L^1(\mathbb{T}) \quad \Rightarrow \quad |f\varrho^{-1}|^q \in L^1(\mathbb{T}).$$

By the substitution $g = |f\tilde{\varrho}|$ this is equivalent to

$$g^p \in L^1(\mathbb{T}) \quad \Rightarrow \quad (g\tilde{\varrho}^{-1}\varrho^{-1})^q \in L^1(\mathbb{T}).$$

In case $p \geq q$ make the substitution $h = g^q$ to conclude that

$$h \in L^{p/q}(\mathbb{T}) \quad \Rightarrow \quad h\tilde{\varrho}^{-q}\varrho^{-q} \in L^1(\mathbb{T}).$$

By the closed graph theorem, the corresponding linear operator is bounded, and this map gives rise to a bounded linear functional on $L^{p/q}(\mathbb{T})$. It follows that $\tilde{\varrho}^{-q}\varrho^{-q} \in L^{p/q}(\mathbb{T})' = L^{p/(p-q)}(\mathbb{T})$, which implies the above.

In case $p < q$, we make a substitution $h = g^p$ to conclude that

$$h \in L^1(\mathbb{T}) \quad \Rightarrow \quad h\tilde{\varrho}^{-p}\varrho^{-p} \in L^{q/p}(\mathbb{T}).$$

Again by the closed graph theorem, the corresponding linear operator must be bounded. There exists $\varepsilon > 0$ such that $E_\varepsilon = \{t \in \mathbb{T} : \tilde{\varrho}(t)^{-p}\varrho(t)^{-p} \geq \varepsilon\}$ has positive measure. Then for each measurable subset $E \subseteq E_\varepsilon$, when taking the characteristic function $h = \chi_E$,

$$\varepsilon\mu(E)^{p/q} = \|\varepsilon\chi_E\|_{q/p} \leq \|\chi_E\tilde{\varrho}^{-p}\varrho^{-p}\|_{q/p} \leq C\|\chi_E\|_1 = C\mu(E).$$

Thus, $0 < \varepsilon/C \leq \mu(E)^{1-p/q}$. Since we can find a sequence of measurable sets $E^{(k)} \subset E_\varepsilon$ with $\mu(E^{(k)}) > 0$ but $\mu(E^{(k)}) \to 0$, a contradiction follows.

Part (b) can be proved similarly, by replacing p with q and ϱ with ϱ^{-1}. $\square$

Based on Lemma 2.1 we will now show that a certain homogeneous Riemann–Hilbert problem has only a trivial solution. A corresponding result holds also for the "adjoint" Riemann–Hilbert problem.

Proposition 2.2. *Let G be an $N \times N$ matrix-valued measurable function on $\mathbb{T}$ satisfying $\tilde{G}^*(t)G(t) = I_N$, and let ϱ satisfy (2.3). Then:*

(a) *If $p \geq q$ and $\varrho^{-1}\tilde{\varrho}^{-1} \in L^r(\mathbb{T})$, then the equation*

$$G(t)f_+(t) = f_-(t) \tag{2.9}$$

with $f_+ \in L_+^p(\mathbb{T};\varrho)^N$ and $f_- \in L_{-,0}^p(\mathbb{T};\varrho)^N$ has only the trivial solution $f_+ = f_- = 0$.

(b) *If $q \geq p$ and $\varrho\tilde{\varrho} \in L^r(\mathbb{T})$, then the equation*

$$G^*(t)h_+(t) = h_-(t) \tag{2.10}$$

with $h_+ \in L_+^q(\mathbb{T};\varrho^{-1})^N$ and $h_- \in L_{-,0}^q(\mathbb{T};\varrho^{-1})^N$ has only the trivial solution $h_+ = h_- = 0$.

Proof. (a): First, notice that the conditions of Lemma 2.1(a) hold. Thus,

$$L^p(\mathbb{T}, \tilde{\varrho}) \subseteq L^q(\mathbb{T}; \varrho^{-1}), \tag{2.11}$$

which is what we are going to use below. Now assume that (2.9) holds. Passing to the complex adjoint and applying the tilde operation yields

$$\tilde{f}_+^*(t)\tilde{G}^*(t) = \tilde{f}_-^*(t).$$

Multiplying the equations together and using $\tilde{G}^*(t)G(t) = I_N$ we obtain

$$\tilde{f}_+^*(t)f_+(t) = \tilde{f}_-^*(t)f_-(t),$$

which is a scalar function since we are multiplying a row with a column vector function. Indeed, using the components of

$$f_\pm(t) = \Big(f_{\pm,1}(t), \ldots, f_{\pm,N}(t)\Big)^T$$

the previous equation reads

$$\sum_{k=1}^N \tilde{f}_{+,k}^*(t)f_{+,k}(t) = \sum_{k=1}^N \tilde{f}_{-,k}^*(t)f_{-,k}(t). \tag{2.12}$$

Because of (2.11) and Hölder's inequality, each of the occuring products is in $L^1(\mathbb{T})$. Furthermore, since $\tilde{f}_{+,k}^* \in L_+^p(\mathbb{T};\tilde{\varrho}) \subseteq L_+^q(\mathbb{T};\varrho^{-1})$, again by (2.11), it follows that each product $\tilde{f}_{+,k}^* f_{+,k}$ and thus the left-hand side of (2.12) is in $L_+^1(\mathbb{T})$. For similar reasons the right-hand belongs to $L^1_{-,0}(\mathbb{T})$. Since $L^1_{+}(\mathbb{T})$ and $L^1_{-,0}(\mathbb{T})$ have a trivial intersection, we conclude that

$$\tilde{f}^*_+(t) f_+(t) = \sum_{k=1}^{N} \tilde{f}^*_{+,k}(t) f_{+,k}(t) = 0. \tag{2.13}$$

There are now two possibilities to finish the proof, both instructive in their own ways. Firstly, one could use the Fourier coefficients $[f_+]_n \in \mathbb{C}^N$ of $f_+(t)$,

$$f_+(t) = \sum_{n=0}^{\infty} [f_+]_n t^n, \qquad |t| = 1,$$

noting that

$$\tilde{f}^*_+(t) = \sum_{n=0}^{\infty} \overline{[f_+]_n}^T t^n, \qquad |t| = 1, \tag{2.14}$$

involves the complex conjugates and the transpose. Thus

$$0 = (\tilde{f}^*_+ f_+)(t) = \sum_{n=0}^{\infty} t^n \sum_{k=0}^{n} \overline{[f_+]_k}^T [f_+]_{n-k},$$

and it follows that for each $n \geq 0$,

$$\sum_{k=0}^{n} \overline{[f_+]_k}^T [f_+]_{n-k} = 0.$$

Consider $n = 0$ to obtain $\overline{[f_+]_0}^T [f_+]_0 = 0$ and thus $[f_+]_0 = 0$. Next consider $n = 2$ to obtain $[f_+]_1 = 0$, then $n = 4$ to get $[f_+]_2 = 0$, and so on. Hence all Fourier coefficients are zero. This implies $f_+(t) = 0$ and thus $f_-(t) = 0$.

An alternative way of reasoning is to realize that both $f_+(z)$ and $\tilde{f}^*_+(z) = \overline{f_+(\bar{z})}^T$ admit analytic continuations into the unit disk $\{ z \in \mathbb{C} : |z| < 1 \}$. In the special case of z being real (and $|z| < 1$) we have $\tilde{f}^*_+(z) = \overline{f_+(z)}^T$, and (2.13) becomes

$$0 = \overline{f_+(z)}^T f_+(z), \text{ for } z \text{ real, } |z| < 1.$$

Since the values of f_+ are vectors, this is a zero sum of non-negative real numbers. Therefore, $f_+(z) = 0$ for z real, and by analytic continuation we obtain the same for all $|z| < 1$. As for a function in $L^1_+(\mathbb{T})$ there is a one-to-one correpondence between its analytic continuation into the unit disk and the boundary values (a.e.) on $\mathbb{T}$, we obtain that $f_+(t) = 0$ for $|t| = 1$ and consequently $f_-(t) = 0$.

(b): We remark that the assumptions and Lemma 2.1 imply that

$$L^q(\mathbb{T}, \tilde{\varrho}^{-1}) \subseteq L^p(\mathbb{T}; \varrho). \tag{2.15}$$

Thus the result can be obtained from part (a) by interchanging p with q and ϱ with ϱ^{-1}. For G^* the corresponding condition $\tilde{G}(t) G^*(t) = I_N$ holds. □

3. Toeplitz operators on the unit circle and real line

Given a Banach space X, we denote by $\mathcal{L}(X)$ the space of all bounded linear operators acting on X. Also, we use the notation

$$\alpha(A) = \dim \ker A \quad \text{and} \quad \beta(A) = \dim X/\overline{\operatorname{Im} A} \ (= \dim \ker A^*)$$

for $A \in \mathcal{L}(X)$. Therein $A^* \in \mathcal{L}(X')$ stands for the adjoint of A.

Toeplitz operators on $L_+^p(\mathbb{T}; \varrho)$. If we assume, in addition to (2.3), that the weight ϱ satisfies the Hunt–Muckenhoupt–Wheeden (or A_p) condition,

$$\sup_I \left(\frac{1}{|I|} \int_I \varrho(t)^p \, |dt| \right)^{1/p} \left(\frac{1}{|I|} \int_I \varrho(t)^{-q} \, |dt| \right)^{1/q} < \infty, \tag{3.1}$$

where the supremum is taken over all subarcs I of $\mathbb{T}$, then it is possible to consider Toeplitz operators on $L_+^p(\mathbb{T})$. In fact, the Hunt–Muckenhoupt–Wheeden condition is necessary and sufficient for the boundedness of Riesz projection

$$P : \sum_{n=-\infty}^{\infty} f_n t^n \mapsto \sum_{n=0}^{\infty} f_n t^n \tag{3.2}$$

on the space $L^p(\mathbb{T}; \varrho)$. Under these conditions the image of P equals the Hardy space $L_+^p(\mathbb{T}; \varrho)$, and the kernel of P equals $L_{-,0}^p(\mathbb{T}; \varrho)$, see [11] or [1] and references therein.

For $G \in L^\infty(\mathbb{T})^{N \times N}$ the block *Toeplitz operator* acting on $L_+^p(\mathbb{T}; \varrho)^N$ is defined by

$$T(G) : f \mapsto P(Gf), \tag{3.3}$$

where Gf stands for the pointwise product on $\mathbb{T}$ of the matrix-valued function G with the vector-valued function f. Under the above assumptions $T(G)$ is a bounded linear operator with norm

$$\|T(G)\|_{\mathcal{L}(L_+^p(\mathbb{T}; \varrho)^N)} \leq C_{p,\varrho} \|G\|_{L^\infty(\mathbb{T})^{N \times N}}, \tag{3.4}$$

where $C_{p,\varrho} = \|P\|_{\mathcal{L}(L^p(\mathbb{T}; \varrho))}$.

Theorem 3.1. *Let $G \in L^\infty(\mathbb{T})^{N \times N}$, and assume that $\widetilde{G}^*(t)G(t) = I_N$. Consider $T(G)$ on $L_+^p(\mathbb{T}; \varrho)^N$, with ϱ satisfying (3.1).*

(a) *If $p \geq q$ and $\varrho^{-1}\tilde{\varrho}^{-1} \in L^r(\mathbb{T})$, then $\alpha(T(G)) = 0$.*
(b) *If $q \geq p$ and $\varrho\tilde{\varrho} \in L^r(\mathbb{T})$, then $\beta(T(G)) = 0$.*

Proof. (a): If the Toeplitz operator $T(G)$ has a non-trivial kernel, then there exists a nonzero $f_+ \in L_+^p(\mathbb{T}; \varrho)$ such that

$$0 = T(G)f_+ = P(Gf_+),$$

which is equivalent to

$$f_- = Gf_+ \in L_{-,0}^p(\mathbb{T}; \varrho).$$

Now Proposition 2.2 implies $f_+ = 0$, which is a contradiction.

(b): The dual space of $L_+^p(\mathbb{T}; \varrho)^N$ can be identified with $L_+^q(\mathbb{T}; \varrho^{-1})^N$ via $\Lambda : L_+^q(\mathbb{T}; \varrho^{-1})^N \to (L_+^p(\mathbb{T}; \varrho)^N)'$,

$$(\Lambda g)(f) = \int_0^{2\pi} \overline{g(e^{ix})}^T f(e^{ix}) \, dx.$$

Under this identification it is easy to see that the adjoint operator of $T(G)$ (acting on $L_+^p(\mathbb{T}; \varrho)^N$) is the operator $T(G^*)$ acting on $L_+^q(\mathbb{T}; \varrho^{-1})^N$. Therefore, one can rely on part (a) and just interchange p with q and ϱ with ϱ^{-1} to obtain the desired result. $\qquad\square$

Factorization. An $N \times N$ matrix-valued measurable function G defined on $\mathbb{T}$ possesses a *factorization* in $L^p(\mathbb{T}; \varrho)$ if it can be written in the form

$$G(t) = G_-(t)\Lambda(t)G_+(t), \qquad t \in \mathbb{T},$$

such that

$$
\begin{aligned}
G_+ &\in L_+^q(\mathbb{T}; \varrho^{-1})^{N \times N}, & G_- &\in L_-^p(\mathbb{T}; \varrho)^{N \times N}, \\
G_+^{-1} &\in L_+^p(\mathbb{T}; \varrho)^{N \times N}, & G_-^{-1} &\in L_-^q(\mathbb{T}; \varrho^{-1})^{N \times N},
\end{aligned}
\tag{3.5}
$$

where

$$\Lambda(t) = \operatorname{diag}[t^{\varkappa_1}, \dots, t^{\varkappa_N}],$$

and $\varkappa_1, \dots, \varkappa_N \in \mathbb{Z}$ are called the *partial indices*. Notice that if G possesses a factorization, then G and G^{-1} belong to $L^1(\mathbb{T})$.

This is a weighted version of the L^p-factorization as defined in [12]. Its existence guarantees that the respective homogeneous Riemann–Hilbert problem has finitely many linearly independent solutions, while the closure $\mathfrak{M}$ of the range $R_G = \{f_- + Gf_+\}$ has finite codimension and, moreover, R_G is "rationally closed", that is, contains all rational vector functions belonging to $\mathfrak{M}$. If $G \in L^\infty(\mathbb{T})^{N \times N}$, then the Toeplitz operator $T(G)$ is Fredholm on $L_+^p(\mathbb{T}; \varrho)^N$ if and only if G is $L^p(\mathbb{T}; \varrho)$ factorable and in addition the operator $G_- P_+ G_-^{-1}$ is bounded in the metric of $L^p(\mathbb{T}; \varrho)^N$.

In the following theorem we assume (2.2), (2.3) and (2.8); however, (3.1) need not hold.

Theorem 3.2. *Let $G \in L^1(\mathbb{T})^{N \times N}$ satisfy $\widetilde{G}^*(t)G(t) = I_N$, and assume in addition that G possesses a factorization in $L^p(\mathbb{T}; \varrho)$.*

(a) *If $p \geq q$ and $\varrho^{-1}\tilde{\varrho}^{-1} \in L^r(\mathbb{T})$, then the partial indices $\varkappa_k \geq 0$.*
(b) *If $q \geq p$ and $\varrho\tilde{\varrho} \in L^r(\mathbb{T})$, then the partial indices $\varkappa_k \leq 0$.*
(c) *If $p = 2$ and both $\varrho\tilde{\varrho}$ and $\varrho^{-1}\tilde{\varrho}^{-1}$ belong to $L^\infty(\mathbb{T})$, then the factorization is canonical, i.e., $\varkappa_k = 0$ and*

$$G(t) = G_-(t)G_+(t). \tag{3.6}$$

(d) *If the factorization is canonical and if the assumptions of (a) or (b) hold, then one can choose the factors to satisfy $\widetilde{G}_-^* G_- = \widetilde{G}_+^* G_+ = I_N$.*

Proof. (a): Assume that for some k, $\varkappa_k < 0$. Then define

$$f_+(t) = G_+^{-1}(t)e_k$$

where e_k is the kth unit vector in $\mathbb{C}^N$. Applying $G(t) = G_-(t)\Lambda(t)G_+(t)$ we obtain

$$G(t)f_+(t) = G_-(t) \cdot t^{\varkappa_k}e_k =: f_-(t).$$

It is now easy to see that $0 \not\equiv f_+ \in L_+^p(\mathbb{T};\varrho)^N$ and $f_- \in L_{-,0}^p(\mathbb{T};\varrho)^N$, whence we get a contradiction to Proposition 2.2(a).

(b): Passing to the complex conjugates in the factorization we see that G^* possesses a factorization in $L^q(\mathbb{T};\varrho^{-1})$,

$$G^*(t) = G_+^*(t)\,\mathrm{diag}[t^{-\varkappa_1},\ldots,t^{-\varkappa_N}]G_-^*(t),$$

with partial indices $-\varkappa_1,\ldots,-\varkappa_N$. Now we can apply the results of part (a), interchanging p with q and replacing ϱ by ϱ^{-1}, or argue similar as in (a) and apply Proposition 2.2(b).

(c): This follows directly from (a) and (b).

(d): From the factorization $G = G_-G_+$ we obtain $\widetilde{G}^* = \widetilde{G}_+^*\widetilde{G}_-^*$ and through inversion and using $G = (\widetilde{G}^*)^{-1}$,

$$G_-G_+ = G = (\widetilde{G}^*)^{-1} = (\widetilde{G}_-^*)^{-1}(\widetilde{G}_+^*)^{-1}.$$

Under the assumptions of (a) we have $L^p(\mathbb{T};\tilde{\varrho}) \subseteq L^q(\mathbb{T};\varrho^{-1})$. Consequently,

$$\widetilde{G}_-^* G_- = (\widetilde{G}_+^*)^{-1} G_+^{-1},$$

with the left and right-hand sides lying in $L_-^1(\mathbb{T})^{N\times N}$ and $L_+^1(\mathbb{T})^{N\times N}$, respectively (compare (3.5)). Thus, each of the products $(\widetilde{G}_+^*)^{-1}G_+^{-1}$ and $\widetilde{G}_-^*G_-$ is identically equal to some $C \in \mathbb{C}^{N\times N}$, which must be nonsingular. In particular, using the analyticity of $G_+(z)$ and $\widetilde{G}_+^*(z) = \overline{G_+(\bar{z})}^T$ for $|z| < 1$, we get

$$C = (G_+(0)^*)^{-1}G_+(0)^{-1},$$

which is therefore positive definite. Letting

$$H_-(t) = G_-(t)G_+(0), \qquad H_+(t) = G_+(0)^{-1}G_+(t),$$

we see that $G = H_-H_+$ is also a factorization, while $\widetilde{H}_-^* H_- = \widetilde{H}_+^* H_+ = I_N$ holds. If the assumptions stated in (b) hold, then we have $L^q(\mathbb{T};\tilde{\varrho}^{-1}) \subseteq L^p(\mathbb{T};\varrho)$ and we consider

$$G_-^{-1}(\widetilde{G}_-^*)^{-1} = G_+\widetilde{G}_+^*,$$

and proceed analogously. $\square$

Examples. We now provide some simple examples illustrating Theorems 3.2 and 3.1, as well as the essentiality of the nesting conditions in order for these theorems to hold.

Both examples feature a scalar piecewise continuous function. For such functions a Fredholmness criterion and index formula were established by Gohberg and Krupnik, see [10] or more recent [4, 5]. Moreover, it is shown in [12] that piecewise continuous functions not satisfying the Fredholmness criterion, do not admit a factorization in the above sense. With these considerations in mind, the statements below are easily established.

Example 3.3. We consider $1 < p < \infty$, $\varrho \equiv 1$, and the following function which has two jump discontinuities at $t = \pm i$. Let $\beta \in \mathbb{R}$ and define

$$\phi(e^{ix}) = \begin{cases} e^{i\pi\beta} & \text{if } -\pi/2 < x < \pi/2 \\ e^{-i\pi\beta} & \text{if } \pi/2 < x < 3\pi/2. \end{cases}$$

This (scalar) function satisfies the condition $\tilde{\phi}^*(t)\phi(t) = 1$. The "size" of the jump discontinuities is given by

$$\frac{1}{2\pi} \arg \frac{\phi(i-0)}{\phi(i+0)} = \beta \quad \text{and} \quad \frac{1}{2\pi} \arg \frac{\phi(-i-0)}{\phi(-i+0)} = -\beta.$$

The function can be written as

$$\phi(t) = (-t/i)^{\beta}(t/i)^{-\beta},$$

and its factorization in $L^p(\mathbb{T})$ can be easily obtained from there. Its specific form will depend on the relation between β and p.

<u>Case 1:</u> If $|\beta| < \min\{\frac{1}{p}, \frac{1}{q}\}$, then a canonical factorization is given by

$$\phi(t) = \left[(1 - i/t)^{-\beta}(1 + i/t)^{\beta}\right] \cdot \left[(1 - t/i)^{\beta}(1 + t/i)^{-\beta}\right].$$

Moreover, $T(\phi)$ is invertible on $L^p(\mathbb{T})$.

<u>Case 2:</u> If $1/p < \beta < 1/q$, then a factorization with $\varkappa_1 = 1$ is given by

$$\phi(t) = \left[i(1 - i/t)^{-\beta+1}(1 + i/t)^{\beta}\right] \cdot t \cdot \left[(1 - t/i)^{\beta-1}(1 + t/i)^{-\beta}\right].$$

The operator $T(\phi)$ is Fredholm on $L^p(\mathbb{T})$ and has a trivial kernel and a cokernel of dimension one.

<u>Case 3:</u> If $1/q < \beta < 1/p$, then a factorization with $\varkappa_1 = -1$ is given by

$$\phi(t) = \left[i(1 - i/t)^{-\beta}(1 + i/t)^{\beta-1}\right] \cdot t^{-1} \cdot \left[(1 - t/i)^{\beta}(1 + t/i)^{-\beta+1}\right].$$

The operator $T(\phi)$ is Fredholm on $L^p(\mathbb{T})$ and has kernel dimension one and a trivial cokernel.

These results are in agreement with Theorems 3.1 and 3.2.

<u>Case 4:</u> If $\beta \in \mathbb{Z} + \{1/p, 1/q\}$, then ϕ possesses no factorization and $T(\phi)$ is not Fredholm. However, by Theorem 3.1, $T(\phi)$ has a trivial kernel in case $p \geq q$ and a trivial cokernel in case $q \geq p$.

These four cases essentially cover all values for β since adding an integer to β changes the function ϕ by at most a sign.

Example 3.4. Let $p = 2$, fix two distinct points $\tau_1, \tau_2 \in \mathbb{T}$ with $\text{Im}(\tau_k) > 0$, and consider the weight

$$\varrho(t) = |1 - \tau_1|^{\alpha_1}|1 - \bar{\tau}_1|^{\alpha_1}|1 - \tau_2|^{\alpha_2}|1 - \bar{\tau}_2|^{\alpha_2}$$

with $|\alpha_k| < 1/2$, which guarantees that ϱ satisfies (2.3) and in fact the A_2 condition. Introduce the function

$$\phi(t) = (-t/\tau_1)^{\beta_1}(-t/\bar{\tau}_1)^{-\beta_1}(-t/\tau_2)^{\beta_2}(-t/\bar{\tau}_2)^{-\beta_2}$$

with $\beta_1, \beta_2 \in \mathbb{R}$. This function satisfies $\tilde{\phi}^*\phi = 1$. We can factor $\phi = \phi_-\phi_+$ with

$$\phi_-(t) = (1 - t/\tau_1)^{-\beta_1+1}(1 - t/\bar{\tau}_1)^{\beta_1}(1 - t/\tau_2)^{-\beta_2}(1 - t/\bar{\tau}_2)^{\beta_2-1}\tau_1/\bar{\tau}_2,$$

$$\phi_+(t) = (1 - t/\tau_1)^{\beta_1-1}(1 - t/\bar{\tau}_1)^{-\beta_1}(1 - t/\tau_2)^{\beta_2}(1 - t/\bar{\tau}_2)^{-\beta_2+1}.$$

This is a *canonical* factorization in $L^p(\mathbb{T}; \varrho)$ if and only if

$$-1/2 + \alpha_1 < \beta_1 - 1 < 1/2 + \alpha_1, \qquad -1/2 + \alpha_2 < \quad \beta_2 \quad < 1/2 + \alpha_2,$$
$$-1/2 + \alpha_1 < \quad -\beta_1 \quad < 1/2 + \alpha_1, \qquad -1/2 + \alpha_2 < -\beta_2 + 1 < 1/2 + \alpha_2.$$

For instance, we can choose the values $\alpha_1 = -1/4$, $\alpha_2 = 1/4$, and $\beta_k \in (1/4, 3/4)$. It is easy to verify that $\tilde{\phi}^*_\pm\phi_\pm \neq 1$, and that this also holds if we modify the factors by some constant. This contrasts the statement of Theorem 3.2(d). Clearly, neither the assumptions of (a) nor (b) hold.

Toeplitz operators on $L^p_+(\mathbb{R}; \mu)$. For a weight μ on the real line $\mathbb{R}$, let the weighted space $L^p(\mathbb{R}; \mu)$ consist of all measurable functions f defined on $\mathbb{R}$ for which

$$\|f\|_{p,\mu} := \left(\int_{-\infty}^\infty |f(x)\mu(x)|^p \, dx \right)^{1/p} < \infty.$$

We will assume $\mu \in L^p(\mathbb{R}; (1 + |x|)^{-1})$ and $\mu^{-1} \in L^q(\mathbb{R}; (1 + |x|)^{-1})$, and furthermore the Hunt–Muckenhoupt–Wheeden (or A_p) condition on $\mathbb{R}$,

$$\sup_I \left(\frac{1}{|I|} \int_I \mu(x)^p \, dx \right)^{1/p} \left(\frac{1}{|I|} \int_I \mu(x)^{-q} \, dx \right)^{1/q} < \infty$$

with the supremum taken over all finite intervals. This condition is equivalent to the boundedness of the projection $P = (I + S)/2$ on $L^p(\mathbb{R}; \mu)$, where S is the singular integral operator on $\mathbb{R}$. In fact, the image and the kernel of P equal $L^p_+(\mathbb{R}; \mu)$ and $L^p_-(\mathbb{R}; \mu)$, respectively, the Hardy spaces of all functions in $L^p(\mathbb{R}; \mu)$ which admit an analytic continuation into the upper/lower complex half-plane (for details, see, e.g., [2, p. 302]).

For $H \in L^\infty(\mathbb{R})^{N \times N}$ the *Toeplitz operator* on $L^p_+(\mathbb{R}; \mu)^N$ is defined by

$$T(H)f = P(Hf).$$

We now have the following result. Therein the notation $\tilde{\mu}(x) = \mu(-x)$, $\tilde{H}(x) = H(-x)$, and $H^*(x) = \overline{H(x)}^T$ is used, along with by now standard (2.2), (2.8).

Theorem 3.5. *Let $H \in L^\infty(\mathbb{R})^{N \times N}$, and assume that $\tilde{H}^*(x)H(x) = I_N$. Consider $T(H)$ on $L^p_+(\mathbb{R}; \mu)^N$.*

(a) *If $p \geq q$ and $\mu^{-1}\tilde{\mu}^{-1} \in L^r(\mathbb{R})$, then $\alpha(T(H)) = 0$.*
(b) *If $q \geq p$ and $\mu\tilde{\mu} \in L^r(\mathbb{R})$, then $\beta(T(H)) = 0$.*

Proof. (a): The result can be proved either in analogy to Theorem 3.1, or by noting that the map Y defined by

$$(Yf)(x) = \frac{2}{i+x} f\left(\frac{i-x}{i+x}\right)$$

is an isometric isomorphism from $L^p(\mathbb{T}; \varrho)$ onto $L^p(\mathbb{R}; \mu)$, where

$$\mu(x) = \frac{2^{1/p-1}}{(1+x^2)^{1/p-1/2}} \varrho\left(\frac{i-x}{i+x}\right).$$

Furthermore, Y maps $L^p_+(\mathbb{T}; \varrho)$ and $L^p_{-,0}(\mathbb{T}; \varrho)$ onto $L^p_+(\mathbb{R}; \mu)$ and $L^p_-(\mathbb{R}; \mu)$, respectively. Moreover, $YT(G)Y^{-1} = T(H)$ with $H(x) = G((i-x)/(i+x))$. Hence the statements about defect numbers of $T(H)$ can be reduced to the corresponding statements for $T(G)$ on the space $L^p_+(\mathbb{T}; \varrho)$. We also remark that ϱ satisfies the A_p condition on $\mathbb{T}$ if and only if μ satisfies the A_p condition on $\mathbb{R}$. Finally, $\tilde{H}^* H = I_N$ implies that $\tilde{G}^* G = I_N$. We now compute that (since $1/p - 1/2 = 1/2(1/p - 1/q)$)

$$\mu(x)^{-1}\tilde{\mu}(x)^{-1} = 2^{-2/q}(1+x^2)^{1/p-1/q}\varrho(\tfrac{i-x}{i+x})^{-1}\tilde{\varrho}(\tfrac{i-x}{i+x})^{-1}.$$

In view of (2.8), it follows that $\mu^{-1}\tilde{\mu}^{-1} \in L^r(\mathbb{R})$ if and only if $\varrho^{-1}\tilde{\varrho}^{-1} \in L^r(\mathbb{T})$. This proves part (a).

(b): This can be proved analogously by passing to the adjoint of $T(H)$, which can be identified with $T(H^*)$ acting of $L^q(\mathbb{R}; \mu^{-1})$. $\square$

4. Toeplitz operators on ℓ^p spaces

In this section, we are going to establish analogous statements for Toeplitz operators on weighted ℓ^p spaces. Toeplitz operators on such spaces have been analyzed, e.g., in [3], although the focus there was to develop a Fredholm theory for piecewise continuous symbols. We also refer to [5] and the references therein.

For $1 < p < \infty$ and a weight function $\varrho : \mathbb{Z} \to \mathbb{R}_+$ define $\ell^p(\varrho)$ as the space of all sequences $x = \{x_n\}_{n=-\infty}^\infty$ such that

$$\|x\|_{p,\varrho} := \left(\sum_{n=-\infty}^\infty |x_n \varrho(n)|^p\right)^{1/p} < \infty.$$

By c_{00} we denote the set of all sequences with finite support, i.e., sequences $\{x_n\}$ for which only finitely many of the x_n's are nonzero. The set c_{00} is dense in the spaces $\ell^p(\varrho)$.

Given two sequences, $x = \{x_n\}_{n=-\infty}^\infty$ and $y = \{y_n\}_{n=-\infty}^\infty$, one can define the convolution

$$z = x * y, \qquad z_n = \sum_{k=-\infty}^\infty x_{n-k}y_k$$

provided that for each n the series defining z_n converges.

Multiplier algebras. Let $M_{p,\varrho}$ stand for the set of all sequences $a = \{a_n\}_{n=-\infty}^{\infty}$ such that

(i) $a * x \in \ell^p(\varrho)$ for each $x \in c_{00}$, and

(ii) $$\|a\|_{M_{p,\varrho}} := \sup_{x \in c_{00}} \frac{\|a * x\|_{p,\varrho}}{\|x\|_{p,\varrho}} < \infty.$$

In this case, the map

$$L(a) : x \mapsto a * x$$

extends via continuity to a bounded linear operator acting on $\ell^p(\varrho)$, called the *convolution operator* with symbol a. Obviously, $\|L(a)\|_{\mathcal{L}(\ell^p(\varrho))} = \|a\|_{M_{p,\varrho}}$.

It will be convenient to use the following notation. First, let

$$U_n : \{x_k\}_{k=-\infty}^{\infty} \mapsto \{x_{k-n}\}_{k=-\infty}^{\infty} \tag{4.1}$$

be the shift operators acting on appropriate spaces of sequences. Denote by e_n the sequence

$$e_n = \{\delta_{k,n}\}_{k=-\infty}^{\infty}, \tag{4.2}$$

where $\delta_{k,n}$ stands for the Kronecker symbol. Note that $U_n a = a * e_n$. Given a weight ϱ, define the weight $\tilde{\varrho}$ by

$$\tilde{\varrho}(n) = \varrho(-n). \tag{4.3}$$

We are now able to state the following basic properties concerning convolutions and convolution operators. Notice in particular that (c) implies that $M_{p,\varrho}$ is indeed an algebra, and thus a Banach algebra with appropriate norm. The unit element in $M_{p,\varrho}$ is e_0.

Proposition 4.1. *Let* (2.2) *hold. Then:*

(a) *We have*

$$M_{p,\varrho} = M_{q,\tilde{\varrho}^{-1}} \subseteq \bigcap_{n \in \mathbb{Z}} U_n \left(\ell^p(\varrho) \cap \ell^q(\tilde{\varrho}^{-1}) \right).$$

(b) *If $a \in M_{p,\varrho}$ and $x \in \ell^p(\varrho)$, then $L(a)x = a * x$ and the series defining the convolution converges absolutely.*

(c) *If $a, b \in M_{p,\varrho}$, then $a * b \in M_{p,\varrho}$ and $L(a * b) = L(a)L(b)$. In particular,*

$$(a * b) * x = a * (b * x), \qquad x \in \ell^p(\varrho).$$

(d) *Suppose that*

$$\sup_{n \in \mathbb{Z}} \left(\frac{\varrho(n+1)}{\varrho(n)} + \frac{\varrho(n-1)}{\varrho(n)} \right) < \infty. \tag{4.4}$$

Then

$$(y * a) * x = y * (a * x)$$

whenever $x \in \ell^p(\varrho)$, $y \in \ell^q(\tilde{\varrho}^{-1})$, and $a \in M_{p,\varrho} = M_{q,\tilde{\varrho}^{-1}}$. In particular, $y * x$ is well defined.

We will need statements (c) and (d) in the proof of the Theorem 4.3 below. Notice that condition (4.4) is equivalent to the boundedness of the shift operators U_n on $\ell^p(\varrho)$. In other words, it means that $c_{00} \subseteq M_{p,\varrho}$.

Proof. (a): From duality, i.e., the identification of the dual space of $\ell^p(\varrho)$ with $\ell^q(\varrho^{-1})$ it follows that $a \in M_{p,\varrho}$ if and only if

$$\sup_{y \in c_{00}} \sup_{x \in c_{00}} \frac{1}{\|y\|_{q,\tilde{\varrho}^{-1}}\|x\|_{p,\varrho}} \left| \sum_k y_{-k} \sum_j a_{k-j}x_j \right| < \infty,$$

which involves only finite sums and is thus equal to

$$\sup_{x \in c_{00}} \sup_{y \in c_{00}} \frac{1}{\|y\|_{q,\tilde{\varrho}^{-1}}\|x\|_{p,\varrho}} \left| \sum_j x_{-j} \sum_k a_{j-k}y_k \right| < \infty.$$

The latter is equivalent to $a \in M_{q,\tilde{\varrho}^{-1}}$. Hence $M_{p,\varrho} = M_{q,\tilde{\varrho}^{-1}}$.

If $a \in M_{p,\varrho}$, then by definition $a * e_n = U_n a$ belongs to $\ell^p(\varrho)$. Since also $a \in M_{q,\tilde{\varrho}^{-1}}$, it follows that $a * e_n = U_n a$ belongs to $\ell^q(\tilde{\varrho}^{-1})$ as well. This concludes the proof of (a).

(b): Let $a \in M_{p,\varrho}$ and $x \in \ell^p(\varrho)$. Since $U_{-n}a \in \ell^q(\tilde{\varrho}^{-1})$, the series

$$\sum_{k=-\infty}^{\infty} |a_{n-k}x_k| \le \left(\sum_{k=-\infty}^{\infty} |a_{n-k}\varrho^{-1}(k)|^q \right)^{1/q} \left(\sum_{k=-\infty}^{\infty} |x_k\varrho(k)|^p \right)^{1/p}$$

is finite. Consequently, the convolution product $a * x$ is well defined and its defining series are absolutely convergent. Moreover,

$$|[a * x]_n| \le \|U_{-n}a\|_{q,\tilde{\varrho}^{-1}}\|x\|_{p,\varrho}. \tag{4.5}$$

Since $L(a)x = a * x$ for $x \in c_{00}$, the estimate (4.5) implies that the equality holds for all $x \in \ell^p(\varrho)$.

(c): Notice first that $a * (b * x)$ is well defined. Also, $a * b$ is well defined due to (b) since $a \in M_{p,\varrho}$ and $b \in M_{p,\varrho} \subseteq \ell^p(\varrho)$. It is easy to see that $(a*b)*x = a*(b*x)$ holds for $x = e_n$ and thus for all $x \in c_{00}$. This is enough to conclude that $a * b$ is a multiplier. Indeed,

$$\|(a * b) * x\|_{p,\varrho} = \|a * (b * x)\|_{p,\varrho} \le \|a\|_{M_{p,\varrho}}\|b * x\|_{p,\varrho} \le \|a\|_{M_{p,\varrho}}\|b\|_{M_{p,\varrho}}\|x\|_{p,\varrho}$$

whenever $x \in c_{00}$.

It follows that $a*b \in M_{p,\varrho}$ and $L(a*b)x = L(a)L(b)x$ for $x \in c_{00}$. Due to the density of c_{00} this holds for all $x \in \ell^p(\varrho)$. Finally, from this and (b) we conclude that $(a * b) * x = a * (b * x)$ for all $x \in \ell^p(\varrho)$.

(d): It is easily seen that

$$e_n * (a * e_m) = (e_n * a) * e_m = U_{n+m}a.$$

Thus $y * (a * x) = (y * a) * x$ for $x, y \in c_{00}$. Due to assumption (4.4), the shift operator U_n is bounded on both $\ell^p(\varrho)$ and $\ell^q(\tilde{\varrho}^{-1})$ for each n. If we consider the nth component in the two products under consideration, we can estimate in analogy to (4.5) as follows:

$$|[y * (a * x)]_n| \le \|U_{-n}y\|_{q,\tilde{\varrho}^{-1}}\|a\|_{M_{p,\varrho}}\|x\|_{p,\varrho}$$

and

$$|[(y * a) * x]_n| \leq \|y\|_{q,\tilde{\varrho}^{-1}} \|a\|_{M_{q,\tilde{\varrho}^{-1}}} \|U_{-n}x\|_{p,\varrho}$$

for all $x \in \ell^p(\varrho)$ and $y \in \ell^q(\tilde{\varrho}^{-1})$. Using these estimates and density, the desired equality follows. Finally, if we take $a = e_0 \in M_{p,\varrho}$, we obtain that $y * x$ is well defined. $\qquad\square$

The discrete analogue of Lemma 2.1 is the following. Notice the slight difference in the conditions.

Lemma 4.2. *Let* (2.2) *and* (2.8) *hold. Then:*

(a) $\ell^p(\tilde{\varrho}) \subseteq \ell^q(\varrho^{-1})$ *if and only if*

$$(p \geq q \text{ and } \varrho^{-1}\tilde{\varrho}^{-1} \in \ell^r) \text{ or } (p < q \text{ and } \varrho^{-1}\tilde{\varrho}^{-1} \in \ell^\infty). \qquad (4.6)$$

(b) $\ell^q(\varrho^{-1}) \subseteq \ell^p(\tilde{\varrho})$ *if and only if*

$$(q \geq p \text{ and } \varrho\tilde{\varrho} \in \ell^r) \text{ or } (q < p \text{ and } \varrho\tilde{\varrho} \in \ell^\infty). \qquad (4.7)$$

Proof. (a) – "if" part: The case $p \geq q$ follows from Hölder's inequality. The case $q > p$ can be reduced to the case $p = q = 2$ due to the inclusions $\ell^p(\tilde{\varrho}) \subseteq \ell^2(\tilde{\varrho})$ and $\ell^2(\varrho^{-1}) \subseteq \ell^q(\varrho^{-1})$.

(a) – "only if" part: The case $p \geq q$ can be settled as in Lemma 2.1. Assume $p < q$. The assumption implies, after making the same kind of substitutions as in Lemma 2.1, that

$$h = \{h_k\} \in \ell^1 \qquad \Longrightarrow \qquad \{h_k\varrho(k)^{-p}\tilde{\varrho}(k)^{-p}\} \in \ell^{q/p}.$$

By the closed graph theorem, this must be a bounded linear map. Now consider $h = e_n$ in order to conclude that

$$\varrho(n)^{-p}\tilde{\varrho}(n)^{-p} = \|\{h_k\varrho(k)^{-p}\tilde{\varrho}(k)^{-p}\}\|_{q/p} \leq C\|\{h_k\}\|_1 = C.$$

This implies that $\varrho(n)^{-1}\tilde{\varrho}(n)^{-1}$ is bounded and thus a sequence in ℓ^∞.

(b): This can be proved by interchanging p with q and ϱ with ϱ^{-1}. $\qquad\square$

Toeplitz operators. Let us first introduce the projection

$$P : \{x_n\}_{n=-\infty}^\infty \in \ell^p(\varrho) \mapsto \{y_n\}_{n=-\infty}^\infty \in \ell^p(\varrho), \quad y_n = \begin{cases} x_n & \text{if } n \geq 0 \\ 0 & \text{if } n < 0 \end{cases}$$

and the spaces

$$\ell_+^p(\varrho) = \left\{ x \in \ell^p(\varrho) : x_n = 0 \text{ for all } n < 0 \right\}$$

and

$$\ell_-^p(\varrho) = \left\{ x \in \ell^p(\varrho) : x_n = 0 \text{ for all } n > 0 \right\}.$$

The image of P equals $\ell_+^p(\varrho)$, and the kernel of P equals $U_{-1}(\ell_-^p(\varrho))$.

For $a \in M_{p,\varrho}^{N \times N}$ we define the *Toeplitz operator* acting on $\ell_+^p(\varrho)^N$ by

$$T(a)x = P(a * x),$$

i.e., $T(a)x = PL(a)x$. Given $x = \{x_n\}$ we define

$$x^T = \{x_n^T\}, \qquad \bar{x} = \{\overline{x_n}\}.$$

The unit element in the matrix version of the multiplier algebra is $e_0 \otimes I_N$.

Theorem 4.3. *Let* $a \in M_{p,\varrho}^{N \times N}$, *let condition* (4.4) *on* ϱ *hold, and assume that*

$$\bar{a}^T * a = e_0 \otimes I_N.$$

Consider $T(a)$ *on* $\ell_+^p(\varrho)^N$.
   (a) *If condition* (4.6) *holds, then* $\alpha(T(a)) = 0$.
   (b) *If condition* (4.7) *holds, then* $\beta(T(a)) = 0$.

Remark. Due to the assumption $a \in M_{p,\varrho}^{N \times N}$, the convolution product $\bar{a}^T * a$ is always well defined. The relation $\bar{a}^T * a = e_0 \otimes I_N$ can be rewritten as

$$\sum_{k=-\infty}^{\infty} (\bar{a}_k)^T a_{n-k} = \delta_{0,n} I_N \qquad \text{for each } n \in \mathbb{Z}.$$

Hence, at least formally, this corresponds to

$$\left(\sum_k \bar{a}_k^T t^k \right) \left(\sum_k a_k t^k \right) = I_N, \quad t \in \mathbb{T},$$

and thus to the condition $\widetilde{G}^*(t) G(t) = I_N$ of the previous section.

Proof. Assume that the kernel of $T(a)$ is non-trivial. Then there exist $x \in \ell_+^p(\varrho)^N$, $x \neq 0$, and $y \in U_{-1}(\ell_-^p(\varrho))^N$ such that

$$a * x = y.$$

Since $a \in M_{p,\varrho}^{N \times N}$ we obviously have $\bar{a}^T \in M_{p,\varrho}^{N \times N}$ as well. Multiplying with $\bar{a}^T$ we obtain

$$\bar{a}^T * y = \bar{a}^T * (a * x) = (\bar{a}^T * a) * x = x$$

due to Proposition 4.1 (c) noting that $x \in \ell^p(\varrho)^N$ and $\bar{a}^T * a = e_0 \oplus I_N$. Now multiply with $\bar{x}^T$ to obtain

$$\bar{x}^T * (\bar{a}^T * y) = \bar{x}^T * x.$$

By Proposition 4.1 (d), this convolution product is well defined and we have

$$\bar{x}^T * (\bar{a}^T * y) = (\bar{x}^T * \bar{a}^T) * y$$

since $\bar{a}^T \in M_{p,\varrho}^{N \times N} = M_{q,\tilde{\varrho}^{-1}}^{N \times N}$ and $y, \bar{x} \in \ell^p(\varrho)^N \subseteq \ell^q(\tilde{\varrho}^{-1})^N$, making now use of the assumption (4.6). From $\bar{x}^T * \bar{a}^T = \bar{y}^T$, which follows from $a * x = y$, we obtain

$$z := \bar{x}^T * x = \bar{y}^T * y.$$

The sequence $z = 0$ because $z_n = 0$ for $n \geq 0$ due to the right-hand side and $z_n = 0$ for $n < 0$ due to the left-hand side. In particular, for $n \geq 0$ we obtain

$$\sum_{k=0}^{n} \overline{x_k}^T x_{n-k} = 0.$$

We recursively consider $n = 0, 2, \ldots$ to conclude $x_n = 0$ for all n. Thus $x = 0$ and $y = 0$.

The proof of part (b) is analogous, by interchanging p with q and ϱ with ϱ^{-1}. Notice that the adjoint of $T(a)$ on $\ell^p_+(\varrho)^N$ can be identified with the operator $T(b)$ acting on $\ell^q_+(\varrho^{-1})^N$ with $b_n = \overline{a_{-n}}^T$. Clearly, $\bar{b}^T * b = e_0 \otimes I_N$ holds as well. Furthermore, $b \in M^{N \times N}_{q, \varrho^{-1}} = M^{N \times N}_{p, \tilde{\varrho}}$. $\qquad\qquad\square$

References

[1] A. Böttcher and Yu.I. Karlovich, *Carleson curves, Muckenhoupt weights, and Toeplitz operators*, Progress in Math., vol. 154, Birkhäuser Verlag, Basel and Boston, 1997.

[2] A. Böttcher, Yu.I. Karlovich, and I.M. Spitkovsky, *Convolution operators and factorization of almost periodic matrix functions*, Operator Theory: Advances and Applications, vol. 131, Birkhäuser Verlag, Basel and Boston, 2002.

[3] A. Böttcher and M. Seybold, *Discrete Wiener–Hopf operators on spaces with Muckenhoupt weight*, Studia Math. **143** (2000), no. 2, 121–144.

[4] A. Böttcher and B. Silbermann, *Introduction to large truncated Toeplitz matrices*, Springer-Verlag, New York, 1999.

[5] ———, *Analysis of Toeplitz operators*, second ed., Springer Monographs in Mathematics, Springer-Verlag, Berlin, 2006, prepared jointly with A. Karlovich.

[6] M.C. Câmara, L. Rodman, and I.M. Spitkovsky, *One sided invertibility of matrices over commutative rings, corona problems, and Toeplitz operators with matrix symbols*, submitted.

[7] K.F. Clancey and I. Gohberg, *Factorization of matrix functions and singular integral operators*, Operator Theory: Advances and Applications, vol. 3, Birkhäuser, Basel and Boston, 1981.

[8] I. Gohberg, M.A. Kaashoek, and I.M. Spitkovsky, *An overview of matrix factorization theory and operator applications*, in: *Factorization and integrable systems* (Faro, 2000), Operator Theory: Advances and Applications, vol. 141, Birkhäuser Verlag, Basel and Boston, 2003, pp. 1–102.

[9] I. Gohberg and M.G. Krein, *Systems of integral equations on a half-line with kernel depending upon the difference of the arguments*, Uspekhi Mat. Nauk **13** (1958), no. 2, 3–72 (in Russian), English translation: Amer. Math. Soc. Transl. **14** (1960), no. 2, 217–287.

[10] I. Gohberg and N. Krupnik, *One-dimensional linear singular integral equations. I*, Operator Theory: Advances and Applications, vol. 53, Birkhäuser Verlag, Basel, 1992, Introduction, translated from the 1979 German translation by B. Luderer and S. Roch and revised by the authors.

[11] R. Hunt, B. Muckenhoupt, and R. Wheeden, *Weighted norm inequalities for the conjugate function and Hilbert transform*, Trans. Amer. Math. Soc. **176** (1973), 227–251.

[12] G.S. Litvinchuk and I.M. Spitkovskii, *Factorization of measurable matrix functions*, Operator Theory: Advances and Applications, vol. 25, Birkhäuser Verlag, Basel, 1987, translated from the Russian by B. Luderer, with a foreword by B. Silbermann.

[13] A.F. Voronin, *On the well-posedness of the Riemann boundary value problem with a matrix coefficient*, Dokl. Akad. Nauk **414** (2007), no. 2, 156–158 (in Russian), English translation: Dokl. Math. **75** (2007), no. 3, 358–360.

[14] _____, *Partial indices of unitary and Hermitian matrix functions*, Sibirsk. Mat. Zh. **51** (2010), no. 5, 1010–1016 (in Russian), English translation: Sib. Math. J. **51** (2010), no. 5, 805–809.

Torsten Ehrhardt
Mathematics Department
University of California
Santa Cruz, CA-95064, USA
e-mail: tehrhard@ucsc.edu

Ilya M. Spitkovsky
Mathematics Department
The College of William and Mary
Williamsburg, VA 23187-8795, USA
e-mail: ilya@math.wm.edu
        imspitkovsky@gmail.com

Operator Theory:
Advances and Applications, Vol. 237, 145–160
© 2013 Springer Basel

Rational Matrix Solutions of a Bezout Type Equation on the Half-plane

A.E. Frazho, M.A. Kaashoek and A.C.M. Ran

Dedicated to Leonia Lerer on the occasion of his 70th birthday, in friendship

Abstract. A state space description is given of all stable rational matrix solutions of a general rational Bezout type equation on the right half-plane.
Included are a state space formula for a particular solution satisfying a certain H^2 minimality condition, a state space formula for the inner function
describing the null space of the multiplication operator corresponding to the
Bezout equation, and a parameterization of all solutions using the particular
solution and this inner function. A state space version of the related Tolokonnikov lemma is also presented.

Mathematics Subject Classification (2010). Primary 47B35, 39B42; Secondary
47A68, 93B28.

Keywords. Bezout equation; stable rational matrix functions; state space representation; algebraic Riccati equation; stabilizing solution, right invertible
multiplication operator; Wiener–Hopf operators.

1. Introduction

In this paper G is a stable rational $m \times p$ matrix function. Here *stable* means that
G is proper, that is, the limit of $G(s)$ as $s \to \infty$ exists, and G has all its poles in
the open left half-plane $\{s \in \mathbb{C} \mid \Re(s) < 0\}$. In other words, G is a rational matrix-
valued H^∞ function, where the latter means that G is analytic and bounded on
the open right half-plane. In this paper p will be larger than m, and thus G will
be a "fat" non-square matrix function. We shall be interested in stable rational
$p \times m$ matrix-valued solutions X of the Bezout type equation

$$G(s)X(s) = I_m, \quad \Re s \geq 0. \tag{1.1}$$

The symbol I_m on the right-hand side denotes the $m \times m$ identity matrix.

Throughout we shall assume that G admits a state space realization of the form

$$G(s) = C(sI_n - A)^{-1}B + D. \tag{1.2}$$

Here A is an $n \times n$ matrix which is assumed to be stable, that is, all the eigenvalues of A are contained in the open left half-plane. Moreover, B, C and D are matrices of appropriate sizes. Our aim is to give necessary and sufficient conditions for the solvability of (1.1), and to give a full description of all stable rational matrix-valued solutions, in terms of the matrices appearing in the realization (1.2). The results we present are the continuous analogs of the main theorems in [6] and [8].

To state the main results we need some additional notation. By P we denote the controllability Gramian associated with the realization (1.2), that is, P is the (unique) solution of the Lyapunov equation

$$AP + PA^* + BB^* = 0. \tag{1.3}$$

Consider the algebraic Riccati equation

$$A^*Q + QA + (C - \Gamma^*Q)^*(DD^*)^{-1}(C - \Gamma^*Q) = 0, \tag{1.4}$$

where Γ is defined by

$$\Gamma = BD^* + PC^*. \tag{1.5}$$

Here it is assumed that D is right invertible, which is a natural condition. Indeed, if (1.1) has a stable rational matrix solution X, then using (1.2) and the fact that X is proper, we see that $DX(\infty) = \lim_{s\to\infty} G(s)X(s) = I_m$. Hence $X(\infty)$ is a right inverse of D, and thus D is right invertible. A solution Q of (1.4) is called the *stabilizing solution* of the algebraic Riccati equation (1.4) if Q is Hermitian and the $n \times n$ matrix A_0 given by

$$A_0 = A - \Gamma(DD^*)^{-1}(C - \Gamma^*Q) \tag{1.6}$$

is stable. If it exists, a stabilizing solution is unique (cf. formula (2.11)). The following is the first main result of this paper.

Theorem 1.1. *There is a stable rational $p \times m$ matrix function X satisfying the equation $G(s)X(s) = I_m$ if and only if the following three conditions hold*

1. *The matrix D is right invertible,*
2. *there exists a stabilizing solution Q of the Riccati equation (1.4), and*
3. *the matrix $I_n - PQ$ is invertible.*

In that case a particular solution of (1.1) is given by

$$\Xi(s) = \left(I_p - C_1(sI_n - A_0)^{-1}(I_n - PQ)^{-1}B\right)D^*(DD^*)^{-1}, \tag{1.7}$$

where A_0 is the stable $n \times n$ matrix given by (1.6) and

$$C_1 = D^*(DD^*)^{-1}(C - \Gamma^*Q) + B^*Q. \tag{1.8}$$

The matrix $D^*(DD^*)^{-1}$ appearing in (1.7) is *Moore–Penrose right inverse* of D. In what follows we shall often denote $D^*(DD^*)^{-1}$ by D^+. Note that $\dim \operatorname{Ker} D = p - m$.

The rational $p \times p$ matrix function appearing in the right-hand side of (1.7) between the brackets will be denoted by Y, that is,

$$Y(s) = I_p - C_1(sI_n - A_0)^{-1}(I_n - PQ)^{-1}B. \tag{1.9}$$

Note that the value of Y at infinity is invertible. Hence $Y(s)^{-1}$ is a well-defined rational matrix function. We shall see that $Y(s)^{-1}$ is again stable. Thus both $Y(s)$ and $Y(s)^{-1}$ are stable rational matrix functions. In other words the entries of both $Y(s)$ and $Y(s)^{-1}$ are H^∞ functions. In this case we say that Y is *invertible outer*.

Among other things the following theorem describes the set of all stable rational solutions to $G(s)X(s) = I_m$.

Theorem 1.2. *There exists a stable rational $p \times m$ matrix function X satisfying $G(s)X(s) = I_m$ if and only if D is right invertible and there exists a stable rational $p \times p$ matrix function Y which is invertible outer and satisfies the equation $G(s)Y(s) = D$. In this case one such Y is given by (1.9) and the inverse of this Y is given by*

$$Y(s)^{-1} = I_p + C_1(I_n - PQ)^{-1}(sI_n - A)^{-1}B. \tag{1.10}$$

Moreover, using this function Y the following holds.

(i) *Let E be any isometry mapping $\mathbb{C}^{p-m}$ into $\mathbb{C}^p$ such that $\operatorname{Im} E = \operatorname{Ker} D$. Then the function*

$$\Theta(s) = Y(s)E = \left(I_p - C_1(sI_n - A_0)^{-1}(I_n - PQ)^{-1}B \right)E \tag{1.11}$$

is a stable rational $p \times (p-m)$ matrix function satisfying $G(s)\Theta(s) = 0$, and Θ is inner, that is, $\Theta(-\bar{s})^*\Theta(s) = I_{p-m}$.

(ii) *If h is any $\mathbb{C}^p$-valued H^2 function satisfying $G(s)h(s) = 0$, then there exists a unique $\mathbb{C}^{(p-m)}$-valued H^2 function w such that $h(s) = \Theta(s)w(s)$. In fact, $w(s) = \Theta(-\bar{s})^*h(s)$.*

(iii) *The set of all stable rational $p \times m$ matrix functions X satisfying $G(s)X(s) = I_m$ is given by*

$$X(s) = \left(I_p - C_1(sI_n - A_0)^{-1}(I_n - PQ)^{-1}B \right) \times$$

$$\times \left(D^*(DD^*)^{-1} + EZ(s) \right), \tag{1.12}$$

where Z is an arbitrary stable rational $(p-m) \times m$ matrix function. Moreover, if X satisfies $G(s)X(s) = I_m$, then Z in (1.12) is given by $Z(s) = E^*Y(s)^{-1}X(s)$.

(iv) *The rational $p \times p$ matrix function*

$$G_{ext}(s) = \begin{bmatrix} G(s) \\ E^*Y(s)^{-1} \end{bmatrix}, \quad \Re s \geq 0, \tag{1.13}$$

is invertible outer and its inverse is given by

$$G_{ext}(s)^{-1} = \begin{bmatrix} \Xi(s) & \Theta(s) \end{bmatrix}, \quad \Re s \geq 0. \tag{1.14}$$

Note that item (ii) tells us that the null space $\operatorname{Ker} M_G$ of the multiplication operator M_G defined by G, mapping H_p^2 into H_m^2, is given by $\operatorname{Ker} M_G = \Theta H_{p-m}^2$. Thus Θ plays the role of the inner function in the Beurling–Lax theorem specified for $\operatorname{Ker} M_G$. Furthermore, (1.12) in item (iii) can be rewritten in the following equivalent form $X(s) = \Xi(s) + \Theta(s)Z(s)$. Using this form of (1.12) we expect our state space formulas also to be useful in deriving rational H^∞ solutions of (1.1) that satisfy an additional H^∞ norm constraint, by reducing the norm constraint problem to a generalized Sarason problem (cf. Section I.7 in [7]). Finally, item (iv) is inspired by Tolokonnikov's lemma (see [18] and [16, Appendix 3]).

The formulas in Theorems 1.1 and 1.2 can be easily converted into a Matlab program to compute Ξ in (1.7), the function Y in (1.9), and Θ in (1.11).

We see Theorems 1.1 and 1.2 as the closed right half-plane analogues of Theorem 1.1 in [6] and Theorem 1.1 in [8], which deal with equation (1.1) in the setting of rational matrix functions analytic in the closed unit disc. Obviously, a way to obtain the set of all stable rational matrix solutions to equation (1.1) is to use the Cayley transform to derive the right half-plane solutions from their analogues in the disc case as given in [8]. However, note that in the present half-plane case there is an additional difficulty: The constant functions are not in L^2, whereas in the disc case they are in L^2. Furthermore, the particular solution Ξ in Theorem 1.1 is not the analogue of the least squares solution in [6]. On the other hand, as we shall show in Section 4, the function Ξ has an interpretation in terms of solutions to a somewhat different minimization problem (see Theorem 4.1).

We take this occasion to mention that Theorem 1.1 in [6] and Theorem 1.1 in [8] have predecessors in the papers [14] and [13]. In particular, see Lemma 4.1 and Theorem 4.2 in [13]. We are grateful to Dr. Sander Wahls for mentioning to us these and several related other references. It is interesting to see the role the Bezout equation plays in solving the engineering problems considered in [14] and [13]. The proofs in [6] and [8] are quite different from those in [14] and [13]; also different Riccati equations are used and different state space formulas are obtained.

There is an extensive literature on the Bezout equation and the related corona equation, see, e.g., the classical papers [4], [9], [18], the books [16], [15], [1], the more recent papers [19], [20], [21], [22], and the references therein. Also, finding rational matrix solutions in state space form for Bezout equations is a classical topic in mathematical system theory; see, e.g., the book [23], and the papers [11], [10]. However, as far as we know the formulas we present here are new and cannot easily be obtained using the methods presented in the classical sources. The interpretation of the special solution (1.7) as a limit of solutions of minimization problems also seems to be new. Moreover, the approach we follow in the present paper and the earlier papers [6, 8] can be extended to a Wiener space setting. In fact, in a Wiener space setting the function Y given by (1.9) appears in a very natural way; see also the comment at the end of Section 2. We plan to return to this in a future paper, also for the discrete case.

The paper consists of four sections, including this introduction. In the second section we present the preliminaries from operator theory used in the proofs, and we explain the role of the Riccati equation (1.4), and prove the necessity of conditions 1, 2, 3 in Theorem 1.1. The third section contains the proofs of Theorem 1.1 and 1.2. In the final section we consider an optimization problem, which helps in identifying Ξ in as a solution with a special minimality property.

2. Operator theory and Riccati equation

In this section we prove the necessity of conditions 1, 2, 3 in Theorem 1.1. Our proof requires some preliminaries from operator theory and uses the Riccati equation (1.4).

Let Ω be any proper rational $k \times r$ matrix function with no pole on the imaginary axis $i\mathbb{R}$. With Ω we associate the Wiener–Hopf operator T_Ω and the Hankel operator H_Ω, both mapping $L_r^2(\mathbb{R}^+)$ into $L_k^2(\mathbb{R}^+)$. These operators are the integral operators defined by

$$(T_\Omega f)(t) = \Omega(\infty)f(t) + \int_0^t \omega(t-\tau)f(\tau)d\tau, \quad t \geq 0, \quad f \in L_r^2(\mathbb{R}^+), \qquad (2.1)$$

$$(H_\Omega f)(t) = \int_0^\infty \omega(t+\tau)f(\tau)d\tau, \qquad\qquad t \geq 0 \quad f \in L_r^2(\mathbb{R}^+). \qquad (2.2)$$

Here ω is the Lebesque integrable (continuous) matrix function on the imaginary axis determined by Ω via the Fourier transform:

$$\Omega(s) = \Omega(\infty) + \int_{-\infty}^\infty e^{-s\tau}\omega(\tau)\,d\tau, \quad s \in i\mathbb{R}.$$

In the sequel we shall freely use the basic theory of Wiener–Hopf and Hankel operators which can be found in Chapters XII and XIII of [12]. Note that in [12] the Fourier transform is taken with respect to the real line instead of the imaginary axis as is done here.

Now let G be the stable rational $p \times m$ function given by (1.2). Then

$$G(s) = D + \int_0^\infty e^{-s\tau}Ce^{\tau A}B\,d\tau, \quad s \in i\mathbb{R}.$$

Hence the Wiener–Hopf operator T_G and the Hankel operator H_G are given by

$$(T_G f)(t) = Df(t) + \int_0^t Ce^{(t-\tau)A}Bf(\tau)d\tau, \quad t \geq 0, \qquad (2.3)$$

$$(H_G f)(t) = \int_0^\infty Ce^{(t+\tau)A}Bf(\tau)d\tau, \qquad t \geq 0. \qquad (2.4)$$

With G we also associate the rational $m \times m$ matrix function R given by $R(s) = G(s)G(-\bar{s})^*$. Note that R is a proper rational $m \times m$ matrix function with no pole on the imaginary axis. By T_R we denote the corresponding Wiener–Hopf operator acting on $L_m^2(\mathbb{R}^+)$. It is well known (see, e.g., formula (24) in Section XII.2 of [12]) that

$$T_R = T_G T_G^* + H_G H_G^*. \tag{2.5}$$

Next assume that the equation $G(s)X(s) = I_m$ has a stable rational matrix solution X. The fact that X is stable implies that X is proper and has no poles on the imaginary axis, and thus T_X is well defined. Furthermore, $T_G T_X = T_{GX}$; see [12, Proposition XIII.1.2]. Since GX is identically equal to the $m \times m$ identity matrix T_{GX} is the identity operator on $L_m^2(\mathbb{R}^+)$, and hence T_X is a right inverse of T_G. The fact T_G that is right invertible, implies that $T_G T_G^*$ is invertible and hence strictly positive. The identity (2.5) then shows that T_R is strictly positive too, and hence is invertible.

In the following proposition we use the algebraic Riccati equation (1.4) to obtain necessary and sufficient conditions for T_R to be invertible in terms of the matrices A, B, and C appearing in the realization (1.2). As in Section 1, we denote by P the controllability Gramian associated with the realization (1.2), that is, P is the solution of the Lyapunov equation (1.3). Finally, Γ is the $n \times m$ matrix defined by (1.5).

Proposition 2.1. *Let* $R(s) = G(s)G(-\bar{s})^*$. *Then the operator* T_R *is invertible if and only if the algebraic Riccati equation*

$$A^*Q + QA + (C - \Gamma^*Q)^* (DD^*)^{-1} (C - \Gamma^*Q) = 0 \tag{2.6}$$

has a stabilizing solution Q, *that is,* Q *is a Hermitian solution of* (2.6) *and the operator* A_0, *defined by*

$$A_0 = A - \Gamma C_0, \quad where \ C_0 = (DD^*)^{-1} (C - \Gamma^*Q), \tag{2.7}$$

is stable.

Proof. The proposition is an immediate consequence of Theorem 14.8 in [3]. To see this, we first show that

$$R(s) = DD^* + C(sI_n - A)^{-1}\Gamma - \Gamma^*(sI_n + A^*)^{-1}C^*. \tag{2.8}$$

This partial fraction expansion for R follows from the Lyapunov equation (1.3), and its immediate consequence

$$-(sI_n - A)^{-1}BB^*(sI_n + A^*)^{-1} = (sI_n - A)^{-1}P - P(sI_n + A^*)^{-1}.$$

By employing $G(s) = D + C(sI_n - A)^{-1}B$ the identity (2.8) then follows from

$$R(s) = G(s)G(-\bar{s})^* = (D + C(sI_n - A)^{-1}B)(D^* - B^*(sI_n + A^*)^{-1}C^*)$$
$$= DD^* + C(sI_n - A)^{-1}BD^* - DB^*(sI_n + A^*)^{-1}C^*$$
$$+ C(sI_n - A)^{-1}PC^* - CP(sI_n + A^*)^{-1}C^*.$$

Using $\Gamma = BD^* + PC^*$, this yields (2.8). Given (2.8) we can apply Theorem 14.8 in [3], replacing J by DD^* and B by Γ, and rewriting the corresponding algebraic Riccati equation in the form (2.6). $\square$

From the partial fraction expansion (2.8) it follows that the action of the Wiener–Hopf operator T_R on $L_m^2(\mathbb{R}^+)$ is given by

$$
\begin{aligned}
(T_R f)(t) = DD^* f(t) &+ \int_0^t C e^{(t-\tau)A} \Gamma f(\tau)\, d\tau \\
&+ \int_t^\infty \Gamma^* e^{-(t-\tau)A^*} C^* f(\tau)\, d\tau. \quad t \geq 0.
\end{aligned}
\tag{2.9}
$$

By W_{obs} and $W_{0,\,\mathrm{obs}}$ we denote the observability operators mapping the state space $\mathbb{C}^n$ into $L_m^2(\mathbb{R}^+)$ defined by

$$
(W_{\mathrm{obs}} x)(t) = C e^{tA} x \quad \text{and} \quad (W_{0,\,\mathrm{obs}} x)(t) = C_0 e^{tA_0} x, \ \text{where } x \in \mathbb{C}^n. \tag{2.10}
$$

Proposition 2.2. *Assume that T_R is invertible, or equivalently, there exists a stabilizing solution Q to the algebraic Riccati equation (2.6). Then this stabilizing solution is uniquely determined by*

$$
Q = W_{\mathrm{obs}}^* T_R^{-1} W_{\mathrm{obs}}. \tag{2.11}
$$

Proof. To establish this, let us first show that Q satisfies the following Lyapunov equation

$$
A^* Q + Q A_0 + C^* C_0 = 0. \tag{2.12}
$$

Recall that $A_0 = A - \Gamma C_0$. Then (2.12) follows from the Riccati equation

$$
\begin{aligned}
0 &= A^* Q + Q A + (C - \Gamma^* Q)^* (DD^*)^{-1} (C - \Gamma^* Q) \\
&= A^* Q + Q (A_0 + \Gamma C_0) + (C - \Gamma^* Q)^* C_0 \\
&= A^* Q + Q A_0 + C^* C_0.
\end{aligned}
$$

Thus (2.12) holds. Because A and A_0 are both stable, the stabilizing solution Q can also be written as

$$
Q = \int_0^\infty e^{tA^*} C^* C_0 e^{tA_0}\, dt = W_{\mathrm{obs}}^* W_{0,\,\mathrm{obs}}. \tag{2.13}
$$

Next we prove that

$$
T_R^{-1} W_{\mathrm{obs}} = W_{0,\,\mathrm{obs}}. \tag{2.14}
$$

This essentially follows from [2], Corollary 6.3. For completeness we provide a proof. It suffices to compute $T_R W_{0,\,\mathrm{obs}}$. To do this, we use (2.9). Fix $x \in \mathbb{C}^n$. From the second identity in (2.10) and the first identiy in (2.7) it follows that

$$
\begin{aligned}
\int_0^t C e^{(t-\tau)A} \Gamma (W_{0,\,\mathrm{obs}} x)(\tau)\, d\tau &= \int_0^t C e^{(t-\tau)A} \Gamma C_0 e^{\tau A_0} x\, d\tau \\
&= \int_0^t C e^{(t-\tau)A} (A - A_0) e^{\tau A_0} x\, d\tau = C e^{tA} \Big(\int_0^t C e^{-\tau A} (A - A_0) e^{\tau A_0}\, d\tau \Big) x \\
&= -C e^{tA} \Big(\int_0^t \frac{d}{d\tau} (e^{-\tau A} e^{\tau A_0})\, d\tau \Big) x = -C e^{tA_0} x + C e^{tA} x.
\end{aligned}
$$

Furthermore, using the Lyapunov identity (2.12) we obtain

$$\int_t^\infty \Gamma^* e^{-(t-\tau)A^*} C^*(W_{0,\,\mathrm{obs}}x)(\tau)\, d\tau = \int_t^\infty \Gamma^* e^{-(t-\tau)A^*} C^* C_0 e^{\tau A_0} x\, d\tau$$

$$= -\int_t^\infty \Gamma^* e^{-(t-\tau)A^*}(A^*Q + QA_0)e^{\tau A_0} x\, d\tau$$

$$= -\Gamma^* e^{-tA^*}\left(\int_t^\infty e^{\tau A^*}(A^*Q + QA_0)e^{\tau A_0}\, d\tau\right)x$$

$$= -\Gamma^* e^{-tA^*}\left(\int_t^\infty \frac{d}{d\tau}(e^{\tau A^*} Q e^{\tau A_0})\, d\tau\right)x = -\Gamma^* e^{-tA^*}\left(-e^{tA^*} Q e^{tA_0}\right)x$$

$$= \Gamma^* Q e^{tA_0} x.$$

Using (2.9) and the second identity in (2.7) we conclude that

$$(T_R W_{0,\,\mathrm{obs}})(t) = DD^* C_0 e^{tA_0} + (-Ce^{tA_0} + Ce^{tA}) + \Gamma^* Q e^{tA_0}$$
$$= (DD^* C_0 + \Gamma^* Q)e^{tA_0} - Ce^{tA_0} + Ce^{tA} = Ce^{tA}.$$

This proves $T_R W_{0,\,\mathrm{obs}} = W_{\mathrm{obs}}$, and hence (2.14) holds. Together (2.13) and (2.14) show that $W_{\mathrm{obs}}^* T_R^{-1} W_{\mathrm{obs}} = W_{\mathrm{obs}}^* W_{0,\,\mathrm{obs}} = Q$. In particular, the stabilizing solution is uniquely determined by (2.11). □

Lemma 2.3. *Assume T_R is invertible. Then $I - H_G^* T_R^{-1} H_G$ is positive. Furthermore, the following are equivalent:*

(i) *T_G is right invertible,*
(ii) *$I - H_G^* T_R^{-1} H_G$ is strictly positive,*
(iii) *$I - H_G^* T_R^{-1} H_G$ is invertible.*

Proof. Rewriting (2.5) as $T_G T_G^* = T_R - H_G H_G^*$, and multiplying the latter identity from the left and from the right by $T_R^{-1/2}$ shows that

$$T_R^{-1/2} T_G T_G^* T_R^{-1/2} = I - T_R^{-1/2} H_G H_G^* T_R^{-1/2}. \tag{2.15}$$

Hence $I - T_R^{-1/2} H_G H_G^* T_R^{-1/2}$ is positive which shows that $H_G^* T_R^{-1/2}$ is a contraction. But then $H_G^* T_R^{-1} H_G = (H_G^* T_R^{-1/2})(H_G^* T_R^{-1/2})^*$ is also a contraction, and thus the operator $I - H_G^* T_R^{-1} H_G$ is positive.

Since $I - H_G^* T_R^{-1} H_G$ is positive, the equivalence of items (ii) and (iii) is trivial. Assume (ii) holds. Then $T_R^{-1/2} H_G$ is a strict contraction, and hence the same holds true for $T_R^{-1/2} H_G H_G^* T_R^{-1/2}$. But then $I - T_R^{-1/2} H_G H_G^* T_R^{-1/2}$ is strictly positive, and (2.15) shows that T_G is right invertible. The converse implication is proved in a similar way. □

Corollary 2.4. *Assume that T_R is invertible, or equivalently, there exists a stabilizing solution Q to the algebraic Riccati equation (2.6). Then the spectral radius of QP is at most one.*

Furthermore, the following are equivalent:

(i) T_G *is right invertible,*
(ii) $r_{\mathrm{spec}}(QP) < 1$,
(iii) $I_n - QP$ *is invertible.*

Proof. Let W_{con} be the controllability operator mapping $L_2^p(\mathbb{R}^+)$ into $\mathbb{C}^n$ defined by

$$W_{\mathrm{con}}h = \int_0^\infty e^{tA}Bh(t)dt, \qquad h \in L_p^2(\mathbb{R}^+).$$

Then $P = W_{\mathrm{con}}W_{\mathrm{con}}^*$ and $H_G = W_{\mathrm{obs}}W_{\mathrm{con}}$. Using these two identities and (2.11), we obtain for the spectral radius of $H_G^* T_R^{-1} H_G$ that

$$
\begin{aligned}
r_{\mathrm{spec}}(H_G^* T_R^{-1} H_G) &= r_{\mathrm{spec}}(W_{\mathrm{con}}^* W_{\mathrm{obs}}^* T_R^{-1} W_{\mathrm{obs}} W_{\mathrm{con}}) \\
&= r_{\mathrm{spec}}(W_{\mathrm{con}}^* Q W_{\mathrm{con}}) \qquad\qquad (2.16) \\
&= r_{\mathrm{spec}}(Q W_{\mathrm{con}} W_{\mathrm{con}}^*) = r_{\mathrm{spec}}(QP).
\end{aligned}
$$

By Lemma 2.3 the operator $I - H_G^* T_R^{-1} H_G$ is positive. Hence the spectral radius of $H_G^* T_R^{-1} H_G$ is at most one, and the preceding calculation shows that $r_{\mathrm{spec}}(QP) \leq 1$. Since $r_{\mathrm{spec}}(QP) \leq 1$, the equivalence of items (ii) and (iii) is trivial. Assume $r_{\mathrm{spec}}(QP) < 1$. Then (2.16) shows that $I - H_G^* T_R^{-1} H_G$ is invertible, and Lemma 2.3 tells us that T_G is right invertible. To prove the converse implication, assume that T_G is right invertible. Then, by Lemma 2.3, the operator $I - H_G^* T_R^{-1} H_G$ is strictly positive. Hence $r_{\mathrm{spec}}(I - H_G^* T_R^{-1} H_G) < 1$, and (2.16) shows that $r_{\mathrm{spec}}(QP) < 1$. $\quad\square$

Necessity of the conditions 1, 2, 3 in Theorem 1.1. Assume that the equation $G(s)X(s) = I_m$ has a stable rational matrix solution X. As was shown in the paragraph preceding Theorem 1.1, this implies that D is right invertible. Thus condition 1 is necessary. Furthermore, in the paragraph directly after (2.5) it was shown that $G(s)X(s) = I_m$ has a stable rational matrix solution also implies that T_G is right invertible and T_R is invertible. Given the latter we can apply Proposition 2.1 to show that condition 2 is necessary. Finally, using Corollary 2.4, we see that T_G is right invertible and T_R is invertible imply that $I_n - PQ$ is invertible, which shows that condition 3 is necessary. $\quad\square$

Comment. The identities appearing in this section can also be used to give an alternative formula for the function Y in (1.9), namely

$$Y(s) = I_p - \int_0^\infty e^{-st}y(t)dt, \ \Re s \geq 0, \quad \text{where} \quad y = T_G^*(T_G T_G^*)^{-1}g. \qquad (2.17)$$

This formula also makes sense in a Wiener space setting. From formula (2.17) for Y it follows that $T_G y = g$, which immediately implies that $G(s)Y(s) = D$. The latter identity will be derived in the next section (see the second paragraph of the proof of Theorem 1.1) using state space computations. We plan to prove (2.17) in a future paper.

3. Proof of the two main theorems

It will be convenient first to prove the two identities given in the following lemma.

Lemma 3.1. *Assume conditions 1, 2, 3 in Theorem 1.1 are satisfied. Then*

$$BC_1 = A(I_n - PQ) - (I_n - PQ)A_0, \tag{3.1}$$

$$DC_1 = C(I_n - PQ). \tag{3.2}$$

Proof. Recall that C_1 and C_0 are respectively defined in (1.8) and (2.7). This implies that

$$C_1 = D^*C_0 + B^*Q. \tag{3.3}$$

To prove the first identity, we use the Lyapunov equation (2.12) with Γ defined in (1.5) to compute

$$\begin{aligned}
BC_1 &= BD^*C_0 + BB^*Q = (\Gamma - PC^*)C_0 - (AP + PA^*)Q \\
&= \Gamma C_0 + PA^*Q + PQA_0 - APQ - PA^*Q \\
&= \Gamma C_0 + PQA_0 - APQ = A - A_0 + PQA_0 - APQ \\
&= A(I_n - PQ) - (I_n - PQ)A_0.
\end{aligned}$$

The second identity follows from

$$\begin{aligned}
DC_1 &= C - \Gamma^*Q + DB^*Q = C - DB^*Q - CPQ + DB^*Q \\
&= C(I_n - PQ).
\end{aligned}$$

Thus both identities are proved. □

Proof of Theorem 1.1. In the previous section we have seen that the conditions 1, 2, 3 in Theorem 1.1 are necessary. Therefore in what follows we assume these three conditions are fullfilled. The latter allows us to introduce the $p \times p$ rational matrix function

$$Y(s) = I_p - C_1(sI_n - A_0)^{-1}(I_n - PQ)^{-1}B. \tag{3.4}$$

Note that Y is stable, because the matrix A_0 which is given by (1.6) is stable. The latter follows from the fact that condition 2 is satisfied. We claim that

$$Y(s)^{-1} = I_p + C_1(I_n - PQ)^{-1}(sI_n - A)^{-1}B. \tag{3.5}$$

Since A is stable, we see that Y is invertible outer. To prove (3.5), we use (3.1). Indeed, using (3.1), we obtain

$$\begin{aligned}
A_0 + (I_n - PQ)^{-1}BC_1 &= A_0 + (I_n - PQ)^{-1}A(I_n - PQ) - A_0 \\
&= (I_n - PQ)^{-1}A(I_n - PQ). \tag{3.6}
\end{aligned}$$

Recall that the inverse of $I_p - \gamma(sI_n - \alpha)^{-1}\beta$ is the state space realization given by $I_p + \gamma(sI_n - (\alpha + \beta\gamma))^{-1}\beta$. Using this for the state space realization for Y in (3.4) with (3.6), we obtain

$$\begin{aligned}
Y(s)^{-1} &= I_p + C_1(sI_n - (I_n - PQ)^{-1}A(I_n - PQ))^{-1}(I_n - PQ)^{-1}B \\
&= I_p + C_1(I_n - PQ)^{-1}(sI_n - A)^{-1}B.
\end{aligned}$$

Hence the inverse of $Y(s)$ is given by (3.5). In particular, Y is an invertible outer function.

Next we show that $G(s)Y(s) = D$. To do this we use (3.2) together with the state space formula for $Y(s)^{-1}$ in (3.5), to obtain

$$DY(s)^{-1} = D + DC_1(I_n - PQ)^{-1}(sI_n - A)^{-1}B$$
$$= D + C(sI_n - A)^{-1}B = G(s).$$

In other words, $G(s) = DY(s)^{-1}$. By multiplying the latter identity from the left by $Y(s)$ we obtain $G(s)Y(s) = D$.

Finally, by comparing (1.7) and (3.4), we see that $\Xi(s) = Y(s)D^*(DD^*)^{-1}$. It follows that

$$G(s)\Xi(s) = G(s)Y(s)D^*(DD^*)^{-1} = DD^*(DD^*)^{-1} = I_m.$$

This completes the proof of Theorem 1.1.   □

Proof of Theorem 1.2. Given the above proof of Theorem 1.1 it remains to prove items (i)–(iv) in Theorem 1.2. We do this in four steps.

STEP 1. First we show that Θ is inner. To do this, recall that

$$\Theta(s) = E + C_1(sI_n - A_0)^{-1}B_i, \quad \text{where } B_i = -(I_n - PQ)^{-1}BE. \tag{3.7}$$

We shall make use of the following Lyapunov equation

$$A_0^*(Q - QPQ) + (Q - QPQ)A_0 + C_1^*C_1 = 0. \tag{3.8}$$

To see this, notice, that (3.1), (3.2) and (3.3) with (2.7) and (2.12) yield

$$C_1^*C_1 = (C_0^*D + QB)C_1$$
$$= C_0^*C(I_n - QP) + QA(I_n - PQ) - Q(I_n - PQ)A_0$$
$$= -(QA + A_0^*Q)(I_n - QP) + QA(I_n - PQ) - Q(I_n - PQ)A_0$$
$$= -A_0^*(Q - QPQ) - (Q - QPQ)A_0.$$

Therefore (3.8) holds. The Lyapunov equation in (3.8) also yields

$$-(sI_n + A_0^*)^{-1}C_1^*C_1(sI_n - A_0)^{-1}$$
$$= (Q - QPQ)(sI_n - A_0)^{-1} - (sI_n + A_0^*)^{-1}(Q - QPQ). \tag{3.9}$$

To see this, simply multiply the previous equation by $sI_n + A_0^*$ on the left and $sI_n - A_0$ on the right.

To show that Θ is an inner function, notice that (3.9) gives

$$
\begin{aligned}
\Theta(-\bar{s})^*\Theta(s) &= \left(E^* - B_i^*(sI_n + A_0^*)^{-1}C_1^*\right)\left(E + C_1(sI_n - A_0)^{-1}B_i\right) \\
&= I_{p-m} + E^*C_1(sI_n - A_0)^{-1}B_i - B_i^*(sI_n + A_0^*)^{-1}C_1^*E \\
&\quad - B_i^*(sI_n + A_0^*)^{-1}C_1^*C_1(sI_n - A_0)^{-1}B_i \\
&= I_{p-m} + E^*C_1(sI_n - A_0)^{-1}B_i - B_i^*(sI_n + A_0^*)^{-1}C_1^*E \\
&\quad + B_i^*(Q - QPQ)(sI_n - A_0)^{-1}B_i \\
&\quad - B_i^*(sI_n + A_0^*)^{-1}(Q - QPQ)B_i \\
&= I_{p-m} + \left(B_i^*(Q - QPQ) + E^*C_1\right)(sI_n - A_0)^{-1}B_i \\
&\quad - B_i^*(sI_n + A_0^*)^{-1}\left(C_1^*E + (Q - QPQ)B_i\right) = I_{p-m}.
\end{aligned}
$$

The last equality follows from the fact that

$$
B_i^*(Q - QPQ) + E^*C_1 = 0. \tag{3.10}
$$

To verify this, observe that

$$
B_i^*(Q - QPQ) = -E^*B^*(I_n - QP)^{-1}(Q - QPQ) = -E^*B^*Q
$$
$$
E^*C_1 = E^*(B^*Q + D^*C_0) = E^*B^*Q.
$$

Hence $B_i^*(Q - QPQ) + E^*C_1 = 0$. Therefore $\Theta(s)$ is an inner function.

STEP 2. It will be convenient first to prove item (iv). Take s in the right half-plane, i.e., $\Re s \geq 0$. Using the definition of Y in (1.9), and the identities (1.7) and (1.11), we see that $\Xi(s) = Y(s)D^+$ and $\Theta(s) = Y(s)E$, and hence

$$
\begin{bmatrix} \Xi(s) & \Theta(s) \end{bmatrix} = Y(s)\begin{bmatrix} D^+ & E \end{bmatrix}.
$$

Next observe that the $p \times p$ matrix $\begin{bmatrix} D^+ & E \end{bmatrix}$ is invertible, and

$$
\begin{bmatrix} D^+ & E \end{bmatrix}\begin{bmatrix} D \\ E^* \end{bmatrix} = I_p.
$$

Thus $\begin{bmatrix} \Xi(s) & \Theta(s) \end{bmatrix}$ is invertible, and

$$
\begin{bmatrix} \Xi(s) & \Theta(s) \end{bmatrix}^{-1} = \begin{bmatrix} D \\ E^* \end{bmatrix}Y(s)^{-1} = \begin{bmatrix} G(s) \\ E^*Y(s)^{-1} \end{bmatrix}.
$$

This proves (1.14). Since the function $\begin{bmatrix} \Xi & \Theta \end{bmatrix}$ is a stable rational $p \times p$ matrix function, we see that the function defined by (1.13) is invertible outer.

STEP 3. In this part we prove item (iii). Let X be given by (1.12). In other words $X(s) = Y(s)(D^*(DD^*)^{-1} + EZ(s))$, where Z is an arbitrary stable rational matrix function of size $(p - m) \times m$. Since $G(s)Y(s) = D$, we see $G(s)X(s) = DD^*(DD^*)^{-1} + DEZ(s)$. But $DD^*(DD^*)^{-1} = I_m$ and $DE = 0$. We conclude that $G(s)X(s) = I_m$, as desired.

Next we deal with the reverse implication. Let X be any stable rational $p \times m$ matrix function satsfying the equation $G(s)X(s) = I_m$. Put $H = X - \Xi$. Then H is a rational matrix-valued function, and $G(s)H(s) = 0$. Notice that

$E^*Y(s)^{-1}\Xi(s) = 0$. Using item (iv) we obtain

$$H(s) = \begin{bmatrix} \Xi(s) & \Theta(s) \end{bmatrix} \begin{bmatrix} G(s) \\ E^*Y(s)^{-1} \end{bmatrix} H(s)$$

$$= \begin{bmatrix} \Xi(s) & \Theta(s) \end{bmatrix} \begin{bmatrix} 0 \\ E^*Y(s)^{-1}H(s) \end{bmatrix} \qquad (3.11)$$

$$= \Theta(s)E^*Y(s)^{-1}H(s) = \Theta(s)E^*Y(s)^{-1}X(s).$$

Thus $H(s) = \Theta(s)Z(s)$, where $Z(s) = E^*Y(s)^{-1}X(s)$. Since Y is invertible outer, the inverse $Y(\cdot)^{-1}$ is a rational $p \times p$ matrix function. Thus Z is a rational matrix function of size $(p - m) \times m$, and X has the desired representation (1.12).

STEP 4. We prove item (ii). Let h be any $\mathbb{C}^p$-valued H^2 function such that $G(s)h(s) = 0$ for $\Re s > 0$. Repeating the first three identities in (3.11) with h in place of H, we see that $h(s) = \Theta(s)w(s)$, where $w(s) = E^*Y(s)^{-1}h(s)$. Since Y is invertible outer, the entries of $Y(\cdot)^{-1}$ are H^∞ functions. Hence the entries of w are H^2 functions. Furthermore, using the fact that Θ is inner, we see that $w(s) = \Theta(-\bar{s})^*h(s)$ for $\Re s > 0$. □

To complete this section, let us establish the following useful (see the next section) identity

$$\Theta(-\bar{s})^*\Xi(s) = -B_i^*(sI_n + A_0^*)^{-1}C_0^*. \qquad (3.12)$$

For convenience, let us set $B_1 = -(I_n - PQ)^{-1}BD^+$. Then (3.12) follows from (3.3), (3.9) and (3.10), that is,

$$\Theta(-\bar{s})^*\Xi(s) = \left(E^* - B_i^*(sI_n + A_0^*)^{-1}C_1^*\right)\left(D^+ + C_1(sI_n - A_0)^{-1}B_1\right)$$

$$= E^*C_1(sI_n - A_0)^{-1}B_1 - B_i^*(sI_n + A_0^*)^{-1}C_1^*D^+$$

$$\quad - B_i^*(sI_n + A_0^*)^{-1}C_1^*C_1(sI_n - A_0)^{-1}B_1$$

$$= \left((E^*C_1 + B_i^*(Q - QPQ))(sI_n - A_0)^{-1}B_1\right.$$

$$\quad - B_i^*(sI_n + A_0^*)^{-1}\left(C_1^*D^+ + (Q - QPQ)B_1\right)$$

$$= -B_i^*(sI_n + A_0^*)^{-1}\left(C_1^*D^+ + (Q - QPQ)B_1\right)$$

$$= -B_i^*(sI_n + A_0^*)^{-1}\left(C_1^* - QB\right)D^*(DD^*)^{-1}$$

$$= -B_i^*(sI_n + A_0^*)^{-1}C_0^*.$$

This establishes (3.12).

4. The minimization problem

Throughout this section G is a stable rational $m \times p$ matrix function, and we assume that G is given by the stable state space representation (1.2). We also assume that T_G is right invertible.

For each $\gamma > 0$ let w_γ be the scalar weight function given by $w_\gamma(s) = (s+\gamma)^{-1}$. Note that for each $X \in H^\infty_{p\times m}$ the function $w_\gamma X$ belongs to $H^2_{p\times m}$. With G and the weight function w_γ we associate the following minimization problem:

$$\inf\left\{\|w_\gamma X\|_2 \mid G(s)X(s) = I_m \ (\Re s > 0) \text{ and } X \in H_{p\times m}^\infty\right\}. \tag{4.1}$$

The problem is to check whether or not the infimum is a minimum, and if so, to find a minimizing function. We shall show in this section that such a minimizing X exists and is unique. It what follows this minimizing function will be denoted by Ξ_γ. The next theorem shows that Ξ_γ is a stable rational matrix function and provides an explicit formula for Ξ_γ.

Theorem 4.1. *For each $\gamma > 0$ there is a unique solution to the optimization problem (4.1), and this solution is given by*

$$\Xi_\gamma(s) = \Xi(s) - \Theta(s)B_i^*(\gamma I - A_0^*)^{-1}C_0^*, \quad \Re s > 0. \tag{4.2}$$

Here Ξ and Θ are the rational matrix functions given by (1.7) and (1.11), respectively, the matrix A_0 is defined by (1.6) and C_0 by (2.7). In particular, we have $\Xi(s) = \lim_{\gamma\to\infty} \Xi_\gamma(s)$.

Proof. Fix $\gamma > 0$. Since for each $X \in H_{p\times m}^\infty$ the function $w_\gamma X$ belongs to $H_{p\times m}^2$, we have

$$\|w_\gamma \Xi_\gamma\|_2 = \inf\left\{\|w_\gamma X\|_2 \mid w_\gamma GX = w_\gamma I_m \text{ and } X \in H_{p\times m}^\infty\right\} \tag{4.3}$$

$$\geq \inf\left\{\|Z\|_2 \mid GZ = w_\gamma I_m \text{ and } Z \in H_{p\times m}^2\right\} = \|Z_\gamma\|_2. \tag{4.4}$$

The last optimization problem is a least squares optimization problem. So the optimal solution Z_γ for the problem (4.4) is unique. We first derive a formula for Z_γ.

From item (ii) in Theorem 1.2 we know that $\operatorname{Ker} T_G = \operatorname{Im} T_\Theta$. By taking the Fourier transform, we see that Z_γ is the unique matrix function in $H_{p\times m}^2$ such that $GZ_\gamma = w_\gamma I_m$ and Z_γ is orthogonal to $\Theta H_{(p-m)\times m}^2$. Using $G\Xi = I_m$, we obtain that all H^2 solutions to $GZ = w_\gamma I_m$ are given by

$$Z = w_\gamma \Xi + \Theta H_{(p-m)\times m}^2.$$

So we are looking for a H^2 function Z_γ such that

$$Z_\gamma = w_\gamma \Xi + \Theta F \quad \text{and} \quad Z_\gamma \perp \Theta H_{(p-m)\times m}^2,$$

where F is a matrix function in $H_{(p-m)\times m}^2$. By exploiting that Θ is inner, we obtain

$$w_\gamma \Theta^* \Xi + F \perp H_{(p-m)\times m}^2.$$

But then (3.12) tells us that the latter is equivalent to

$$-w_\gamma B_i^*(sI_n + A_0^*)^{-1}C_0^* + F \perp H_{(p-m)\times m}^2. \tag{4.5}$$

However, $-w_\gamma(s)B_i^*(sI_n+A_0^*)^{-1}C_0^*$ admits a partial fraction expansion of the form

$$-w_\gamma(s)B_i^*(sI_n + A_0^*)^{-1}C_0^* = (s+\gamma)^{-1}B_i^*(\gamma I_n - A_0^*)^{-1}C_0^* + \Omega^*(s),$$

where Ω is in $H_{(p-m)\times m}^2$. Using this in the orthogonality relation (4.5), we see that $F(s) = -w_\gamma(s)B_i^*(\gamma I_n - A_0^*)^{-1}C_0^*$. In other words,

$$Z_\gamma(s) = w_\gamma(s)\Xi(s) - w_\gamma(s)\Theta(s)B_i^*(\gamma I_n - A_0^*)^{-1}C_0^*. \tag{4.6}$$

Next put $\Xi_\gamma(s) = (s + \gamma)Z_\gamma(s)$. Then (4.6) implies that Ξ_γ is given by (4.2). Hence Ξ_γ is a stable rational matrix function. In particular, Ξ_γ belongs to $H^\infty_{p \times m}$. Furthermore, it follows that the inequality on the left-hand side of (4.4) is an equality. We conclude that Ξ_γ given by (4.2) is the unique solution to the minimization problem (4.1). $\qquad\square$

References

[1] M. Bakonyi and H.J. Woerdeman, *Matrix completions, moments, and sums of Hermitian squares*, Princeton University Press, 2010.

[2] H. Bart, I. Gohberg, M.A. Kaashoek, and A.C.M. Ran, *Factorization of matrix and operator functions: the state space method*, OT **178**, Birkhäuser Verlag, Basel, 2008.

[3] H. Bart, I. Gohberg, M.A. Kaashoek, and A.C.M. Ran, *A state space approach to canonical factorization: convolution equations and mathematical systems*, OT **200**, Birkhäuser Verlag, Basel, 2010.

[4] L. Carlson, Interpolation by bounded analytic functions and the corona problem, *Ann. Math.* **76** (1962), 547–559.

[5] A.E. Frazho, M.A. Kaashoek, and A.C.M. Ran, The non-symmetric discrete algebraic Riccati equation and canonical factorization of rational matrix functions on the unit circle, *Integral Equations and Operator Theory* **66** (2010), 215–229.

[6] A.E. Frazho, M.A. Kaashoek and A.C.M. Ran, Right invertible multiplication operators and stable rational matrix solutions to an associate Bezout equation, I. the least squares solution. *Integral Equations and Operator Theory* **70** (2011), 395–418.

[7] C. Foias, A. Frazho, I. Gohberg, and M.A. Kaashoek, *Metric constrained interpolation, commutant lifting and systems*, Birkhäuser Verlag, Basel, 1998;

[8] A.E. Frazho, M.A. Kaashoek and A.C.M. Ran, Right invertible multiplication operators and stable rational matrix solutions to an associate Bezout equation, II: Description of all solutions, *Operators and Matrices* **6** (2012), 833–857.

[9] P. Fuhrmann, On the corona theorem and its applications to spectral problems in Hilbert space, *Trans. Amer. Math. Soc.* **132** (1968), 55–66.

[10] P.A. Fuhrmann and R. Ober, On coprime factorizations, in: *Operator Theory and its Applications. The Tsuyoshi Ando Anniversary Volume.* OT **62**, Birkhäuser Verlag, Basel, 1993, pp. 39–75.

[11] T. Georgiou and M.C. Smith, Optimal robustness in the gap metric, *IEEE Trans. Automatic Control* **35** (1990), 673–686.

[12] I. Gohberg, S. Goldberg, and M.A. Kaashoek, *Classes of Linear Operators*, Volume I, OT **49**, Birkhäuser Verlag, Basel, 1990.

[13] G. Gu and E.F. Badran, Optimal design for channel equalization via the filterbank approach, *IEEE Trans.Signal Proc.* **52** (2004), 536–545.

[14] G. Gu and L. Li, Worst-case design for optimal channel equalization in filterbank transceivers, *IEEE Trans. Signal Proc.* **51** (2003), 2424–2435.

[15] J.W. Helton, *Operator Theory, analytic functions, matrices and electrical engineering*, Regional Conference Series in Mathematics **68**, Amer. Math. Soc. Providence RI, 1987.

[16] N.K. Nikol'skii, *Treatise on the shift operator*, Grundlehren **273**, Springer Verlag, Berlin 1986.

[17] V.V. Peller, *Hankel Operators and their Applications*, Springer Monographs in Mathematics, Springer 2003.

[18] V.A. Tolokonnikov, Estimates in Carlson's corona theorem. Ideals of the algebra H^∞, the problem of Szekefalvi-Nagy, *Zap. Naučn. Sem. Leningrad. Otdel. Mat. Inst. Steklov.* (LOMI) **113** (1981), 178–1981 (Russian).

[19] S. Treil, Lower bounds in the matrix corona theorem and the codimension one conjecture, *GAFA* **14** (2004), 1118–1133.

[20] S. Treil and B.D. Wick, The matrix-valued H^p corona problem in the disk and polydisk, *J. Funct. Anal.* **226** (2005), 138–172.

[21] T.T. Trent and X. Zhang, A matricial corona theorem, *Proc. Amer. Math. Soc.* **134** (2006), 2549–2558.

[22] T.T. Trent, An algorithm for the corona solutions on $H^\infty(D)$, *Integral Equations and Operator Theory* **59** (2007), 421–435.

[23] M. Vidyasagar, *Control system synthesis: a factorization approach*, The MIT Press, Cambridge, MA, 1985.

A.E. Frazho
Department of Aeronautics and Astronautics
Purdue University
West Lafayette, IN 47907, USA
e-mail: frazho@ecn.purdue.edu

M.A. Kaashoek
Department of Mathematics
Faculty of Sciences
VU University
De Boelelaan 1081a
NL-1081 HV Amsterdam, The Netherlands
e-mail: ma.kaashoek@vu.nl

A.C.M. Ran
Department of Mathematics
Faculty of Sciences
VU University
De Boelelaan 1081a
NL-1081 HV Amsterdam, The Netherlands

and

Unit for BMI, North-West University
Potchefstroom, South Africa
e-mail: a.c.m.ran@vu.nl

Operator Theory:
Advances and Applications, Vol. 237, 161–187

Inverting Structured Operators Related to Toeplitz Plus Hankel Operators

M.A. Kaashoek and F. van Schagen

Dedicated to our dear friend Leonia Lerer, on the occasion of his 70th birthday.

Abstract. In this paper the Ellis–Gohberg–Lay theorem on inversion of certain Toeplitz plus Hankel operators is derived as a corollary of an abstract inversion theorem for a certain class of structured operators. The main results also cover the inversion theorems considered in [6].

Mathematics Subject Classification (2010). Primary 47A62, 47B35; Secondary 47A50, 15A09, 65F05.

Keywords. Inversion, structured operators, structured matrices, Toeplitz operators, Hankel operators, Stein equation.

1. Introduction

To introduce the main results of this paper we first present an example. Let g and h be matrix functions of sizes $p \times q$ and $q \times p$, respectively, with entries in the Wiener algebra on the unit circle (see Section XXIX.2 in [7]). With g and h we associate Hankel operators G and H, where $G : \ell_+^2(\mathbb{C}^q) \to \ell_+^2(\mathbb{C}^p)$ and $H : \ell_+^2(\mathbb{C}^p) \to \ell_+^2(\mathbb{C}^q)$, as follows:

$$
G = \begin{bmatrix} g_0 & g_1 & g_2 & \cdots \\ g_1 & g_2 & g_3 & \cdots \\ g_2 & g_3 & g_4 & \\ \vdots & \vdots & & \ddots \end{bmatrix} \quad \text{and} \quad H = \begin{bmatrix} h_0 & h_1 & h_2 & \cdots \\ h_1 & h_2 & h_3 & \cdots \\ h_2 & h_3 & h_4 & \\ \vdots & \vdots & & \ddots \end{bmatrix}. \tag{1.1}
$$

Here g_j and h_j, $j = 0, 1, 2, \ldots$, are the Fourier coefficients corresponding to the analytic parts of g and h, respectively. Furthermore, we have two invertible block Toeplitz operators R and V acting on $\ell_+^2(\mathbb{C}^p)$ and $\ell_+^2(\mathbb{C}^q)$, respectively. As for G and H the entries of the matrix functions defining R and V belong to the Wiener algebra on the unit circle. We are interested in inverting the operator $R - GV^{-1}H$. For this purpose we need linear maps

$$a_1, a_2 : \mathbb{C}^p \to \ell^1_+(\mathbb{C}^p), \quad d_1, d_2 : \mathbb{C}^q \to \ell^1_+(\mathbb{C}^q) \tag{1.2}$$

satisfying the equations

$$(R - GV^{-1}H)a_1 = \varepsilon_p, \quad (V - HR^{-1}G)d_1 = \varepsilon_q, \tag{1.3}$$

$$a_2^*(R - GV^{-1}H) = \varepsilon_p^*, \quad d_2^*(V - HR^{-1}G) = \varepsilon_q^*. \tag{1.4}$$

Here for each positive integer r the symbol ε_r denotes the canonical embedding of $\mathbb{C}^r$ into $\ell^2_+(\mathbb{C}^r)$, that is, $\varepsilon_r u = \mathrm{col}\,[\delta_{j,0}u]_{j=0}^\infty$, where $u \in \mathbb{C}^r$ and $\delta_{j,k}$ is the Kronecker delta. Note that $\varepsilon_r \varepsilon_r^* = I - S_r S_r^*$, where S_r is the forward shift on $\ell^2_+(\mathbb{C}^r)$. Since $\ell^1_+(\mathbb{C}^p)$ and $\ell^1_+(\mathbb{C}^q)$ are contained in $\ell^2_+(\mathbb{C}^p)$ and $\ell^2_+(\mathbb{C}^q)$, respectively, the products $a_{j0} = \varepsilon_p^* a_j$ and $d_{j0} = \varepsilon_q^* d_j$, $j = 1, 2$, are well defined.

For any linear map x from $\mathbb{C}^m$ into $\ell^1_+(\mathbb{C}^m)$ we denote by T_x the block lower triangular Toeplitz operator acting on $\ell^2_+(\mathbb{C}^m)$ whose first column is equal to x, and for an $m \times m$ matrix u the symbol Δ_u denotes the block diagonal operator on $\ell^2_+(\mathbb{C}^m)$ with all diagonal entries equal to u.

Theorem 1.1. *Assume there exist linear maps a_1, a_2 and d_1, d_2 as in (1.2) satisfying equations (1.3) and (1.4). Then $a_{10} = a_{20}^*$ and $d_{10} = d_{20}^*$. Furthermore, assume that at least one of the matrices a_{10} and d_{10} is invertible. Then both the matrices a_{10} and d_{10} are invertible, and the operators $R - GV^{-1}H$ and $V - HR^{-1}G$ are invertible. Moreover,*

$$(R - GV^{-1}H)^{-1} = T_{a_1}\Delta_{a_{10}^{-1}}T_{a_2}^* - S_p T_{b_1}\Delta_{d_{10}^{-1}}T_{b_2}^* S_p^*, \tag{1.5}$$

$$(V - HR^{-1}G)^{-1} = T_{d_1}\Delta_{d_{10}^{-1}}T_{d_2}^* - S_q T_{c_1}\Delta_{a_{10}^{-1}}T_{c_2}^* S_q^*. \tag{1.6}$$

Here

$$b_1 = -R^{-1}Gd_1, \; c_1 = -V^{-1}Ha_1, \; b_2^* = -d_2^*HR^{-1}, \; c_2^* = -a_2^*GV^{-1}. \tag{1.7}$$

Note that invertibility of $R - GV^{-1}H$ (or $V - HR^{-1}G$) is equivalent to invertibility of the operator T given by

$$T = \begin{bmatrix} R & G \\ H & V \end{bmatrix} : \begin{bmatrix} \ell^2_+(\mathbb{C}^p) \\ \ell^2_+(\mathbb{C}^q) \end{bmatrix} \to \begin{bmatrix} \ell^2_+(\mathbb{C}^p) \\ \ell^2_+(\mathbb{C}^q) \end{bmatrix}. \tag{1.8}$$

Moreover, in that case

$$T^{-1} = \begin{bmatrix} (R - GV^{-1}H)^{-1} & -R^{-1}G(V - HR^{-1}G)^{-1} \\ -V^{-1}H(R - GV^{-1}H)^{-1} & (V - HR^{-1}G)^{-1} \end{bmatrix}.$$

Using this connection one sees that for the selfadjoint case the above theorem is equivalent to Theorem 3.1 in [5] which is an infinite-dimensional generalization of the Gohberg–Heinig inversion theorem from [8]. In a somewhat less explicit form the above theorem also appears in Section 5 of [6].

With the term "Toeplitz plus Hankel operator" used in the title we have in mind operators of the type (1.8).

In the present paper we put Theorem 1.1 in a more general setting. More precisely we derive Theorem 1.1 as a corollary of the two abstract inversion theorems presented below, the alternative versions of these theorems arising from Remark 1.4, and the auxiliary result Lemma 2.2.

To state our main results we need some additional notation. Consider the Hilbert space operators

$$A_1 : \mathcal{X}_1 \to \mathcal{X}_1, \quad B_1 : \mathcal{U} \to \mathcal{X}_1, \quad C_1 : \mathcal{X}_1 \to \mathcal{Y}, \tag{1.9}$$

$$A_2 : \mathcal{X}_2 \to \mathcal{X}_2, \quad B_2 : \mathcal{Y} \to \mathcal{X}_2, \quad C_2 : \mathcal{X}_2 \to \mathcal{U}. \tag{1.10}$$

Throughout we assume that $P : \mathcal{X}_2 \to \mathcal{X}_1$ and $Q : \mathcal{X}_1 \to \mathcal{X}_2$ are operators satisfying the following Stein equations

$$P - A_1 P A_2 = B_1 C_2, \quad Q - A_2 Q A_1 = B_2 C_1. \tag{1.11}$$

If the identities in (1.11) are satisfied, we shall refer to the set of operators $(A_1, B_1, C_1; A_2, B_2, C_2)$ as a *data set associated* with the pair (P, Q). We summarize the structure of the data set associated with (P, Q) with the following diagram:

$$\begin{array}{ccccccc}
& B_1 & & A_1 & & C_1 & \\
\mathcal{U} & \longrightarrow & \mathcal{X}_1 & \longrightarrow & \mathcal{X}_1 & \longrightarrow & \mathcal{Y} \\
& & P \uparrow & & Q \downarrow & & \\
\mathcal{U} & \longleftarrow & \mathcal{X}_2 & \longleftarrow & \mathcal{X}_2 & \longleftarrow & \mathcal{Y}. \\
& C_2 & & A_2 & & B_2 &
\end{array} \tag{1.12}$$

We shall be interested in inverting the operators $I_{\mathcal{X}_1} - PQ$ using solutions of the following four equations

$$(I_{\mathcal{X}_1} - PQ)X = B_1, \quad (I_{\mathcal{X}_2} - QP)W = B_2, \tag{1.13}$$

$$Y(I_{\mathcal{X}_1} - PQ) = C_1, \quad Z(I_{\mathcal{X}_2} - QP) = C_2. \tag{1.14}$$

Here the unknowns are operators

$$X : \mathcal{U} \to \mathcal{X}_1, \quad Y : \mathcal{X}_1 \to \mathcal{Y}, \quad W : \mathcal{Y} \to \mathcal{X}_2, \quad Z : \mathcal{X}_2 \to \mathcal{U}. \tag{1.15}$$

Theorem 1.2. *Let the operators X, Y, Z, and W in (1.15) be solutions of the equations (1.13) and (1.14). Then*

$$I_{\mathcal{Y}} + Y P B_2 = I_{\mathcal{Y}} + C_1 P W, \quad I_{\mathcal{U}} + C_2 Q X = I_{\mathcal{U}} + Z Q B_1. \tag{1.16}$$

Assume in addition that $I_{\mathcal{X}_1} - PQ$ is invertible. Then $I_{\mathcal{X}_1} - A_1 P A_2 Q$ is invertible if and only if at least one of the two operators $I_{\mathcal{U}} + C_2 Q X$ and $I_{\mathcal{Y}} + Y P B_2$ is invertible. In that case both $I_{\mathcal{U}} + C_2 Q X$ and $I_{\mathcal{Y}} + Y P B_2$ are invertible and

$$\begin{aligned}
&Q(I_{\mathcal{X}_1} - PQ)^{-1} - A_2 Q(I_{\mathcal{X}_1} - PQ)^{-1} A_1 \\
&= W(I_{\mathcal{Y}} + Y P B_2)^{-1} Y - A_2 Q X(I_{\mathcal{U}} + C_2 Q X)^{-1} Z Q A_1.
\end{aligned} \tag{1.17}$$

Theorem 1.3. *Assume that there exist operators Y and Z as in (1.15) such that the identities in (1.14) are satisfied. If the associate operator $I_{\mathcal{Y}} + Y P B_2$ is invertible, then $\operatorname{Ker}(I_{\mathcal{X}_1} - PQ) \subset \bigcap_{\nu=0}^{\infty} \operatorname{Ker} C_2 A_2^{\nu} Q$. Moreover, if the operator $I_{\mathcal{X}_1} - PQ$ is Fredholm of index zero and $\bigcap_{\nu=0}^{\infty} \operatorname{Ker} C_2 A_2^{\nu} = \{0\}$, then $I_{\mathcal{X}_1} - PQ$ is invertible.*

Remark 1.4. *Notice that there is a lot of symmetry in the diagram* (1.12) *with respect to the roles of P and Q, the indices 1 and 2, and the spaces $\mathcal{U}$ and $\mathcal{Y}$. Therefore, in* Theorems 1.2 *and* 1.3 *one may simultaneously interchange these pairs and obtain analoguous results. Also one can use duality which, for instance, interchanges the roles of B_j and C_j, $j = 1, 2$. To be more precise, let $(A_1, B_1, C_1;$ $A_2, B_2, C_2)$ be a data set associated with (P, Q), then $(A_2, B_2, C_2; A_1, B_1, C_1)$ is a data set associated with (Q, P) and $(A_2^*, C_2^*, B_2^*; A_1^*, C_1^*, B_1^*)$ is a data set associated with (P^*, Q^*). These invariances under symmetry and duality yield alternative versions of* Theorem 1.2 *and* Theorem 1.3; *see* Theorem 2.3 *and* Theorems 3.6, 3.7 *and* 3.8. *The latter three theorems also play a role in the proof of* Theorem 1.1.

To illustrate the preceding Theorems 1.2 and 1.3 we briefly sketch how Theorem 1.1 can be obtained as a corollary of these two theorems. For simplicity we take here $R = I$ and $V = I$. First we make a special choice of the data in (1.9)–(1.11), as follows:

$$P = GH \text{ on } \mathcal{X}_1 = \ell_+^2(\mathbb{C}^p) \text{ and } Q \text{ is the identity operator on } \mathcal{X}_2 = \ell_+^2(\mathbb{C}^p),$$

$$A_1 = S_p^*, \quad B_1 = G\varepsilon_q : \mathbb{C}^q \to \ell_+^2(\mathbb{C}^p), \quad C_1 = \varepsilon_p^* : \ell_+^2(\mathbb{C}^p) \to \mathbb{C}^p,$$

$$A_2 = S_p, \quad B_2 = \varepsilon_p : \mathbb{C}^p \to \ell_+^2(\mathbb{C}^p), \quad C_2 = \varepsilon_q^* H : \ell_+^2(\mathbb{C}^p) \to \mathbb{C}^q.$$

Using that G and H are Hankel operators, we see that

$$P - A_1 P A_2 = GH - S_p^* GH S_p = GH - GS_q S_q^* H$$
$$= G(I - S_q S_q^*)H = G\varepsilon_q \varepsilon_q^* H = B_1 C_2.$$

Furthermore, $Q - A_2 Q A_1 = I - S_p S_p^* = \varepsilon_p \varepsilon_p^* = B_2 C_1$. Thus the corresponding Stein equations (1.11) are satisfied. Furthermore, since $P = GH$ and $Q = I$, we have $I - PQ = I - GH$ and the left-hand side of (1.17) becomes

$$(I - GH)^{-1} - S_p(I - GH)^{-1}S_p^*.$$

Next one shows that in this case, the equations (1.13) and (1.14) are equivalent to the equations (1.3) and (1.4). Hence for this case (1.16) yields $a_{10} = a_{20}^*$ and $d_{10} = d_{20}^*$. It is then a matter of direct checking to prove that the invertibility of $I - GH$ and $I - HG$ follows from Theorem 1.3, and that formulas (1.5) and (1.6) can be obtained from (1.17). For further details see the full proof of Theorem 1.1 in Section 4.

The setting described by the formulas (1.9)–(1.14) is of particular interest when the data are matrices. To illustrate this, assume that G and H in (1.1) are of finite rank and $R = I$ and $V = I$. In that case we know from mathematical systems theory (see, e.g., [3, Chapter 7]) that the defining functions g and h are rational matrix functions and admit stable finite-dimensional realizations:

$$g(z) = C_1(I_{n_1} - zA_1)^{-1}B_1 \quad \text{and} \quad h(z) = C_2(I_{n_2} - zA_2)^{-1}B_2. \tag{1.18}$$

Here *stable* means that A_1 and A_2 have their eigenvalues in the open unit disc. From (1.18) it follows that the entries g_j and h_j in G and H are given by $g_j =$

$C_1 A_1^j B_1$ and $h_j = C_2 A_2^j B_2$ for $j = 0, 1, 2, \ldots$. Hence $G = \Gamma_1 \Lambda_1$ and $H = \Gamma_2 \Lambda_2$ with

$$\Lambda_j = \begin{bmatrix} B_j & A_j B_j & A_j^2 B_j & \cdots \end{bmatrix}, \quad \Gamma_j = \begin{bmatrix} C_j \\ C_j A_j \\ C_j A_j^2 \\ \vdots \end{bmatrix}. \tag{1.19}$$

Furthermore, (1.11) holds with

$$P = \sum_{\nu=0}^{\infty} A_1^\nu B_1 C_2 A_2^\nu = \Lambda_1 \Gamma_2 \quad \text{and} \quad Q = \sum_{\nu=0}^{\infty} A_2^\nu B_2 C_1 A_1^\nu = \Lambda_2 \Gamma_1. \tag{1.20}$$

In particular, one sees that $I - GH = I - \Gamma_1 \Lambda_1 \Gamma_2 \Lambda_2$ is invertible if and only if $I - PQ = I - \Lambda_1 \Gamma_2 \Lambda_2 \Gamma_1$ is invertible. Moreover, the inverse of $I - GH$ can be expressed in terms of the inverse of $I - PQ$ and vice versa (cf. the remark preceding Lemma 2.2).

Now assume, as in Theorem 1.1, that there exist linear maps a_1, a_2, d_1, d_2 as in (1.2) satisfying equations (1.3) and (1.4) (with R and V identity operators) such that a_{10} or d_{10} is invertible. As we have seen, this implies that $I - GH$ is invertible. To compute $(I - GH)^{-1}$ we apply Theorem 1.2 with P, Q, and the associate data $\{A_1, B_1, C_1; A_2, B_2, C_2\}$ as in (1.20). In this case equations (1.13) and (1.14) are just matrix equations, and identity (1.17) leads to the following inversion formula:

$$(I - GH)^{-1} = I + H_{b_1} \Delta_{d_{10}^{-1}} H_{b_2}^* - S_p^* H_{a_1} \Delta_{a_{10}^{-1}} H_{a_2}^* S_p. \tag{1.21}$$

Here b_1 and b_2 are defined by the first and third identity in (1.7), and H_x denotes the Hankel operator with first column equal to x. Moreover, the Hankel operators appearing in (1.21) are all of finite rank. The inversion formula (1.21) is the discrete analogue of the inversion formula in [10, Theorem 0.1]. The full proof of (1.21) will be presented in Subsection 4.2.

We shall also show that the inversion formulas for $R - GV^{-1}H$ and $V - HR^{-1}G$ in Theorem 1.1 can be replaced by formulas analogous to (1.21), that is, with the Toeplitz operators in (1.5) and (1.6) being replaced by Hankel operators. See Theorem 4.3 for the precise formulation.

Theorems 1.2 and 1.3 also apply to other inversion problems than the ones related to Theorem 1.1. As an illustration we derive Theorem 2.1 in [6] as a corollary of Theorems 1.2 and 1.3. This implies that all examples in [6] are also covered by Theorems 1.2 and 1.3 above. Whether or not the main inversion theorems in Section 2 of [6] also imply Theorems 1.2 and 1.3 remains an open question.

The operators considered in this paper belong to the area of structured matrices and operators which include Toeplitz and Hankel operators, Vandermonde, Cauchy and Pick matrices, resultants and Bezoutians, controllability and observability operators, and many other classes of matrices and operators. The literature on the subject is vast. Here we only mention [14], the review paper [12], and the books [11], [13], [15], [16]. We see our Theorems 1.2 and 1.3 as a contribution to this rich field of research.

This paper consists of six sections including the present introduction. The proofs of Theorems 1.2 and 1.3 are given in Sections 2 and 3, respectively. In Section 4 we return to Theorem 1.1, and complete the sketch of the proof given above. In this section we also prove (1.21) and derive formulas for $(R - GV^{-1}H)^{-1}$ and $(V - HR^{-1}G)^{-1}$ analogous to (1.21). In Section 5 we derive Theorem 2.1 in [6] as a corollary of Theorems 1.2 and 1.3 above. In general, with P and Q given, one can find different sets of operators $(A_1, B_1, C_1; A_2, B_2, C_2)$ such the equations (1.9), (1.10), and (1.11) are satisfied. These different choices of the data set $(A_1, B_1, C_1; A_2, B_2, C_2)$ often lead to different versions of formula (1.17), and it can happen that for some choice of the data set formula (1.17) leads to a formula for $Q(I - PQ)^{-1}$ while for another choice it does not. This phenomenon is illustrated on finite block Toeplitz matrices in the final section.

2. Proof of Theorem 1.2

In this section we prove Theorem 1.2. For simplicity, in this section, any identity operator will be denoted by I, that is, in what follows we shall omit the subscript indicating on which space the identity operator acts. Furthermore, we shall freely use the operators appearing in (1.9), (1.10), and (1.11). We begin with two lemmas.

Lemma 2.1. *If the second identity of* (1.13) *and the first of* (1.14) *are satisfied, then* $I + YPB_2 = I + C_1PW$. *Analogously, if the first identity of* (1.13) *and the second of* (1.14) *are satisfied, then* $I + C_2QX = I + ZQB_1$.

Proof. From the second identity of (1.13) and the first of (1.14) we get

$$I + YPB_2 = I + YP(I - QP)W = I + Y(I - PQ)PW = I + C_1PW.$$

The other two identities give

$$I + C_2QX = I + Z(I_\mathcal{X} - QP)QX = I + ZQ(I - PQ)X = I + ZQB_1. \qquad \square$$

In the proof of the following lemma we use a few times the classical result that given two operators F_1 and F_2 the invertibility of $I + F_1F_2$ is equivalent to the invertibility of $I + F_2F_1$. Moreover, in that case the inverse of $I + F_1F_2$ is given by $(I + F_1F_2)^{-1} = I - F_1(I + F_2F_1)^{-1}F_2$ (see [1], first paragraph on page 30).

Lemma 2.2. *Assume that* $I - PQ$ *is invertible, and let* X *and* Y *be the operators defined by the first equations of* (1.13) *and* (1.14). *Then the following are equivalent:*

(i) $I + C_2QX$ *is invertible,*
(ii) $I + YPB_2$ *is invertible,*
(iii) $I - A_1PA_2Q$ *is invertible.*

Furthermore, in that case $I - PA_2QA_1$ is also invertible and

$$(I + C_2QX)^{-1} = I - C_2Q(I - A_1PA_2Q)^{-1}B_1, \qquad (2.1)$$

$$(I + YPB_2)^{-1} = I - C_1(I - PA_2QA_1)^{-1}PB_2. \qquad (2.2)$$

Proof. Since $X = (I - PQ)^{-1}B_1$, we have

$$I + C_2QX = I + C_2Q(I - PQ)^{-1}B_1.$$

It follows that $I + C_2QX$ is invertible if and only if

$$I + B_1C_2Q(I - PQ)^{-1} = (I - (PQ - B_1C_2Q))(I - PQ)^{-1}$$

is invertible. By the first identity in (1.11), we have $PQ - B_1C_2Q = A_1PA_2Q$. This proves the equivalence of (i) and (iii). Moreover in that case

$$
\begin{aligned}
(I + C_2QX)^{-1} &= \left(I + C_2Q(I - PQ)^{-1}B_1\right)^{-1} \\
&= I - C_2Q(I - PQ)^{-1}\left(I + B_1C_2Q(I - PQ)^{-1}\right)^{-1}B_1 \\
&= I - C_2Q\left(I - A_1PA_2Q\right)^{-1}B_1.
\end{aligned}
$$

This proves identity (2.1).

Since $I - A_1PA_2Q$ is invertible if and only if $I - PA_2QA_1$ is invertible, the equivalence of (ii) and (iii), and the identity (2.2) can be proved using an appropriate modification of the arguments employed in the previous paragraph.

$\square$

Proof of Theorem 1.2. Assuming that $I - PQ$ is invertible and given Lemmas 2.1 and 2.2, it remains to prove the identity (1.17). We will show that

$$W(I + YPB_2)^{-1}Y = Q(I - PQ)^{-1} - (I - A_2QA_1P)^{-1}A_2QA_1, \quad (2.3)$$

$$A_2QX(I + C_2QX)^{-1}ZQA_1 = A_2Q(I - PQ)^{-1}A_1 - (I - A_2QA_1P)^{-1}A_2QA_1. \qquad (2.4)$$

Subtracting these two identities (1.17) appears. In deriving (2.3) and (2.4) we shall use a few times that

$$B_1C_2Q = (I - A_1PA_2Q) - (I - PQ), \qquad (2.5)$$

$$B_2C_1P = (I - A_2QA_1P) - (I - QP). \qquad (2.6)$$

These identities follow from the ones in (1.11).

Let us now prove (2.3). Using (2.2) and (2.6), a standard computation (cf. the state space formulas in Theorem 2.4 in [2]) yields

$$
\begin{aligned}
W(I + YPB_2)^{-1} &= (I - QP)^{-1}B_2\left(I - C_1(I - PA_2QA_1)^{-1}PB_2\right) \\
&= (I - QP)^{-1}B_2\left(I - C_1P(I - A_2QA_1P)^{-1}B_2\right) \\
&= (I - A_2QA_1P)^{-1}B_2.
\end{aligned}
$$

We proceed with

$$W(I + YPB_2)^{-1}Y = (I - A_2QA_1P)^{-1}B_2C_1(I - PQ)^{-1}$$
$$= (I - A_2QA_1P)^{-1}B_2C_1\Big(I + PQ(I - PQ)^{-1}\Big)$$
$$= (I - A_2QA_1P)^{-1}B_2C_1$$
$$+ (I - A_2QA_1P)^{-1}(B_2C_1P)(I - QP)^{-1}Q.$$

Again using (2.6), it follows that

$$W(I + YPB_2)^{-1}Y = (I - A_2QA_1P)^{-1}B_2C_1$$
$$+ (I - QP)^{-1}Q - (I - A_2QA_1P)^{-1}Q.$$

According to the second identity in (1.11) we have $B_2C_1 - Q = -A_2QA_1$, and hence the above calculations yield (2.3).

Next we prove (2.4). Using (2.1) and (2.5) (cf. the state space formulas in Theorem 2.4 in [2]) we have

$$A_2QX(I + C_2QX)^{-1} = A_2Q(I - PQ)^{-1}B_1\Big(I - C_2Q(I - A_1PA_2Q)^{-1}B_1\Big)$$
$$= A_2Q(I - A_1PA_2Q)^{-1}B_1.$$

Hence

$$A_2QX(I + C_2QX)^{-1}Z = A_2Q(I - A_1PA_2Q)^{-1}B_1C_2(I - QP)^{-1}$$
$$= A_2Q(I - A_1PA_2Q)^{-1}B_1C_2$$
$$+ A_2Q(I - A_1PA_2Q)^{-1}B_1C_2QP(I - QP)^{-1}$$
$$= A_2Q(I - A_1PA_2Q)^{-1}B_1C_2$$
$$+ A_2Q(I - A_1PA_2Q)^{-1}B_1C_2Q(I - PQ)^{-1}P.$$

Using (2.5) we see that

$$A_2QX(I + C_2QX)^{-1}Z = A_2Q(I - A_1PA_2Q)^{-1}B_1C_2$$
$$- A_2Q(I - A_1PA_2Q)^{-1}P + A_2Q(I - PQ)^{-1}P$$
$$= -A_2Q(I - A_1PA_2Q)^{-1}A_1PA_2 + A_2QP(I - QP)^{-1}.$$

Now rewrite

$$A_2Q(I - A_1PA_2Q)^{-1}A_1PA_2 = A_2QA_1P(I - A_2QA_1P)^{-1}A_2$$
$$= \Big(I - (I - A_2QA_1P)\Big)(I - A_2QA_1P)^{-1}A_2$$
$$= (I - A_2QA_1P)^{-1}A_2 - A_2.$$

Similarly

$$A_2Q(I - PQ)^{-1}P = A_2QP(I - QP)^{-1}$$
$$= A_2\Big(I - (I - QP)\Big)(I - QP)^{-1} = A_2(I - QP)^{-1} - A_2.$$

It follows that

$$A_2 Q X (I + C_2 Q X)^{-1} Z = -(I - A_2 Q A_1 P)^{-1} A_2 + A_2 (I - Q P)^{-1}.$$

Hence

$$A_2 Q X (I + C_2 Q X)^{-1} Z Q A_1 = -(I - A_2 Q A_1 P)^{-1} A_2 Q A_1 + A_2 (I - Q P)^{-1} Q A_1$$
$$= -(I - A_2 Q A_1 P)^{-1} A_2 Q A_1 + A_2 Q (I - P Q)^{-1} A_1.$$

Thus (2.4) holds, and the proof is complete. $\qquad\square$

Using Remark 1.4 about interchanging the roles of P and Q one obtains the following alternative of Theorem 1.2.

Theorem 2.3. *Let the operators* X, Y, Z, *and* W *in* (1.15) *be solutions of the equations* (1.13) *and* (1.14)*. Then*

$$I_\mathcal{Y} + Y P B_2 = I_\mathcal{Y} + C_1 P W, \quad I_\mathcal{U} + C_2 Q X = I_\mathcal{U} + Z Q B_1.$$

Assume in addition that $I_{\mathcal{X}_2} - Q P$ *is invertible. Then* $I_{\mathcal{X}_2} - A_2 Q A_1 P$ *is invertible if and only if at least one of the two operators* $I_\mathcal{Y} + C_1 P W$ *and* $I_\mathcal{U} + Z Q B_1$ *is invertible. In that case both* $I_\mathcal{Y} + C_1 P W$ *and* $I_\mathcal{U} + Z Q B_1$ *are invertible and*

$$P (I_{\mathcal{X}_2} - Q P)^{-1} - A_1 P (I_{\mathcal{X}_2} - Q P)^{-1} A_2$$
$$= X (I_\mathcal{U} + Z Q B_1)^{-1} Z - A_1 P W (I_\mathcal{Y} + C_1 P W)^{-1} Y P A_2.$$

3. Conditions of invertibility, proof of Theorem 1.3

The next four results extend and sharpen Theorem 1.3. Hence in order to prove Theorem 1.3 it suffices to prove the four results presented below.

Proposition 3.1. *Assume that there exist operators* Y *and* Z *as in* (1.15) *such that* (1.14) *is satisfied, and let the associated operator* $I_\mathcal{Y} + Y P B_2$ *be invertible. Then*

$$(I_{\mathcal{X}_1} - P Q) x = 0 \quad \Rightarrow \quad C_2 A_2^\nu Q x = 0, \quad (\nu = 0, 1, 2, \ldots).$$

Proposition 3.2. *Assume that there exist operators* X *and* W *as in* (1.15) *such that* (1.13) *is satisfied, and let the associated operator* $I_\mathcal{U} + C_2 Q X$ *be invertible. Then*

$$y (I_{\mathcal{X}_1} - P Q) = 0 \quad \Rightarrow \quad y P A_2^\nu B_2 = 0 \quad (\nu = 0, 1, 2, \ldots).$$

Corollary 3.3. *Assume that the operator* $I_{\mathcal{X}_1} - P Q$ *is Fredholm of index zero and that* $\bigcap_{\nu=0}^\infty \operatorname{Ker} C_2 A_2^\nu = \{0\}$*. Furthermore, assume that there exist operators* Y *and* Z *as in* (1.15) *such that* (1.14) *is satisfied. Then* $I_\mathcal{Y} + Y P B_2$ *is invertible implies that* $I_{\mathcal{X}_1} - P Q$ *is invertible.*

Corollary 3.4. *Assume that the operator* $I_{\mathcal{X}_1} - P Q$ *is Fredholm of index zero and that* $\bigcap_{\nu=0}^\infty \operatorname{Ker} B_2^* A_2^{*\nu} = \{0\}$*. Furthermore, assume that there exist operators* X *and* W *as in* (1.15) *such that* (1.13) *is satisfied. Then* $I_\mathcal{U} + C_2 Q X$ *is invertible implies that* $I_{\mathcal{X}_1} - P Q$ *is invertible.*

Using Remark 1.4 we see that it suffices to prove Proposition 3.1 and Corollary 3.3.

Proof of Proposition 3.1. Assume that the operator $I_y + YPB_2$ is invertible, and let $(I_{\mathcal{X}_1} - PQ)x = 0$. Then $C_1 x = Y(I_{\mathcal{X}_1} - PQ)x = 0$ and

$$C_2 Qx = Z(I_{\mathcal{X}_1} - QP)Qx = ZQ(I_{\mathcal{X}_1} - PQ)x = 0.$$

Also

$$(I_{\mathcal{X}_1} - A_1 P A_2 Q)x = \big(I_{\mathcal{X}_1} - (P - B_1 C_2)Q\big)x = (I_{\mathcal{X}_1} - PQ + B_1 C_2 Q)x$$
$$= (I_{\mathcal{X}_1} - PQ)x + B_1 C_2 Qx = 0,$$

and

$$(I_{\mathcal{X}_1} - P A_2 Q A_1)x = \big(I_{\mathcal{X}_1} - P(Q - B_2 C_1)\big)x = (I_{\mathcal{X}_1} - PQ + PB_2 C_1)x$$
$$= (I_{\mathcal{X}_1} - PQ)x + PB_2 C_1 x = 0.$$

Next observe that

$$(I_y + YPB_2)C_1 = C_1 + YPB_2 C_1 = C_1 + YP(Q - A_2 Q A_1)$$
$$= C_1 + YPQ - YP A_2 Q A_1 = Y - YP A_2 Q A_1.$$

We see that

$$(I_y + YPB_2)C_1 P A_2 Qx = YP A_2 Qx - YP A_2 Q A_1 P A_2 Qx$$
$$= YP A_2 Q(I_{\mathcal{X}_1} - A_1 P A_2 Q)x = 0.$$

By assumption $I_y + YPB_2$ is invertible. Hence $C_1 P A_2 Qx = 0$. Therefore

$$0 = B_2 C_1 P A_2 Qx = (Q - A_2 Q A_1)P A_2 Qx = QP A_2 Qx - (A_2 Q)A_1 P A_2 Qx$$
$$= QP A_2 Qx - A_2 Qx = -(I_{\mathcal{X}_2} - QP)A_2 Qx.$$

We conclude that

$$(I_{\mathcal{X}_1} - PQ)x = 0 \quad \Rightarrow \quad (I_{\mathcal{X}_2} - QP)A_2 Qx = 0. \tag{3.1}$$

Next we prove that

$$(I_{\mathcal{X}_1} - PQ)x = 0 \quad \Rightarrow \quad (I_{\mathcal{X}_2} - QP)A_2^\nu Qx = 0, \quad \nu = 0, 1, 2, \dots. \tag{3.2}$$

For $\nu = 0$ we have $(I_{\mathcal{X}_2} - QP)Qx = Q(I_{\mathcal{X}_2} - PQ)x = 0$. We proceed by induction. Assume that the right-hand side of (3.2) holds for $\nu = k \geq 0$. Then

$$(I_{\mathcal{X}_1} - PQ)P A_2^k Qx = P(I_{\mathcal{X}_2} - QP)A_2^k Qx = 0.$$

Thus $(I_{\mathcal{X}_1} - PQ)\tilde{x} = 0$ where $\tilde{x} = P A_2^k Qx$. Now apply (3.1) with $\tilde{x}$ replacing x. It follows that

$$0 = (I_{\mathcal{X}_2} - QP)A_2 Q\tilde{x} = (I_{\mathcal{X}_2} - QP)A_2 QP A_2^k Qx$$
$$= -(I_{\mathcal{X}_2} - QP)A_2(I_{\mathcal{X}_2} - QP)A_2^k Qx + (I_{\mathcal{X}_2} - QP)A_2 A_2^k Qx$$
$$= (I_{\mathcal{X}_2} - QP)A_2^{k+1} Qx.$$

Thus by induction (3.2) holds. Using (3.2) we see that

$$C_2 A_2^\nu Q x = Z(I_{\mathcal{X}_2} - QP)A_2^\nu Q x = 0, \quad \nu = 0, 1, 2, \dots.$$

We proved the proposition.   □

Proof of Corollary 3.3. Let the operator $I_{\mathcal{Y}} + YPB_2$ be invertible, and assume that $(I_{\mathcal{X}_1} - PQ)x = 0$. According to Proposition 3.1 this implies that the vector Qx belongs to $\bigcap_{\nu=0}^{\infty} \operatorname{Ker} C_2 A_2^\nu$. By our hypotheses the latter space consists of the zero vector only. So $Qx = 0$. But then $x = PQx = 0$. Since $I_{\mathcal{X}_1} - PQ$ is Fredholm of index zero and $\operatorname{Ker}(I_{\mathcal{X}_1} - PQ) = 0$, it follows that the operator $I_{\mathcal{X}_1} - PQ$ is invertible.   □

Remark 3.5. *Assume that $\bigcap_{\nu=0}^{\infty} \operatorname{Ker} C_2 A_2^\nu = \{0\}$ and $I_{\mathcal{X}_1} - PQ$ is Fredholm of index zero. Assume also that Y and Z exist satisfying (1.14). Then Proposition 3.1 and Lemma 2.2 imply that the operator $I_{\mathcal{Y}} + YPB_2$ is invertible if and only if the operators $I_{\mathcal{X}_1} - PQ$ and $I_{\mathcal{X}_1} - A_1 P A_2 Q$ are both invertible.*

We conclude this section with three alternatives of Theorem 1.3. They follow directly from Theorem 1.3 by using the symmetry and duality arguments mentioned in Remark 1.4.

Theorem 3.6. *Assume that there exist operators Y and Z as in (1.15) such that the identities in (1.14) are satisfied. If the associated operator $I_{\mathcal{U}} + ZQB_1$ is invertible, then $\operatorname{Ker}(I_{\mathcal{X}_2} - QP) \subset \bigcap_{\nu=0}^{\infty} \operatorname{Ker} C_1 A_1^\nu P$. Moreover, if the operator $I_{\mathcal{X}_2} - QP$ is Fredholm of index zero and $\bigcap_{\nu=0}^{\infty} \operatorname{Ker} C_1 A_1^\nu = \{0\}$, then $I_{\mathcal{X}_2} - QP$ is invertible.*

Theorem 3.7. *Assume that there exist operators X and W as in (1.15) such that the identities in (1.13) are satisfied. If the associate operator $I_{\mathcal{U}} + C_2 QX$ is invertible, then $\operatorname{Im} PA_2^\nu B_2 \subset \operatorname{Im}(I_{\mathcal{X}_1} - PQ)$ for $\nu = 0, 1, 2, \dots$. Moreover, if $I_{\mathcal{X}_1} - PQ$ is Fredholm of index zero and $\operatorname{span}\{\operatorname{Im} A_2^\nu B_2 \mid \nu = 0, 1, 2, \dots\}$ is dense in $\mathcal{X}_2$, then $I_{\mathcal{X}_1} - PQ$ is invertible.*

Theorem 3.8. *Assume that there exist operators Y and Z as in (1.15) such that the identities in (1.14) are satisfied. If the associate operator $I_{\mathcal{Y}} + C_1 PW$ is invertible, then $\operatorname{Im} QA_1^\nu B_1 \subset \operatorname{Im}(I_{\mathcal{X}_2} - QP)$ for $\nu = 0, 1, 2, \dots$. Moreover, if $I_{\mathcal{X}_2} - QP$ is Fredholm of index zero and $\operatorname{span}\{\operatorname{Im} A_1^\nu B_1 \mid \nu = 0, 1, 2, \dots\}$ is dense in $\mathcal{X}_1$, then $I_{\mathcal{X}_2} - QP$ is invertible.*

4. Toeplitz plus Hankel operators

This section consists of three subsections. In the first we prove Theorem 1.1. In the second subsection we derive the identity (1.21), and in the third we derive formulas for $(R - GV^{-1}H)$ and $(V - HR^{-1}G)^{-1}$ analogous to the one in (1.21). We begin with a general remark.

Remark. *Since the entries of the matrix functions defining the operators G, H, R, V all belong to the Wiener algebra on the unit circle, it follows that for arbitrary* *linear maps* $x : \mathbb{C}^s \to \ell^1_+(\mathbb{C}^p)$ *and* $y : \mathbb{C}^s \to \ell^1_+(\mathbb{C}^q)$ *one has*

$$Hx : \mathbb{C}^s \to \ell^1_+(\mathbb{C}^q), \quad Rx : \mathbb{C}^s \to \ell^1_+(\mathbb{C}^p), \quad R^{-1}x : \mathbb{C}^s \to \ell^1_+(\mathbb{C}^p),$$
$$Gy : \mathbb{C}^s \to \ell^1_+(\mathbb{C}^p), \quad Vy : \mathbb{C}^s \to \ell^1_+(\mathbb{C}^q), \quad V^{-1}y : \mathbb{C}^s \to \ell^1_+(\mathbb{C}^q).$$

In particular, the linear maps b_1, b_2 *and* c_1, c_2 *defined by* (1.7) *have their values in* $\ell^1_+(\mathbb{C}^p)$ *and* $\ell^1_+(\mathbb{C}^q)$, *respectively.*

4.1. Proof of Theorem 1.1

In this subsection we prove Theorem 1.1. In order to do this we apply Theorems 1.2 and 1.3 with a special choice of P and Q and the associated data set, namely

$$P = I_{\mathcal{X}_1} - (R - GV^{-1}H), \quad Q = I_{\mathcal{X}_2}, \tag{4.1}$$
$$A_1 = S_p^*, \quad B_1 = GV^{-1}\varepsilon_q(\varepsilon_q^*V^{-1}\varepsilon_q)^{-1}, \quad C_1 = \varepsilon_p^*, \tag{4.2}$$
$$A_2 = S_p, \quad B_2 = \varepsilon_p, \quad C_2 = \varepsilon_q^*V^{-1}H. \tag{4.3}$$

Here $\mathcal{X}_1 = \mathcal{X}_2 = \ell^2_+(\mathbb{C}^p)$, and S_p is the forward shift on $\ell^2_+(\mathbb{C}^p)$. Note that

$$I - PQ = R - GV^{-1}H.$$

Since R is the invertible and G (and H) are compact operators, we see that $I - PQ$ is a Fredholm operator of index zero.

To see that for the above operators the identities in (1.11) are valid we first recall a useful equality. Note that $\varepsilon_q^*V^{-1}\varepsilon_q$ is the entry in the left upper corner of the block matrix representing V^{-1}. Since V is an invertible Toeplitz operator, the matrix $\varepsilon_q^*V^{-1}\varepsilon_q$ is invertible and

$$V^{-1} - S_qV^{-1}S_q^* = (V^{-1}\varepsilon_q)(\varepsilon_q^*V^{-1}\varepsilon_q)^{-1}(\varepsilon_q^*V^{-1}). \tag{4.4}$$

This result is well known and follows using a standard Schur complement argument (see, e.g., [6, Section 4]).

Next we deal with the Stein equations (1.11). Using that G and H are Hankel operators, that R is a Toeplitz operator, and that V^{-1} satisfies (4.4) we see that

$$\begin{aligned} P - A_1PA_2 &= I - (R - GV^{-1}H) - S_p^*S_p + S_p^*(R - GV^{-1}H)S_p \\ &= -R + S_p^*RS_p + GV^{-1}H - S_p^*GV^{-1}HS_p \\ &= G(V^{-1} - S_qV^{-1}S_q^*)H \\ &= G(V^{-1}\varepsilon_q)(\varepsilon_q^*V^{-1}\varepsilon_q)^{-1}(\varepsilon_q^*V^{-1})H = B_1C_2. \end{aligned}$$

Also

$$Q - A_2QA_1 = I - S_pS_p^* = \varepsilon_p\varepsilon_p^* = B_2C_1.$$

Thus equations (1.11) are satisfied. Furthermore, since $I - PQ = R - GV^{-1}H$, the left-hand side of (1.17) becomes

$$(R - GV^{-1}H)^{-1} - S_p(R - GV^{-1}H)^{-1}S_p^*.$$

Finally we see that

$$
\begin{aligned}
I - A_1 P A_2 Q &= I - S_p^* \left(I - (R - GV^{-1}H) \right) S_p \\
&= I - S_p^* S_p + (S_p^* R S_p - S_p^* GV^{-1} H S_p) \qquad (4.5) \\
&= R - G_1 V^{-1} H_1,
\end{aligned}
$$

where $G_1 = S_p^* G$ and $H_1 = S_q^* H$.

As a next step towards the proof of Theorem 1.1 it will be convenient first to prove the following proposition. In what follows we freely use the terminology and notation introduced in Theorem 1.1 and in the paragraphs preceding Theorem 1.1.

Proposition 4.1. *The following five conditions are equivalent.*

(1) *Equation (1.3) has solutions a_1 and d_1 and at least one of the matrices a_{10} and d_{10} is invertible.*

(2) *Equation (1.3) has solutions a_1 and d_1 and both matrices a_{10} and d_{10} are invertible.*

(3) *Equation (1.4) has solutions a_2^* and d_2^* and at least one of the matrices a_{20}^* and d_{20}^* is invertible.*

(4) *Equation (1.4) has solutions a_2^* and d_2^* and both matrices a_{20}^* and d_{20}^* are invertible.*

(5) *The operators $R - GV^{-1}H$ and $R - G_1 V^{-1} H_1$, where $G_1 = S_p^* G$ and $H_1 = S_q^* H$, are invertible.*

Moreover in that case

$$
\begin{aligned}
&(R - GV^{-1}H)^{-1} - S_p (R - GV^{-1}H)^{-1} S_p^* \\
&\quad = a_1 a_{10}^{-1} a_2^* - S_p R^{-1} G d_1 d_{10}^{-1} d_2^* H R^{-1} S_p^*,
\end{aligned} \qquad (4.6)
$$

$$
\begin{aligned}
&(V - HR^{-1}G)^{-1} - S_q (V - HR^{-1}G)^{-1} S_q^* \\
&\quad = d_1 d_{10}^{-1} d_2^* - S_q V^{-1} H a_1 a_{10}^{-1} a_2^* G V^{-1} S_q^*.
\end{aligned} \qquad (4.7)
$$

Proof. We split the proof into seven parts. The first part has an auxiliary character. The equivalence of the five conditions is proved in Parts 2–6. In the final part we prove formulas (4.6) and (4.7). Throughout we use the operators defined by (4.1), (4.2), and (4.3).

PART 1. We shall present operators satisfying (1.13) and (1.14). When a_1 and d_1 are linear maps satisfying equation (1.3), we put

$$
X = R^{-1} G d_1 (\varepsilon_q^* V^{-1} \varepsilon_q)^{-1}, \qquad W = a_1. \qquad (4.8)
$$

We claim that with P and Q defined as in the first paragraph of this subsection, the operators X and W in (4.8) satisfy (1.13). Indeed, using (1.3), we have

$$
\begin{aligned}
(I - PQ)X &= (R - GV^{-1}H) R^{-1} G d_1 (\varepsilon_q^* V^{-1} \varepsilon_q)^{-1} \\
&= GV^{-1}(V - HR^{-1}G) d_1 (\varepsilon_q^* V^{-1} \varepsilon_q)^{-1} \\
&= GV^{-1} \varepsilon_q (\varepsilon_q^* V^{-1} \varepsilon_q)^{-1} = B_1, \\
(I - QP)W &= (R - GV^{-1}H) a_1 = \varepsilon_p = B_2.
\end{aligned}
$$

Hence (1.13) is satisfied. Furthermore, (1.3) gives

$$
\begin{aligned}
I + C_1 PW &= I + \varepsilon_p^* P a_1 = I + \varepsilon_p^* (I - (R - GV^{-1}H)) a_1 \\
&= \varepsilon_p^* a_1 + I - \varepsilon_p^* \varepsilon_p = a_{10},
\end{aligned}
$$

$$
\begin{aligned}
I + C_2 QX &= I + \varepsilon_q^* V^{-1} H R^{-1} G d_1 (\varepsilon_q^* V^{-1} \varepsilon_q)^{-1} \\
&= I - \varepsilon_q^* V^{-1} (V - H R^{-1} G) d_1 (\varepsilon_q^* V^{-1} \varepsilon_q)^{-1} + \varepsilon_q^* d_1 (\varepsilon_q^* V^{-1} \varepsilon_q)^{-1} \\
&= I - \varepsilon_q^* V^{-1} \varepsilon_q (\varepsilon_q^* V^{-1} \varepsilon_q)^{-1} + \varepsilon_q^* d_1 (\varepsilon_q^* V^{-1} \varepsilon_q)^{-1} \\
&= d_{10} (\varepsilon_q^* V^{-1} \varepsilon_q)^{-1}.
\end{aligned}
$$

Similar results hold for a_2 in place of a_1 and d_2 in place of d_1. Indeed, when a_2 and d_2 are linear maps satisfying equation (1.4), we put

$$
Y = a_2^*, \quad Z = d_2^* H R^{-1}. \tag{4.9}
$$

These operators satisfy (1.14). Indeed, using (1.4), we have

$$
\begin{aligned}
Z(I - QP) &= d_2^* H R^{-1} (R - GV^{-1}H) \\
&= d_2^* (V - H R^{-1} G) V^{-1} H = \varepsilon_q^* V^{-1} H = C_2, \\
Y(I - PQ) &= a_2^* (R - GV^{-1}H) = \varepsilon_p^* = C_1.
\end{aligned}
$$

Hence (1.14) is satisfied. Furthermore, (1.4) gives

$$
\begin{aligned}
I + YPB_2 &= I + a_2^* (I - (R - GV^{-1}H)) \varepsilon_p \\
&= I - a_2^* (R - GV^{-1}H) \varepsilon_p + a_2^* \varepsilon_p = a_2^* \varepsilon_p = a_{20}, \\
I + ZQB_1 &= I + d_2^* H R^{-1} G V^{-1} \varepsilon_q (\varepsilon_q^* V^{-1} \varepsilon_q)^{-1} \\
&= I + d_2^* \varepsilon_q (\varepsilon_q^* V^{-1} \varepsilon_q)^{-1} - d_2^* (V - H R^{-1} G) V^{-1} \varepsilon_q (\varepsilon_q^* V^{-1} \varepsilon_q)^{-1} \\
&= I + d_2^* \varepsilon_q (\varepsilon_q^* V^{-1} \varepsilon_q)^{-1} - \varepsilon_q^* V^{-1} \varepsilon_q (\varepsilon_q^* V^{-1} \varepsilon_q)^{-1} = d_{20} (\varepsilon_q^* V^{-1} \varepsilon_q)^{-1}.
\end{aligned}
$$

PART 2. In this part we show that condition (5) implies conditions (1)–(4). So assume (5) is satisfied. Then there exists linear maps as in (1.2) satisfying equations (1.3) and (1.4). Also $I - PQ$ is invertible, and using (4.5) we see that the same holds true for $I - A_1 P A_2 Q$. From Lemma 2.2 we conclude that $I + C_2 QX$ and $I + YPB_2$ are invertible. Hence d_{10} and a_{20}^* are invertible. Now use Theorem 1.2 and notice that the equalities in (1.16) imply that $I + ZQB_1$ and $I + C_1 PW$ also are invertible. We conclude that d_{20}^* and a_{10} are invertible, $a_{10} = a_{20}^*$ and $d_{10} = d_{20}^*$. So indeed condition (5) implies conditions (1)–(4).

Next we make a remark that will help to simplify the remaining parts of the proof. Assume that $I - PQ$ is invertible. Then the equations (1.13) and (1.14) are uniquely solvable, and with these solutions the equalities in (1.16) hold true. Moreover Lemma 2.2 shows that if one of the four operators in (1.16) is invertible, then $I - A_1 P A_2 Q$ is invertible and hence condition (5) is satisfied. Conclusion: in order to finish the proof of the equivalence of the five conditions (1)–(5) we only have to show that (1) and (3) each separately imply that $I - PQ$ is invertible. (Trivially, condition (2) implies (1) and (4) implies (3).)

PART 3. In this part we show that condition (1) with a_{10} invertible implies that $I - PQ$ is invertible. Define X and W by (4.8). As we have seen in the first part of the proof, $I + C_1 PW = a_{10}$, and hence $I + C_1 PW$ is invertible. Assume that $y(I - QP) = 0$. According to Theorem 3.8 it follows that $yQA_1^n B_1 = 0$ for $n = 0, 1, 2, \ldots$. So for all nonnegative integers n we obtain that $y(S_p^*)^n GV^{-1}\varepsilon_q = 0$. Here we use that $\varepsilon_q^* V^{-1}\varepsilon_q$ is invertible. Since V is invertible, V factors as $V = V_- V_+$, where V_+ and V_- are invertible Toeplitz operators, V_+ and V_+^{-1} are both lower triangular, and V_- and V_-^{-1} are both upper triangular (see [9] or [2, Theorem 1.2]). Then

$$0 = y(S_p^*)^n GV^{-1}\varepsilon_q = (yG)S_q^n V^{-1}\varepsilon_q = (yG)S_q^n V_+^{-1} V_-^{-1}\varepsilon_q = (yG)V_+^{-1} S_q^n \varepsilon_q v_{-0},$$

where v_{-0} is the invertible $(1,1)$-entry of V_-^{-1}. Hence $(yGV_+^{-1})S_q^n \varepsilon_p = 0$ for $n = 0, 1, 2, \ldots$. So we obtain $yGV_+^{-1} = 0$, and $yG = 0$. Since $I - PQ = R - GV^{-1}H$ we have $y(R - GV^{-1}H) = 0$, and $yG = 0$ implies that $yR = 0$. But R is invertible, and therefore $y = 0$. So $I - PQ$ has a dense range. Since $I - PQ$ is a Fredholm operator of index zero, it follows that $I - PQ$ is invertible.

PART 4. In this part we show that condition (1) with d_{10} invertible implies that $I - PQ$ is invertible. Define X and W by (4.8). As we have seen in the first part of the proof, $I + C_2 QX = d_{10}$, and hence the operator $I + C_2 QX$ is invertible. Assume that $y(I - PQ) = 0$. From Theorem 3.7 we see that $0 = yPA_2^n B_2 = yPS_p^n \varepsilon_p$ for $n = 0, 1, 2, \ldots$. It follows that $yP = 0$, and therefore $y = y - yPQ = y(I - PQ) = 0$. So $I - PQ$ has a dense range. Since $I - PQ$ is a Fredholm operator of index zero, it follows that $I - PQ$ is invertible.

PART 5. In this part we show that condition (3) with a_{20}^* invertible implies that $I - PQ$ is invertible. Define Y and Z by (4.9). As we have seen in the first part of the proof, $I + YPB_2 = a_{20}^*$, and hence $I + YPB_2$ is invertible. From Theorem 1.3 we conclude that

$$\operatorname{Ker}(I - PQ) \subset \bigcap_{\nu=0}^{\infty} \operatorname{Ker} C_2 A_2^\nu Q.$$

Assume that $(I - PQ)x = 0$. Then, by the previous identity $C_2 A_2^\nu Qx = C_2 A_2^\nu x = 0$. Using the definitions of A_2 and C_2, we obtain $\varepsilon_p^* V^{-1} HS_p^n x = 0$ for $n = 0, 1, 2, \ldots$. As above in Part 3 write $V = V_- V_+$. Then

$$0 = \varepsilon_p^* V_+^{-1} V_-^{-1} HS_p^n x = \varepsilon_p^* V_+^{-1} V_-^{-1} (S_q^*)^n Hx = v_{+0} \varepsilon_p^* (S_q^*)^n V_-^{-1} Hx.$$

We see that $Hx = 0$. But then $Rx = 0$ and $x = 0$. As above we conclude that $R - GV^{-1}H$ is invertible. So $I - PQ$ is invertible.

PART 6. In this part we show that condition (3) with d_{20}^* invertible that $I - PQ$ is invertible. Define Y and Z by (4.9). As we have seen in the first part of the proof $I + ZQB_1 = d_{20}^*$, and hence $I + ZQB_1$ is invertible. Assume that $(I - PQ)x = 0$. According to Theorem 3.6 we have that $C_1 A_1^n Px = 0$ for $n = 0, 1, 2, \ldots$. So $\varepsilon_p S_p^n Px = 0$ for all n. But then $Px = 0$, and we conclude that $x = 0$. Hence $I - PQ$ is invertible.

PART 7. Finally we apply Theorem 1.2 to show that the inverse of $R - GV^{-1}H$ is given by (1.17) whenever one of the conditions (1)–(5) is satisfied. Formula (1.17) translates to (4.6), and hence

$$(R - GV^{-1}H)^{-1} - S_p(R - GV^{-1}H)^{-1}S_p^*$$
$$= a_1 a_{10}^{-1} a_2^* - S_p R^{-1} G d_1 d_{10}^{-1} d_2^* H R^{-1} S_p^*.$$

The identity (4.7) one obtains by just switching the roles of R and V, G and H, p and q, a_{j0} and d_{j0}. $\qquad\square$

We now are ready to prove Theorem 1.1. Recall that for a linear map $\alpha : \mathbb{C}^m \to \ell_+^2(\mathbb{C}^m)$ we denote by T_α the $m \times m$ block lower triangular Toeplitz operator with first column equal to α. For an $m \times m$ matrix u the symbol Δ_u denotes the block diagonal operator on $\ell_+^2(\mathbb{C}^m)$ with all diagonal entries equal to u.

Proof of Theorem 1.1. First we prove (1.5). Use Proposition 4.1 to derive the identity (4.6). By multiplying this identity $n - 1$ times from the left by S_p and $n - 1$ times from the right by S_p^*, and adding the resulting identities one gets (also using (1.7))

$$(R - GV^{-1}H)^{-1} - S_p^n(R - GV^{-1}H)^{-1}(S_p^*)^n$$
$$= \sum_{k=0}^{n-1} S_p^k a_1 a_{10}^{-1} a_2^*(S_p^*)^k - S_p \left(\sum_{k=0}^{n-1} S_p^k b_1 d_{10}^{-1} b_2^*(S_p^*)^k \right) S_p^*.$$

Since for any h we have $\lim_{n \to \infty} (S_p^*)^n h = 0$, the left-hand side converges pointwise to $(R - GV^{-1}H)^{-1}$. Notice that

$$\sum_{k=0}^{n-1} S_p^k a_1 a_{10}^{-1} a_2^*(S_p^*)^k = T_{a_1} \Delta_{a_{10}^{-1}} \Pi_n T_{a_2}^*,$$

where $\Pi_n = I - S_p^n(S_p^*)^n$, and $\lim_{n \to \infty} T_{a_1} \Delta_{a_{10}^{-1}} \Pi_n T_{a_2}^* = T_{a_1} \Delta_{a_{10}^{-1}} T_{a_2}^*$. Similarly, using that $b_1, b_2 : \mathbb{C}^q \to \ell_+^1(\mathbb{C}^p)$, one finds

$$\sum_{n=0}^{\infty} S_p^n b_1 d_{10}^{-1} b_2^*(S_p^*)^n = T_{b_1} \Delta_{d_{10}^{-1}} T_{b_2}^*.$$

So we get

$$(R - GV^{-1}H)^{-1} = T_{a_1} \Delta_{a_{10}^{-1}} T_{a_2}^* - S_p T_{b_1} \Delta_{d_{10}^{-1}} T_{b_2}^* S_p^*.$$

We proved (1.5). Formula (1.6) one obtains in a similar way from (4.7). $\qquad\square$

It is interesting to specify Theorem 1.1 for the case when $G = 0$, $H = 0$ and $V = I$. Then $R - GV^{-1}H = R$. Recall that R is assumed to be invertible. The hypotheses $G = 0$, $H = 0$ and $V = I$ imply that (1.3) and (1.4) reduce to

$$a_1 = R^{-1}\varepsilon_p, \quad d_1 = \varepsilon_q, \quad a_2^* = \varepsilon_p^* R^{-1}, \quad d_2^* = \varepsilon_q^*.$$

In particular, a_1 is the first column of R^{-1}, a_2^* is the first row of R^{-1}, and $a_{10} = a_{20}^*$ is the $(1, 1)$ entry of R^{-1}. Since we assume that the entries of the matrix function defining R belong to the Wiener algebra on the circle, the classical result from [9] then tells us that a_{10} is invertible and

$$R^{-1} = T_{a_1} \Delta_{a_{10}^{-1}} T_{a_2}^*. \tag{4.10}$$

The above formula for R^{-1} is precisely (1.5) for the case when G and H are zero. Indeed, when G and H are zero, then (1.7) tells us that the matrices b_1 and b_2 are zero. But in that case (1.5) reduces to (4.10).

4.2. Finite rank Hankel operators

In this subsection we return to the case when G and H are of finite rank and $R = I_p$ and $V = I_q$. We shall derive the identity (1.21). To do this we use the data appearing in (1.18), (1.19) and (1.20).

Proof of (1.21). Assume there exist linear maps a_1, a_2, d_1, d_2 as in (1.2) satisfying equations (1.3) and (1.4) (with R and V identity operators) such that a_{10} or d_{10} is invertible. As Theorem 1.1 tells us, this implies that $I - GH$ is invertible. We intend to apply Theorem 1.2 with the data as in (1.18), (1.19) and (1.20). First one checks that the operators

$$X = \Lambda_1 a_1, \quad Y = d_2^* \Gamma_1, \quad W = \Lambda_2 d_1, \quad Z = a_2^* \Gamma_2 \tag{4.11}$$

satisfy the identities (1.13) and (1.14). This allows us to prove the following identities:

$$I + C_2 QX = a_{10}, \quad I + C_1 PW = d_{10}, \tag{4.12}$$
$$I + ZQB_1 = a_{20}^*, \quad I + YPB_2 = d_{20}^*. \tag{4.13}$$

To see this let us establish the first identity in (4.12):

$$I + C_2 QX = I + \varepsilon_p^* \Gamma_2 \Lambda_2 \Gamma_1 \Lambda_1 a_1 = I + \varepsilon_p^* GH a_1$$
$$= I + \varepsilon_p^*(a_1 - (I - GH)a_1) = a_{10}.$$

The other identities in (4.12) and (4.13) are proved in a similar way. It follows (see Lemma 2.1) that $d_{10} = d_{20}^*$ and $a_{10} = a_{20}^*$. Since we assume that one of the matrices a_{10} and d_{10} is invertible, both are invertible, and we get from (1.17), (4.11), (4.12), and (4.13) that

$$Q(I - PQ)^{-1} - A_2 Q(I - PQ)^{-1} A_1$$
$$= \Lambda_2 d_1 d_{10}^{-1} d_2^* \Gamma_1 - A_2 \Lambda_2 \Gamma_1 \Lambda_1 a_1 a_{10}^{-1} a_2^* \Gamma_2 \Lambda_2 \Gamma_1 A_1$$
$$= \Lambda_2 d_1 d_{10}^{-1} d_2^* \Gamma_1 - A_2 \Lambda_2 H a_1 a_{10}^{-1} a_2^* G \Gamma_1 A_1.$$

The fact that both A_1 and A_2 are stable then yields:

$$Q(I - PQ)^{-1} = \sum_{\nu=0}^{\infty} A_2^\nu \Lambda_2 d_1 d_{10}^{-1} d_2^* \Gamma_1 A_1^\nu - \sum_{\nu=0}^{\infty} A_2^{\nu+1} \Lambda_2 H a_1 a_{10}^{-1} a_2^* G \Gamma_1 A_1^{\nu+1}.$$

Using $(I - GH)^{-1} = I + \Gamma_2 Q (I - PQ)^{-1} \Lambda_1$, we find

$$(I - GH)^{-1} = I + \sum_{\nu=0}^{\infty} \Gamma_2 A_2^{\nu} \Lambda_2 d_1 d_{10}^{-1} d_2^* \Gamma_1 A_1^{\nu} \Lambda_1$$

$$- \sum_{\nu=0}^{\infty} \Gamma_2 A_2^{\nu+1} \Lambda_2 H a_1 a_{10}^{-1} a_2^* G \Gamma_1 A_1^{\nu+1} \Lambda_1.$$

Next put $b_1 = -Gd_1$ and $b_2^* = -d_2^* H$; cf. the identities in (1.7). Then

$$\Gamma_2 A_2^{\nu} \Lambda_2 d_1 = (S_p^*)^{\nu}(-b_1), \quad d_2^* \Gamma_1 A_1^{\nu} \Lambda_1 = -b_2^* S_p^{\nu} \quad (\nu \geq 0).$$

Also use

$$\Gamma_2 A_2^{\nu+1} \Lambda_2 H a_1 = (S_p^*)^{\nu+1} G H a_1 = (S_p^*)^{\nu+1}(a_1 - \varepsilon_p) = (S_p^*)^{\nu+1} a_1,$$

$$a_2^* G \Gamma_1 A_1^{\nu+1} \Lambda_1 = a_2^* G H S_p^{\nu+1} = (a_2^* - \varepsilon_p^*) S_p^{\nu+1} = a_2^* S_p^{\nu+1}.$$

In this way we obtain

$$(I - GH)^{-1} = I + \sum_{\nu=0}^{\infty} (S_p^*)^{\nu} b_1 d_{10}^{-1} b_2^* S_p^{\nu} - \sum_{\nu=0}^{\infty} (S_p^*)^{\nu+1} a_1 a_{10}^{-1} a_2^* S_p^{\nu+1}$$

$$= I + H_{b_1} \Delta_{d_{10}^{-1}} H_{b_2}^* - S_p^* H_{a_1} \Delta_{a_{10}^{-1}} H_{a_2}^* S_p. \qquad (4.14)$$

This proves (1.21). $\qquad\qquad\qquad\qquad\qquad\qquad\qquad\qquad\qquad\qquad\qquad\qquad\square$

Remark. Notice that all the Hankel operators in (4.14) are of finite rank. For example, the identity $H_{b_1} = -\Gamma_2 \begin{bmatrix} W & A_2 W & A_2^2 W & \cdots \end{bmatrix}$ implies that the rank of H_{b_1} is at most the order of A_2.

Formula (1.21) remains true in the more general case when G and H are of finite rank. To see this, assume there exist linear maps a_1, a_2, d_1, d_2 as in (1.2) satisfying equations (1.3) and (1.4) (with R and V identity operators) such that a_{10} or d_{10} is invertible. Hence, by Theorem 1.1, the operator $I - GH$ is invertible To derive the analogue of the formula (1.21) for this case we apply Theorem 1.2. Put

$$P = H, \quad Q = G,$$
$$A_1 = S_q^*, \quad B_1 = H\varepsilon_p, \quad C_1 = \varepsilon_p^*,$$
$$A_2 = S_p^*, \quad B_2 = G\varepsilon_q, \quad C_2 = \varepsilon_q^*.$$

The corresponding Stein equations (1.11) are satisfied, and

$$X = H a_1, \quad W = G d_1, \quad Y = d_2^*, \quad \text{and } Z = a_2^*$$

solve the equations (1.13) and (1.14). Notice that the equalities (4.12), (4.13) hold true. Furthermore, $d_{10} = d_{20}^*$ and $a_{10} = a_{20}^*$. Theorem 1.2 yields

$$G(I - HG)^{-1} - S_p^* G(I - HG)^{-1} S_q^* = G d_1 d_{10}^{-1} d_2^* - S_p^* G H a_1 a_{10}^{-1} a_2^* G S_q^*.$$

Since $S_p^* G H a_1 = S_p^*(a_1 - \varepsilon_p) = S_p^* a_1$ and $G d_1 = -b_1$ and $a_2^* G = -c_2^*$, we get

$$G(I - HG)^{-1} - S_p^* G(I - HG)^{-1} S_q^* = -b_1 d_{10}^{-1} d_2^* + S_p^* a_1 a_{10}^{-1} c_2^* S_q^*.$$

Using the same reasoning as in the proof of Theorem 1.1 we obtain

$$G(I - HG)^{-1} = -H_{b_1}\Delta_{d_{10}}^{-1}T_{d_2}^* + S_p^*H_{a_1}\Delta_{a_{10}}^{-1}T_{c_2}^*S_q^*. \tag{4.15}$$

To derive from (4.15) a formula for $(I - GH)^{-1}$, we use the identity

$$(I - GH)^{-1} = I + G(I - HG)^{-1}H. \tag{4.16}$$

Thus we have to multiply (4.15) from the right by H. For this purpose we use the following lemma.

Lemma 4.2. *Let* $x : \mathbb{C}^r \to \ell_+^1(\mathbb{C}^r)$, *and let* $K : \ell_+^1(\mathbb{C}^r) \to \ell_+^1(\mathbb{C}^s)$ *be a Hankel operator of which the defining matrix function has entries in the Wiener algebra. Then*

$$KT_x = H_{Kx} : \ell_+^1(\mathbb{C}^r) \to \ell_+^1(\mathbb{C}^s).$$

Here T_x *is the lower triangular Toeplitz operator with first column* x *and* H_{Kx} *is the Hankel operator with with first column* Kx.

Proof. Note that $Kx : \mathbb{C}^r \to \ell_+^1(\mathbb{C}^s)$. The jth column of KT_x is $KS_r^j x$ and the jth column of H_{Kx} is $(S_s^*)^j Kx$. These are equal since $KS_r^j = (S_s^*)^j K$. $\square$

From this lemma we see that $T_{d_2}^*H = H_{H^*d_2}^*$. With as before $b_2^* = -d_2^*H$ it follows that $T_{d_2}^*H = -H_{b_2}^*$. Next we consider $T_{c_2}^*S_q^*H$. The dual of this operator is $H^*S_qT_{c_2} = H^*T_{S_qc_2} = H_{H^*S_qc_2}$. Note that

$$c_2^*S_q^*H = -a_2^*GHS_p = (-a_2^* + \varepsilon_p^*)S_p = -a_2^*S_p,$$

to see that $H^*S_qT_{c_2} = -H_{S_p^*a_2} = -S_p^*H_{a_2}$. We get $T_{c_2}^*S_q^*H = -H_{a_2}^*S_p$, and therefore (4.15) and (4.16) together yield

$$(I - GH)^{-1} = I + H_{b_1}\Delta_{d_{10}}^{-1}H_{b_2}^* - S_p^*H_{a_1}\Delta_{a_{10}}^{-1}H_{a_2}^*S_p, \tag{4.17}$$

which is (1.21) for the present case.

Remark. Using (4.16) the identity (4.17) also follows from (1.6) with $R = I$ and $V = I$. Indeed, using (4.16) and (1.6) we obtain

$$(I - GH)^{-1} = I + GT_{d_1}\Delta_{d_{10}}^{-1}T_{d_2}^*H - GS_qT_{c_1}\Delta_{a_{10}}^{-1}T_{c_2}^*S_q^*H.$$

Applying Lemma 4.2 yields the identities:

$$GT_{d_1} = -H_{b_1}, \ GS_qT_{c_1} = -S_p^*H_{a_1}, \ T_{d_2}^*H = -H_{b_2}^*, \ T_{c_2}^*S_q^*H = -H_{a_2}^*S_p.$$

Using these identities one obtains (4.17).

In the next section we shall prove the analogue of formula (4.17) in the general setting of Theorem 1.1.

4.3. Hankel type formulas for the inverse of Toeplitz plus Hankel

In this section we derive formulas for the inverses of $R - GV^{-1}H$ and $V - HR^{-1}G$ that generalize formula (1.21) (and (4.17)). We begin with some notation.

Let φ be the defining function of R, and ψ the one of V. Since R and V are invertible, $\det \varphi$ and $\det \psi$ have no zero on the unit circle, and hence φ^{-1} and ψ^{-1} are well defined matrix functions. Moreover, the entries of φ^{-1} and ψ^{-1} belong to the Wiener algebra on the unit circle. We denote by $R^{\times}$ and $V^{\times}$ the block Toeplitz operators defined by φ^{-1} and ψ^{-1}, respectively.

Theorem 4.3. *Assume there exist linear maps a_1, a_2 and d_1, d_2 as in (1.2) satisfying equations (1.3) and (1.4). Then $a_{10} = a_{20}^*$ and $d_{10} = d_{20}^*$. Furthermore, assume that at least one of the matrices a_{10} and d_{10} is invertible. Then both the matrices a_{10} and d_{10} are invertible, and the operators $R - GV^{-1}H$ and $V - HR^{-1}G$ are invertible. Moreover,*

$$(R - GV^{-1}H)^{-1} = R^{\times} + H_{b_1}\Delta_{d_{10}}^{-1}H_{b_2}^* - S_p^* H_{a_1}\Delta_{a_{10}}^{-1}H_{a_2}^* S_p, \qquad (4.18)$$

$$(V - HR^{-1}G)^{-1} = V^{\times} + H_{c_1}\Delta_{a_{10}}^{-1}H_{c_2}^* - S_q^* H_{d_1}\Delta_{d_{10}}^{-1}H_{d_2}^* S_q. \qquad (4.19)$$

Here

$$b_1 = -R^{-1}Gd_1, \ c_1 = -V^{-1}Ha_1, \ b_2^* = -d_2^*HR^{-1}, \ c_2^* = -a_2^*GV^{-1}.$$

Proof. As in the proof of Theorem 1.1, we conclude from Proposition 4.1 that the identity (4.6) holds. Multiply this identity from the left by S_p^* and from the right by $-S_p$. This yields (cf. the first identity in (5.16) of [6]):

$$(R - GV^{-1}H)^{-1} - S_p^*(R - GV^{-1}H)^{-1}S_p = b_1 d_{10}^{-1}b_2^* - S_p^* a_1 a_{10}^{-1} a_2^* S_p. \qquad (4.20)$$

Next, by $n - 1$ times repeatedly multiplying (4.20) from the left by S_p^* and from the right by S_p, and adding the resulting identities, we get

$$(R - GV^{-1}H)^{-1} - (S_p^*)^n(R - GV^{-1}H)^{-1}S_p^n$$

$$= \sum_{k=0}^{n-1}(S_p^*)^k b_1 d_{10}^{-1}b_2^* S_p^k - \sum_{k=0}^{n-1}(S_p^*)^k S_p^* a_1 a_{10}^{-1} a_2^* S_p S_p^k. \qquad (4.21)$$

In order to determine for the terms in (4.21) the limits for n going to infinity, we will deal with each of these terms separately.

We begin with the first term on the right-hand side. Let Π_n be the projection of $\ell_+^2(\mathbb{C}^p)$ mapping $(x_0, x_1, x_2, \ldots)$ onto $(x_0, \ldots, x_{n-1}, 0, 0, \ldots)$. Then

$$\sum_{k=0}^{n-1}(S_p^*)^k b_1 d_{10}^{-1}b_2^* S_p^k = H_{b_1}\Delta_{d_{10}}^{-1}\Pi_n H_{b_2}^*.$$

Since $H_{b_2}^*$ is compact, $\lim_{n \to \infty} \Pi_n H_{b_2}^* = H_{b_2}^*$, with convergence in the operator norm, and hence

$$\sum_{k=0}^{\infty}(S_p^*)^k b_1 d_{10}^{-1}b_2^* S_p^k = H_{b_1}\Delta_{d_{10}}^{-1}H_{b_2}^*. \qquad (4.22)$$

Similarly one derives that

$$\sum_{k=0}^{\infty}(S_p^*)^k S_p^* a_1 a_{10}^{-1} a_2^* S_p S_p^k = S_p^* H_{a_1}\Delta_{a_{10}}^{-1}H_{a_2}^* S_p. \tag{4.23}$$

Next we proceed with the left-hand side of the identity (4.21). First remark that the convergence of the right-hand side yields the existence of the limit $\lim_{n\to\infty}(S_p^*)^n(R-GV^{-1}H)^{-1}S_p^n$. We will prove

$$\lim_{n\to\infty}(S_p^*)^n(R-GV^{-1}H)^{-1}S_p^n = R^{\times}. \tag{4.24}$$

The proof of (4.24) will be based on the following two observations: (a) the operator $(R-GV^{-1}H)^{-1}-R^{-1}$ is compact, and (b) the operator $R^{-1}-R^{\times}$ is compact. Note that (a) follows from

$$(R-GV^{-1}H)^{-1} = R^{-1}+R^{-1}GV^{-1}H(R-GV^{-1}H)^{-1}.$$

Indeed, using the latter identity and G (or H) is compact, we get item (a). To get item (b), we use that R is invertible, and hence the defining function φ of R admits a canonical factorizations, $\varphi = \varphi_-\varphi_+$. Recall that $R^{\times}$ is the block Toeplitz operator defined by $\varphi^{-1} = \varphi_+^{-1}\varphi_-^{-1}$ and $R^{-1} = R_+^{\times}R_-^{\times}$, where $R_+^{\times}$ and $R_-^{\times}$ are the block Toeplitz operators defined by φ_+^{-1} and φ_-^{-1}, respectively. But then, using a standard identity (see formula (4) in [7, Section XXIII]), we see that $R^{\times}-R^{-1}$ is the product of two compact Hankel operators, which proves item (b).

Given items (a) and (b) we see that $(R-GV^{-1}H)^{-1} = R^{\times}+K$, where K is a compact operator. Since $(S_p^*)^n \to 0$ pointwise, the compactness of K implies that $(S_p^*)^n K \to 0$ in operator norm, and thus $(S_p^*)^n K S_p^n \to 0$ in operator norm as $n\to\infty$. The fact that $R^{\times}$ is a block Toeplitz operator is equivalent to $S_p^* R^{\times} S_p = R^{\times}$, and hence $(S_p^*)^n R^{\times} S_p^n = R^{\times}$ for each n. We conclude that the left-hand side of (4.24) is equal to

$$\lim_{n\to\infty}(S_p^*)^n R^{\times} S_p^n + \lim_{n\to\infty}(S_p^*)^n K S_p^n = R^{\times}.$$

This proves (4.24). Combining the limits (4.22), (4.23), and (4.24) we obtain (4.18). The proof of (4.19) can be done in exactly the same manner. □

5. New proof of the main inversion theorem in [6]

In this section we show how Theorem 2.1 in [6] can be obtained from Theorem 1.2.

We begin with some preliminaries. Let $\mathcal{X}$ be a Hilbert space with two orthogonal direct sum decompositions:

$$\mathcal{X} = \mathcal{U}_1 \oplus \mathcal{Y}_1 = \mathcal{U}_2 \oplus \mathcal{Y}_2. \tag{5.1}$$

On $\mathcal{X}$ we have two operators A and K such that relative to these decompositions

$$A = \begin{bmatrix} \alpha_1 & 0 \\ 0 & \alpha_2 \end{bmatrix} : \begin{bmatrix} \mathcal{U}_1 \\ \mathcal{Y}_1 \end{bmatrix} \to \begin{bmatrix} \mathcal{U}_2 \\ \mathcal{Y}_2 \end{bmatrix}, \quad \text{where } \alpha_2 \text{ is invertible;} \tag{5.2}$$

$$K = \begin{bmatrix} \kappa_1 & 0 \\ 0 & \kappa_2 \end{bmatrix} : \begin{bmatrix} \mathcal{U}_2 \\ \mathcal{Y}_2 \end{bmatrix} \to \begin{bmatrix} \mathcal{U}_1 \\ \mathcal{Y}_1 \end{bmatrix}, \quad \text{where } \kappa_2 \text{ is invertible and } \kappa_2 = \alpha_2^{-1}. \tag{5.3}$$

In what follows we denote by $\pi_{\mathcal{H}}$ the orthogonal projection of $\mathcal{X}$ onto the subspace $\mathcal{H}$, viewed as an operator from $\mathcal{X}$ to $\mathcal{H}$. Furthermore, $\tau_{\mathcal{H}}$ denotes the canonical embedding of $\mathcal{H}$ into $\mathcal{X}$, that is, $\tau_{\mathcal{H}} = \pi_{\mathcal{H}}^*$. The following result is Theorem 2.1 in [6].

Theorem 5.1. *Let T be an invertible operator on $\mathcal{X}$ and let A and K be as in* (5.2) *and* (5.3), *respectively. Assume that*

$$\pi_{\mathcal{Y}_1}(T - KTA)\tau_{\mathcal{Y}_1} = 0. \tag{5.4}$$

Consider the operators defined by

$$\Xi = T^{-1}\tau_{\mathcal{U}_2} : \mathcal{U}_2 \to \mathcal{X}, \quad \Psi = T^{-1}\tau_{\mathcal{U}_1} : \mathcal{U}_1 \to \mathcal{X},$$

$$\Upsilon = \pi_{\mathcal{U}_2}T^{-1} : \mathcal{X} \to \mathcal{U}_2, \quad \Omega = \pi_{\mathcal{U}_1}T^{-1} : \mathcal{X} \to \mathcal{U}_1.$$

Furthermore, put $\xi_0 = \pi_{\mathcal{U}_2}\Xi$ and $\psi_0 = \pi_{\mathcal{U}_1}\Psi$. Then ξ_0 is invertible if and only if ψ_0 is invertible, and in this case the inverse of T satisfies the identity

$$T^{-1} - AT^{-1}K = \Xi\xi_0^{-1}\Upsilon - A\Psi\psi_0^{-1}\Omega K. \tag{5.5}$$

First we state and prove a preliminary lemma.

Lemma 5.2. *Let T be an invertible operator on $\mathcal{X}$ and let A and K be as in* (5.2) *and* (5.3), *respectively. Let*

$$A_0 = \begin{bmatrix} 0 & 0 \\ 0 & \alpha_2 \end{bmatrix} : \begin{bmatrix} \mathcal{U}_1 \\ \mathcal{Y}_1 \end{bmatrix} \to \begin{bmatrix} \mathcal{U}_2 \\ \mathcal{Y}_2 \end{bmatrix}, \quad K_0 = \begin{bmatrix} 0 & 0 \\ 0 & \kappa_2 \end{bmatrix} : \begin{bmatrix} \mathcal{U}_2 \\ \mathcal{Y}_2 \end{bmatrix} \to \begin{bmatrix} \mathcal{U}_1 \\ \mathcal{Y}_1 \end{bmatrix}.$$

Then (5.5) *holds true if and only if*

$$T^{-1} - A_0T^{-1}K_0 = \Xi\xi_0^{-1}\Upsilon - A_0\Psi\psi_0^{-1}\Omega K_0.$$

Proof. It is sufficient to prove that

$$AT^{-1}K - A_0T^{-1}K_0 = A\Psi\psi_0^{-1}\Omega K - A_0\Psi\psi_0^{-1}\Omega K_0.$$

Write

$$AT^{-1}K - A_0T^{-1}K_0 = (A - A_0)T^{-1}K + A_0T^{-1}(K - K_0),$$

$$A\Psi\psi_0^{-1}\Omega K - A_0\Psi\psi_0^{-1}\Omega K_0 = (A - A_0)\Psi\psi_0^{-1}\Omega K + A_0\Psi\psi_0^{-1}\Omega(K - K_0).$$

So it suffices to prove

$$(A - A_0)T^{-1}K = (A - A_0)\Psi\psi_0^{-1}\Omega K \tag{5.6}$$

and

$$A_0T^{-1}(K - K_0) = A_0\Psi\psi_0^{-1}\Omega(K - K_0). \tag{5.7}$$

Write

$$T^{-1} = \begin{bmatrix} \psi_0 & \tau_{12} \\ \tau_{21} & \tau_{22} \end{bmatrix} : \begin{bmatrix} \mathcal{U}_1 \\ \mathcal{Y}_1 \end{bmatrix} \to \begin{bmatrix} \mathcal{U}_1 \\ \mathcal{Y}_1 \end{bmatrix}.$$

Then

$$\Psi = T^{-1}\tau_{\mathcal{U}_1} = \begin{bmatrix} \psi_0 \\ \tau_{21} \end{bmatrix}, \quad \text{and} \quad \Omega = \pi_{\mathcal{U}_1}T^{-1} = \begin{bmatrix} \psi_0 & \tau_{12} \end{bmatrix}.$$

A straightforward computation with these matrix representations reveals that (5.6) and (5.7) hold true. $\square$

Proof of Theorem 5.1. In view of Lemma 5.2 we may assume that $\alpha_1 = 0$ and $\kappa_1 = 0$. Let $\mathcal{X}_1 = \mathcal{X}_2 = \mathcal{X}$, $\mathcal{Y} = \mathcal{U}_2$, $\mathcal{U} = \mathcal{U}_1 \oplus \mathcal{U}_1$ and

$$P = I_{\mathcal{X}} - T, \quad Q = I_{\mathcal{X}},$$
$$A_1 = K, \quad B_1 = \begin{bmatrix} \tau_{\mathcal{U}_1} & -T\tau_{\mathcal{U}_1} + \tau_{\mathcal{U}_1}(I_{\mathcal{U}_1} + \pi_{\mathcal{U}_1}T\tau_{\mathcal{U}_1}) \end{bmatrix}, \quad C_1 = \pi_{\mathcal{U}_2},$$
$$A_2 = A, \quad B_2 = \tau_{\mathcal{U}_2}, \quad C_2 = \begin{bmatrix} -\pi_{\mathcal{U}_1}T \\ \pi_{\mathcal{U}_1} \end{bmatrix}.$$

First we present a few simple auxiliary identities that will play a role in the sequel:

$$\pi_{\mathcal{U}_1}A_1 = \pi_{\mathcal{U}_1}K = 0, \qquad A_2\tau_{\mathcal{U}_1} = A\tau_{\mathcal{U}_1} = 0,$$
$$I_{\mathcal{X}} - A_2A_1 = I_{\mathcal{X}} - AK = \tau_{\mathcal{U}_2}\pi_{\mathcal{U}_2}, \qquad I_{\mathcal{X}} = \tau_{\mathcal{Y}_1}\pi_{\mathcal{Y}_1} + \tau_{\mathcal{U}_1}\pi_{\mathcal{U}_1}.$$

Using these equalities and (5.4), we obtain

$$T - KTA = (\tau_{\mathcal{Y}_1}\pi_{\mathcal{Y}_1} + \tau_{\mathcal{U}_1}\pi_{\mathcal{U}_1})(T - KTA)(\tau_{\mathcal{Y}_1}\pi_{\mathcal{Y}_1} + \tau_{\mathcal{U}_1}\pi_{\mathcal{U}_1})$$
$$= \tau_{\mathcal{U}_1}\pi_{\mathcal{U}_1}T + T\tau_{\mathcal{U}_1}\pi_{\mathcal{U}_1} - \tau_{\mathcal{U}_1}\pi_{\mathcal{U}_1}T\tau_{\mathcal{U}_1}\pi_{\mathcal{U}_1}.$$

So it follows that

$$P - A_1PA_2 = I_{\mathcal{X}} - KA - (T - KTA)$$
$$= \tau_{\mathcal{U}_1}\pi_{\mathcal{U}_1} - \tau_{\mathcal{U}_1}\pi_{\mathcal{U}_1}T - T\tau_{\mathcal{U}_1}\pi_{\mathcal{U}_1} + \tau_{\mathcal{U}_1}\pi_{\mathcal{U}_1}T\tau_{\mathcal{U}_1}\pi_{\mathcal{U}_1} = B_1C_2.$$

Furthermore,

$$Q - A_2QA_1 = I_{\mathcal{X}} - AK = \tau_{\mathcal{U}_2}\pi_{\mathcal{U}_2} = B_2C_1.$$

Next let us define

$$X = \begin{bmatrix} \Psi & -\tau_{\mathcal{U}_1} + \Psi(I_{\mathcal{U}_1} + \pi_{\mathcal{U}_1}T\tau_{\mathcal{U}_1}) \end{bmatrix}, \quad Y = \Upsilon$$
$$W = \Xi, \quad Z = \begin{bmatrix} -\pi_{\mathcal{U}_1} \\ \Omega \end{bmatrix}.$$

Then

$$TX = \begin{bmatrix} \tau_{\mathcal{U}_1} & -T\tau_{\mathcal{U}_1} + \tau_{\mathcal{U}_1}(I_{\mathcal{U}_1} + \pi_{\mathcal{U}_1}T\tau_{\mathcal{U}_1}) \end{bmatrix} = B_1,$$
$$YT = C_1, \quad TW = B_2, \quad ZT = \begin{bmatrix} -\pi_{\mathcal{U}_1}T \\ \pi_{\mathcal{U}_1} \end{bmatrix} = C_2.$$

We proceed with $I_{\mathcal{U}} + C_2 QX$ and $I_{\mathcal{Y}} + YPB_2$. First

$$I_{\mathcal{Y}} + YPB_2 = I_{\mathcal{U}_2} + Y(I_{\mathcal{X}} - T)\tau_{\mathcal{U}_2} = I_{\mathcal{U}_2} + Y\tau_{\mathcal{U}_2} - YT\tau_{\mathcal{U}_2}$$
$$= I_{\mathcal{U}_2} + \pi_{\mathcal{U}_2} T^{-1}\tau_{\mathcal{U}_2} - \pi_{\mathcal{U}_2}\tau_{\mathcal{U}_2} = \xi_0.$$

Next

$$I_{\mathcal{U}} + C_2 QX = \begin{bmatrix} I_{\mathcal{U}_1} & 0 \\ 0 & I_{\mathcal{U}_1} \end{bmatrix} + \begin{bmatrix} -\pi_{\mathcal{U}_1} T \\ \pi_{\mathcal{U}_1} \end{bmatrix} \begin{bmatrix} \Psi & -\tau_{\mathcal{U}_1} + \Psi(I_{\mathcal{U}_1} + \pi_{\mathcal{U}_1} T\tau_{\mathcal{U}_1}) \end{bmatrix}.$$

We compute the four entries of this 2×2 matrix separately. The $(1,1)$-entry is $I_{\mathcal{U}_1} - \pi_{\mathcal{U}_1} T\Psi = I_{\mathcal{U}_1} - \pi_{\mathcal{U}_1}\tau_{\mathcal{U}_1} = 0$. The $(1,2)$-entry is

$$\pi_{\mathcal{U}_1} T\tau_{\mathcal{U}_1} - \pi_{\mathcal{U}_1} T\Psi(I_{\mathcal{U}_1} + \pi_{\mathcal{U}_1} T\tau_{\mathcal{U}_1}) = -I_{\mathcal{U}_1}.$$

For the $(2,1)$-entry $\pi_{\mathcal{U}_1}\Psi = \pi_{\mathcal{U}_1} T^{-1}\tau_{\mathcal{U}_1} = \psi_0$. Finally the $(2,2)$-entry is given by

$$I_{\mathcal{U}_1} - \pi_{\mathcal{U}_1}\tau_{\mathcal{U}_1} + \pi_{\mathcal{U}_1}\Psi(I_{\mathcal{U}_1} + \pi_{\mathcal{U}_1} T\tau_{\mathcal{U}_1}) = \pi_{\mathcal{U}_1}\Psi(I_{\mathcal{U}_1} + \pi_{\mathcal{U}_1} T\tau_{\mathcal{U}_1}).$$

We have

$$I_{\mathcal{U}} + C_2 QX = \begin{bmatrix} 0 & I_{\mathcal{U}_1} \\ \psi_0 & \psi_0(I_{\mathcal{U}_1} + \pi_{\mathcal{U}_1} T\tau_{\mathcal{U}_1}) \end{bmatrix},$$

which is invertible if and only if ψ_0 is invertible. Our assumption was that ξ_0 or ψ_0 is invertible. So we have that $I_{\mathcal{U}} + C_2 QX$ or $I_{\mathcal{U}_2} + YPB_2$ is invertible and hence, according to Theorem 1.2 both are invertible and formula (1.17) holds true. The left-hand side of this formula is $T^{-1} - AT^{-1}K$ since we have that $Q(I_{\mathcal{X}} - PQ)^{-1} = T^{-1}$. To finish the proof we have to check that the right-hand side of (1.17) gives the right-hand side of (5.5). For the first term this is immediate from the above-established equalities $W = \Xi$, $I_{\mathcal{U}_2} + YPB_2 = \xi_0$ and $Y = \Upsilon$. For the second term first notice that

$$A_2 QX(I_{\mathcal{U}} + C_2 QX)^{-1} ZQA_1$$

$$= A \begin{bmatrix} \Psi & -\tau_{\mathcal{U}_1} + \Psi(I_{\mathcal{U}_1} + \pi_{\mathcal{U}_1} T\tau_{\mathcal{U}_1}) \end{bmatrix} \begin{bmatrix} -(I_{\mathcal{U}_1} + \pi_{\mathcal{U}_1} T\tau_{\mathcal{U}_1}) & \psi_0^{-1} \\ I_{\mathcal{U}_1} & 0 \end{bmatrix} \begin{bmatrix} -\pi_{\mathcal{U}_1} \\ \Omega \end{bmatrix} K.$$

Use $A\tau_{\mathcal{U}_1} = 0$ and $\pi_{\mathcal{U}_1} K = 0$ to see that this is equal to $A\Psi\psi_0^{-1}\Omega K$. We proved formula (5.5). □

6. Examples

In this section T is an invertible $n \times n$ block Toeplitz matrix with blocks of size $p \times p$, and we take $P = I_{\mathcal{X}} - T$ and $Q = I_{\mathcal{X}}$, where $\mathcal{X} = (\mathbb{C}^p)^n$. With these P and Q we shall associate two different sets of matrices $\{A_1, B_1, C_1; A_2, B_2, C_2\}$ such the equations (1.9), (1.10), and (1.11) are satisfied. For both choices we compute (1.17). We shall see that for the first choice (1.17) cannot be used to obtain a formula for T^{-1}, while for the second choice (1.17) leads to the Gohberg–Heinig formula [8] for T^{-1}.

First example. To introduce our first choice for the set $(A_1, B_1, C_1; A_2, B_2, C_2)$ associated with P and Q we need some auxiliary operators. We define the forward block shift $N : (\mathbb{C}^p)^n \to (\mathbb{C}^p)^n$ and the two embedding operators $\varepsilon : \mathbb{C}^p \to (\mathbb{C}^p)^n$ and $\eta : \mathbb{C}^p \to (\mathbb{C}^p)^n$ by

$$
N \begin{bmatrix} x_0 \\ x_1 \\ \vdots \\ x_{n-1} \end{bmatrix} = \begin{bmatrix} 0 & 0 & \cdots & 0 \\ I_p & 0 & & 0 \\ & \ddots & \ddots & \vdots \\ & & I_p & 0 \end{bmatrix} \begin{bmatrix} x_0 \\ x_1 \\ \vdots \\ x_{n-1} \end{bmatrix}, \quad \varepsilon x = \begin{bmatrix} x \\ 0 \\ \vdots \\ 0 \end{bmatrix}, \quad \eta x = \begin{bmatrix} 0 \\ \vdots \\ 0 \\ x \end{bmatrix},
$$

for each $x \in \mathbb{C}^p$. Let $\mathcal{X} = \mathcal{X}_1 = \mathcal{X}_2 = (\mathbb{C}^p)^n$, $\mathcal{U} = (\mathbb{C}^p)^2$ and $\mathcal{Y} = \mathbb{C}^p$. Put $A = N + \varepsilon\eta^*$ and

$$
A_1 = A, \quad B_1 = \begin{bmatrix} -T\varepsilon + A_1 T\eta & -\varepsilon \end{bmatrix}, \quad C_1 = 0,
$$
$$
A_2 = A^* = A^{-1}, \quad B_2 = 0, \quad C_2 = \begin{bmatrix} \varepsilon^* \\ \varepsilon^* T - \eta^* T A_2 \end{bmatrix}.
$$

Then $Q - A_2 Q A_1 = B_2 C_1$ and $P - A_1 P A_2 = B_1 C_2$. The latter equality one can check as follows. First note that $T - N T N^* = T\varepsilon\varepsilon^* + \varepsilon\varepsilon^* T - \varepsilon\varepsilon^* T\varepsilon\varepsilon^*$. It follows that

$$
\begin{aligned}
-(P - A_1 P A_2) &= T - A_1 T A_2 = T - (N + \varepsilon\eta^*) T (N^* + \eta\varepsilon^*) \\
&= T\varepsilon\varepsilon^* + \varepsilon\varepsilon^* T - \varepsilon\varepsilon^* T\varepsilon\varepsilon^* - \varepsilon\eta^* T A_2 - N T\eta\varepsilon^* \\
&= T\varepsilon\varepsilon^* + \varepsilon\varepsilon^* T - \varepsilon\eta^* T\eta\varepsilon^* - \varepsilon\eta^* T A_2 - N T\eta\varepsilon^* \\
&= T\varepsilon\varepsilon^* + \varepsilon\varepsilon^* T - A_1 T\eta\varepsilon^* - \varepsilon\eta^* T A_2 \\
&= -B_1 C_2.
\end{aligned}
$$

Recall that T is assumed to be invertible. Define x, w, z^* and y^* by

$$
Tx = \varepsilon, \quad Tw = A_1 T\eta, \quad z^* T = \varepsilon^*, \quad y^* T = \eta^* T A_2.
$$

Then

$$
X = \begin{bmatrix} -\varepsilon + w & -x \end{bmatrix}, \quad Y = 0, \quad W = 0, \quad Z = \begin{bmatrix} z^* \\ \varepsilon^* - y^* \end{bmatrix}
$$

solve (1.13) and (1.14) and

$$
I_{\mathcal{U}} + C_2 X = \begin{bmatrix} \varepsilon^* w & -\varepsilon^* x \\ (\varepsilon^* - y^*)(Tw - T\varepsilon) & y^* \varepsilon \end{bmatrix}
$$

Next notice that $I_{\mathcal{Y}} + Y P W = I_{\mathcal{Y}}$ is invertible. Thus Theorem 1.2 tells us that $I_{\mathcal{U}} + C_2 X$ is invertible and

$$
T^{-1} - A_2 T^{-1} A_1 = A_2 X (I + C_2 X)^{-1} Z A_1. \tag{6.8}
$$

Note that in this case A_1 and A_2 are not stable, and it is not clear how one can use (6.8) to derive a formula for T^{-1}.

Second example. The spaces $\mathcal{X}$, $\mathcal{X}_1$, $\mathcal{X}_2$, $\mathcal{U}$, and $\mathcal{Y}$ are as in the previous example. We choose

$$A_1 = N^*, \quad B_1 = \begin{bmatrix} \eta & -T\eta + \eta t_0 \end{bmatrix}, \quad C_1 = \varepsilon^*,$$

$$A_2 = N, \quad B_2 = \varepsilon, \quad C_2 = \begin{bmatrix} \eta^* - \eta^* T \\ \eta^* \end{bmatrix},$$

where t_0 is the left upper entry of T. Then the equations in (1.11) are satisfied. Using T is invertible, define x, y, w and z by

$$Tx = \varepsilon, \quad Tz = \eta, \quad y^*T = \varepsilon^*, \quad w^*T = \eta^*.$$

To satisfy (1.13) and (1.14) put

$$W = x, \quad Y = y^*, \quad X = \begin{bmatrix} z & -\eta + zt_0 \end{bmatrix}, \quad Z = \begin{bmatrix} w^* - \eta^* \\ w^* \end{bmatrix}.$$

Then $I_{\mathcal{Y}} + YPB_2 = y^*\varepsilon_p$ and

$$I_{\mathcal{U}} + C_2 QX = \begin{bmatrix} \eta^*z & -I + \eta^*zt_0 \\ \eta^*z & \eta^*zt_0 \end{bmatrix}.$$

Now write for short $y_0 = y^*\varepsilon_p$ and $z_m = \eta^*z$. So $I_{\mathcal{U}} + C_2 QX$ is invertible if and only if z_m is. A simple computation gives in this case that

$$T^{-1} - NT^{-1}N^* = xy_0^{-1}y^* - Nzz_m^{-1}w^*N^*. \tag{6.9}$$

This is a well-known formula for the inverse of T from [8]. In this case $N^n = 0$ and $(N^*)^n = 0$. Hence one can easily derive from (6.9) a closed expression for T^{-1}.

References

[1] H. Bart, I. Gohberg, M.A. Kaashoek, and A.C.M. Ran, *Factorization of matrix and operator functions: the state space approach*, OT **178**, Birkhäuser Verlag, Basel, 2008.

[2] H. Bart, I. Gohberg, M.A. Kaashoek, and A.C.M. Ran, *A state space approach to canonical factorization with applications*, OT **200**, Birkhäuser Verlag, Basel, 2010.

[3] M.J. Corless and A.E. Frazho, *Linear systems and control*, Marcel Dekker, Inc., New York, 2003.

[4] R.L. Ellis and I. Gohberg, *Orthogonal systems and convolution operators*, OT **140**, Birkhäuser Verlag, Basel, 2003.

[5] R.L. Ellis, I. Gohberg, and D.C. Lay, Infinite analogues of block Toeplitz matrices and related orthogonal functions, *Integral Equations and Operator Theory* **22** (1995), 375–419.

[6] A.E. Frazho and M.A. Kaashoek, A contractive operator view on an inversion formula of Gohberg–Heinig, in: *Topics in Operator Theory I. Operators, matrices and analytic functions*, OT **202**, Birkhäuser Verlag, Basel, 2010, pp. 223–252.

[7] I. Gohberg, S. Goldberg, and M.A. Kaashoek, *Classes of Linear Operators*, Volume II, OT **63**, Birkhäuser Verlag, Basel, 1993.

[8] I. Gohberg, G. Heinig, The inversion of finite Toeplitz matrices consisting of elements of a non-commutative algebra, *Rev. Roum. Math. Pures et Appl.* **20** (1974), 623–663 (in Russian); English transl. in: *Convolution Equations and Singular Integral Operators*, (eds. L. Lerer, V. Olshevsky, I.M. Spitkovsky), OT **206**, Birkhäuser Verlag, Basel, 2010, pp. 7–46.

[9] I.C. Gohberg and M.G. Krein, Systems of integral equations with kernels depending on the difference of arguments, *Uspekhi Math. Nauk* **13** 2(80) (1958), 3–72 (Russian); English Transl., *Amer. Math. Soc. Transl. (Series 2)* **14** (1960), 217–287.

[10] G.J. Groenewald and M.A. Kaashoek, A Gohberg–Heinig type inversion formula involving Hankel operators,in: *Interpolation, Schur functions and moment problems*, OT **165**, Birkhäuser Verlag, Basel, 2005, pp. 291–302.

[11] G. Heinig and K. Rost, *Algebraic methods for Toeplitz-like matrices and operators*, Akademie-Verlag, Berlin, 1984.

[12] T. Kailath and A.H. Sayed, Displacement structure: Theory and applications, *SIAM Rev.* **37** (1995), 297–386.

[13] T. Kailath and A.H. Sayed (editors), *Fast Reliable Algorithms for Matrices with Structure*, SIAM, Philadelphia, 1999.

[14] I. Koltracht, B.A. Kon, and L. Lerer, Inversion of structured operators, *Integral equations and Operator Theory* **20** (1994), 410–448.

[15] V. Olshevsky (editor), *Structured matrices in mathematics, Computer Science, and Engineering*, Contempary Math. Series **280, 281**, Amer. Math. Soc. 2001.

[16] V.Y. Pan, *Structured matrices and polynomials*, Birkhäuser Boston, 2001.

M.A. Kaashoek and F. van Schagen
Department of Mathematics
Faculty of Sciences
VU University
De Boelelaan 1081a
NL-1081 HV Amsterdam, The Netherlands
e-mail: ma.kaashoek@vu.nl
          f.van.schagen@vu.nl

Operator Theory:
Advances and Applications, Vol. 237, 189–196
© 2013 Springer Basel

On the Sign Characteristics of Selfadjoint Matrix Polynomials

Peter Lancaster and Ion Zaballa

Dedicated to Leonid Lerer on the occasion of his seventieth birthday.

Abstract. An important role is played in the spectral analysis of selfadjoint matrix polynomials by the so-called "sign characteristics" associated with real eigenvalues. In this paper the ordering of the real eigenvalues by their sign characteristics is clarified. In particular, the roles played by the signature of the leading and trailing polynomial coefficients are discussed.

Mathematics Subject Classification (2010). 15A21, 47B15.

Keywords. Matrix polynomial. Sign characteristics.

1. Introduction

Let $L_0, L_1, \ldots, L_\ell \in \mathbb{C}^{n \times n}$. We consider *matrix polynomials*:

$$L(\lambda) := L_\ell \lambda^\ell + L_{\ell-1} \lambda^{\ell-1} + \cdots + L_0, \quad \lambda \in \mathbb{C}, \quad \det L_\ell \neq 0. \quad (1)$$

Such a polynomial is said to be *selfadjoint* if the coefficients are either complex Hermitian or, in particular, real and symmetric. The *eigenvalues* of L are the zeros of $\det L(\lambda)$, and the *eigenfunctions* are the real analytic functions on $\mathbb{R}$ formed by the zeros of $\det L(\lambda)$; say $\mu_1(\lambda), \mu_2(\lambda), \ldots, \mu_n(\lambda)$ (in some order to be decided). In this way the eigenvalues can also be characterized as the zeros of the eigenfunctions and, if λ_0 is an eigenvalue of $L(\lambda)$, then $\dim \operatorname{Ker} L(\lambda_0)$ is exactly the number of eigenfunctions that annihilate at λ_0.

The notion of "sign characteristic" associated with a real eigenvalue plays an important role in the spectral analysis and perturbation theory of selfadjoint matrix polynomials; see [1], [2], and [3], for example. In particular, it should be

The first author was supported in part by the Natural Sciences and Engineering Research Council of Canada.

The second author was supported in part by MICINN MTM2010-19356-C02-01, EJ GIC10/169-IT-361-10 and UPV/EHU UFI11/52.

noted that, for convenience, and because of many applications, it was assumed in [3] that $L_\ell > 0$. Here, the more general case of *nonsingular* L_ℓ prevails as in (the more comprehensive) references [1] and [2]. In particular, we will need the following fundamental result (Theorem 3.7 of [2] and Theorem 6.10 of [1]).

Theorem 1.1. *Let $L(\lambda)$ be an $n \times n$ selfadjoint matrix polynomial with nonsingular leading coefficient and let $\mu_1(\lambda)$, ..., $\mu_n(\lambda)$ be real analytic functions of real λ for which $\det(\mu_j I_n - L(\lambda)) = 0$, $j = 1, \ldots, n$. Let $\lambda_1 < \cdots < \lambda_r$ be the different real eigenvalues of $L(\lambda)$. For each j write*

$$\mu_j(\lambda) = (\lambda - \lambda_i)^{m_{ij}} \nu_{ij}(\lambda), \qquad i = 1, \ldots, r,$$

where $\nu_{ij}(\lambda_i) \neq 0$ is real. Then the non-zero numbers among $m_{i1}, \ldots, m_{in}$ are the partial multiplicities *of $L(\lambda)$ associated with λ_i.*

The sign of $\nu_{ij}(\lambda_i)$ (for $m_{ij} \neq 0$) is the sign characteristic *attached to the elementary divisors $(\lambda - \lambda_i)^{m_{ij}}$ of $L(\lambda)$.*

Note that this statement provides definitions of *partial multiplicities* and *sign characteristics* and these are associated with each elementary divisor. (The reader is referred to [1] and [2] for more comprehensive dicussion.) In particular, if $L(\lambda)$ is semisimple (that is, $m_{ij} = 1$ for all i and j) then λ_i is said to be of positive or negative type according as the sign characteristic attached to the corresponding elementary divisor $(\lambda - \lambda_i)$ is positive or negative, respectively.

2. Admissible sign characteristics

Our first objectives are to provide characterizations of admissible sign characteristics for polynomials $L(\lambda)$ with either *positive definite leading coefficient L_ℓ*, or *positive definite trailing coefficient L_0*.

The first result relates the inertia of the leading coefficient L_ℓ to the asymptotic behaviour of the eigenfunctions.

Theorem 2.1. *Let $(\pi, n - \pi, 0)$ be the inertia of L_ℓ, and let $\lambda_{\max}$ be the largest real eigenvalue of $L(\lambda)$. Then there are π indices $\{i_1, \ldots, i_\pi\} \subseteq \{1, \ldots, n\}$ such that for all $\lambda > \lambda_{\max}$,*

$$\mu_j(\lambda) > 0 \quad if \quad j \in \{i_1, \ldots, i_\pi\} \quad and \quad \mu_j(\lambda) < 0 \quad if \quad j \notin \{i_1, \ldots, i_\pi\}.$$

Proof. For $j = 1, \ldots, n$ the zeros of $\mu_j(\lambda)$ are real eigenvalues of the polynomial matrix $L(\lambda)$ and, since this matrix has at most $n\ell$ real eigenvalues (counting with multiplicities), the number of real zeros of $\mu_j(\lambda)$ is finite. Then for $\lambda > \lambda_{\max}$ either $\mu_j(\lambda) > 0$ or $\mu_j(\lambda) < 0$ for $j = 1, \ldots, n$.

On the other hand, for any fixed real λ, the real number $\mu(\lambda)$ is an eigenvalue of the selfadjoint (real or complex) matrix $L(\lambda)$. Let $\lambda_1(L_\ell) \geq \cdots \geq \lambda_n(L_\ell)$ denote the eigenvalues of L_ℓ with

$$\lambda_1(L_\ell) \geq \cdots \geq \lambda_\pi(L_\ell) > 0 > \lambda_{\pi+1}(L_\ell) \geq \cdots \geq \lambda_n(L_\ell)$$

We will use the "Weyl inequalities" for the eigenvalues of the sum of two symmetric or Hermitian matrices (see [4, Th. 4.3.1], for example). If H_1, H_2 are $n \times n$ Hermitian or symmetric matrices then

$$\lambda_j(H_1) + \lambda_n(H_2) \leq \lambda_j(H_1 + H_2) \leq \lambda_j(H_1) + \lambda_1(H_2)$$

where the eigenvalues of H_1, H_2 and $H_1 + H_2$ are arranged in non-increasing order.

Let λ_0 be a real number. Applying the left-hand Weyl inequality repeatedly to the symmetric matrix $L(\lambda_0) = (L_\ell \lambda_0^\ell + \cdots + L_1 \lambda_0) + (L_0)$ we have

$$\lambda_j(L(\lambda_0)) \geq \lambda_j(L_\ell \lambda_0^\ell + \cdots + L_1 \lambda_0) + \lambda_n(L_0)$$
$$\geq \lambda_j(L_\ell \lambda_0^\ell + \cdots + L_2 \lambda_0^2) + \lambda_0 \lambda_n(L_1) + \lambda_n(L_0)$$
$$\geq \cdots$$
$$\geq \lambda_0^\ell \lambda_j(L_\ell) + \lambda_0^{\ell-1} \lambda_n(L_{n-1}) + \cdots + \lambda_n(L_0)$$

Thus, if $m = \min\{\lambda_n(L_{n-1}), \ldots, \lambda_n(L_0)\}$, then for $\lambda_0 > 1$,

$$\lambda_j(L(\lambda_0)) \geq \lambda_0^\ell \lambda_j(L_\ell) + (\lambda_0^{\ell-1} + \cdots + \lambda_0 + 1)m$$
$$\geq \lambda_0^\ell \lambda_j(L_\ell) + \frac{\lambda_0^\ell - 1}{\lambda_0 - 1}m. \tag{2}$$

Assume now that $\lambda_j(L_\ell) > 0$. If $m \geq 0$ then $\lambda_j(L(\lambda_0)) > 0$ for $\lambda_0 > 1$. Also, if $m < 0$ then for $\lambda_0 > 1 - \frac{m}{\lambda_j(L_\ell)}$ we have

$$\lambda_0^\ell(\lambda_0 - 1) > -\frac{m}{\lambda_j(L_\ell)}\lambda_0^\ell > -\frac{m}{\lambda_j(L_\ell)}\lambda_0^\ell + \frac{m}{\lambda_j(L_\ell)},$$

whence

$$\lambda_0^\ell(\lambda_0 - 1) > -\frac{m}{\lambda_j(L_\ell)}(\lambda_0^\ell - 1),$$

and so

$$\lambda_j(L_\ell)\lambda_0^\ell > -m\frac{\lambda_0^\ell - 1}{\lambda_0 - 1}.$$

Using this in (2) we find that $\lambda_j(L(\lambda_0)) > 0$ for $\lambda_0 > 1 + \frac{|m|}{\lambda_j(L_\ell)}$.

Next, we use the right-hand Weyl inequality ($\lambda_j(H_1 + H_2) \leq \lambda_j(H_1) + \lambda_1(H_2)$) to show in a similar way that

$$\lambda_j(L(\lambda_0)) \leq \lambda_0^\ell \lambda_j(L_\ell) + \lambda_0^{\ell-1} \lambda_1(L_{n-1}) + \cdots + \lambda_1(L_0),$$
$$\leq \lambda_0^\ell \lambda_j(L_\ell) + \frac{\lambda_0^\ell - 1}{\lambda_0 - 1}M,$$

with $M = \max\{\lambda_1(L_0), \ldots, \lambda_1(L_{\ell-1})\}$.

If we assume that $\lambda_j(L_\ell) < 0$ then, as above, $M < 0$ implies $\lambda_j(L(\lambda_0)) < 0$ for $\lambda_0 > 1$. Similarly, $M > 0$ implies that $\lambda_j(L(\lambda_0)) < 0$ for $\lambda_0 > 1 - \frac{M}{\lambda_j(L_\ell)}$. Hence, for $j = \pi + 1, \ldots, n$, $\lambda_j(L(\lambda_0)) < 0$ for $\lambda_0 > 1 - \frac{|M|}{\lambda_j(L_\ell)}$.

Bearing in mind that $\lambda_i(L_\ell) \geq \lambda_\pi(L_\ell)$, $i = 1, \ldots, \pi$ and $\lambda_i(L_\ell) \leq \lambda_{\pi+1}(L_\ell)$ for $i = \pi+1, \ldots, n$, we conclude that $\lambda_j(L(\lambda_0)) > 0$ for $j = 1, \ldots, \pi$ and $\lambda_j(L(\lambda_0)) < 0$ for $j = \pi+1, \ldots, n$ whenever

$$\lambda_0 > \max\left\{1 + \frac{|m|}{\lambda_\pi(L_\ell)}, \ 1 - \frac{|M|}{\lambda_{\pi+1}(L_\ell)}\right\}. \tag{3}$$

But the eigenvalues of $L(\lambda_0)$ are $\mu_1(\lambda_0), \ldots, \mu_n(\lambda_0)$. Then, for λ_0 satisfying (3), there are π indices $\{i_1, \ldots, i_\pi\} \subseteq \{1, \ldots, n\}$ such that, if

$$\{j_1, \ldots, j_{n-\pi}\} = \{1, \ldots, n\} \backslash \{i_1, \ldots, i_\pi\},$$

then $\mu_{i_k}(\lambda_0) > 0$, $k = 1, \ldots, \pi$, and $\mu_{j_k}(\lambda_0) < 0$, $k = 1, \ldots, n-\pi$. The theorem follows using the fact that $\mu_j(\lambda)$ is either positive or negative for $\lambda > \lambda_{\max}$. $\square$

In a similar way, the behaviour of the eigenfunctions of $L(\lambda)$ near zero is closely related to the inertia of the trailing coefficient, L_0.

Theorem 2.2. *Let* $L(\lambda) = L_\ell\lambda^\ell + L_{\ell-1}\lambda^{\ell-1} + \cdots + L_0$ *be an* $n \times n$ *selfadjoint matrix polynomial with* $\det L_\ell \neq 0$. *Let* $\mu_1(\lambda)$, $\ldots$, $\mu_n(\lambda)$ *be the eigenfunctions of* $L(\lambda)$ *and let* (π, ν, δ) *be the inertia of* L_0. *Let* λ_z *be the positive real eigenvalue of* $L(\lambda)$ *closest to zero. Then there are* π *indices* $\{i_1, \ldots, i_\pi\} \subseteq \{1, \ldots, n\}$ *and* ν *indices* $\{j_1, \ldots j_\nu\} \subseteq \{1, \ldots, n\} \backslash \{\{i_1, \ldots, i_\pi\}$ *such that for* $0 < \lambda < \lambda_z$, $\mu_k(\lambda) > 0$ *if* $k \in \{i_1, \ldots, i_\pi\}$ *and* $\mu_k(\lambda) < 0$ *if* $k \in \{j_1, \ldots, j_\nu\}$.

Proof. The proof follows the same lines as that of Theorem 2.1. First, for $j = 1, \ldots, n$, $\mu_j(\lambda)$ is either positive or negative for λ between any two consecutive real eigenvalues of $L(\lambda)$. In particular, each eigenfunction has constant sign in $(0, \lambda_z)$. Let the eigenvalues of L_0 be

$$\lambda_1(L_0) \geq \cdots \geq \lambda_\pi(L_0) > 0 > \lambda_{\pi+\delta+1}(L_0) \geq \cdots \geq \lambda_n(L_0),$$

and let λ_0 be a positive real number. Then $\lambda_j(L_0) > 0$ for $j = 1, \ldots, \pi$ and, applying successively the left-hand Weyl inequalities to $L(\lambda_0)$, we obtain

$$\lambda_j(L(\lambda_0)) \geq \lambda_n(L_\ell\lambda_0^\ell + \cdots + L_1\lambda_0) + \lambda_j(L_0),$$
$$\geq \cdots$$
$$\geq (\lambda_0^\ell \lambda_n(L_\ell) + \cdots + \lambda_0\lambda_n(L_1)) + \lambda_j(L_0),$$
$$\geq m(\lambda_0 + \cdots + \lambda_0^\ell) + \lambda_j(L_0), \tag{4}$$
$$= m\lambda_0 \frac{1 - \lambda_0^\ell}{1 - \lambda_0} + \lambda_j(L_0),$$

where $m = \min\{\lambda_n(L_1), \ldots, \lambda_n(L_\ell)\}$.

Now, if $0 < \lambda_0 < 1 - \frac{|m|}{\lambda_j(L_0)+|m|}$ then $\lambda_j(L(\lambda_0)) > 0$. In fact, if $m \geq 0$ then $0 < \lambda_0 < 1$ and it is plain that $\lambda_j(L(\lambda_0)) \geq \lambda_j(L_0) > 0$. And if $m < 0$ then

$$1 + \frac{m}{\lambda_j(L_0) - m} > \lambda_0 \quad \Rightarrow \quad \frac{\lambda_j(L_0)}{\lambda_j(L_0) - m} > \lambda_0 \quad \Rightarrow \quad \frac{-\lambda_j(L_0)/m}{1 - \lambda_j(L_0)/m} > \lambda_0$$

$$\Rightarrow -\frac{\lambda_j(L_0)}{m} > \lambda_0 \left(1 - \frac{\lambda_j(L_0)}{m}\right) \quad \Rightarrow \quad (1 - \lambda_0)\left(-\frac{\lambda_j(L_0)}{m}\right) > \lambda_0$$

$$\Rightarrow -\frac{\lambda_j(L_0)}{m} > \lambda_0 \frac{1}{1-\lambda_0} > \lambda_0 \frac{1-\lambda_0^\ell}{1-\lambda_0} \quad \Rightarrow \quad \lambda_j(L_0) > -m\lambda_0 \frac{1-\lambda_0^\ell}{1-\lambda_0}.$$

It follows from (4) that $\lambda_j(L(\lambda_0)) \geq m\lambda_0 \frac{1-\lambda_0^\ell}{1-\lambda_0} + \lambda_j(L_0) > 0$.

Similarly, if $j = \pi + \delta + 1, \dots, n$ then $\lambda_j(L_0) < 0$ and we can apply the right-hand Weyl inequalities to show that

$$\lambda_j(L(\lambda_0)) \leq M\lambda_0 \frac{1-\lambda_0^\ell}{1-\lambda_0} + \lambda_j(L_0),$$

where $M = \max\{\lambda_1(L_1), \dots, \lambda_1(L_n)\}$. As in the previous case, if $0 \leq \lambda_0 \leq 1 + \frac{|M|}{\lambda_j(L_0)-|M|}$ then $\lambda_j(L(\lambda_0)) < 0$.

Therefore, for $\lambda_0 > 0$ close enough to zero, $\lambda_j(L(\lambda_0)) > 0$ or $\lambda_j(L(\lambda_0)) < 0$ according as $\lambda_j(L_0) > 0$ or $\lambda_j(L_0) < 0$, respectively. Since $\mu_k(\lambda_0)$ is an eigenvalue of $L(\lambda_0)$ and $\mu_k(\lambda)$ does not change sign in $(0, \lambda_z)$ there must be π eigenfunctions taking positive values in $(0, \lambda_z)$ and ν eigenfunctions taking negative values in the same open interval. $\qquad\square$

3. The semisimple case

With the help of Theorems 2.1 and 2.2 we can establish a necessary condition on the sign characteristics of the real eigenvalues of *semisimple* selfadjoint matrix polynomials with positive definite leading and/or trailing coefficient. This leads to a proof of the following result which was stated in [5] – without proof.

Theorem 3.1. *Let $L(\lambda)$ be an $n \times n$ semisimple selfadjoint matrix polynomial with $L_\ell > 0$ and maximal and minimal real eigenvalues $\lambda_{\max}$ and $\lambda_{\min}$, respectively. For any $\alpha \leq \lambda_{\max}$, let $p(\alpha)$ denote the number of real eigenvalues (counting multiplicities) of $L(\lambda)$ of positive type in $(\alpha, +\infty)$ and $n(\alpha)$ denote the number of real eigenvalues (counting multiplicites) of $L(\lambda)$ of negative type in $[\alpha, +\infty)$. Then*

$$n(\alpha) \leq p(\alpha) \text{ for all } \alpha \in [\lambda_{\min}, \lambda_{\max}]. \tag{5}$$

In particular, this theorem says that, in the semisimple case, if $L_\ell > 0$ then for each real eigenvalue of negative type there is at least one larger real eigenvalue of positive type.

Proof. Let $\alpha \in \mathbb{R}$ be such that $\lambda_{\min} \leq \alpha \leq \lambda_{\max}$ and let $\lambda_0 \in [\alpha, \lambda_{\max}]$ be an eigenvalue of $L(\lambda)$. Since $L(\lambda)$ is semisimple the algebraic multiplicity of λ_0

coincides with its geometric multiplicity. Bearing in mind that $\dim \operatorname{Ker} L(\lambda_0)$ is the number of eigenfunctions that have λ_0 as a zero, we can associate an eigenfunction (perhaps in more than one way) with each eigenvalue.

Let $\mu_0(\lambda)$ be the eigenfunction associated with λ_0. Then, according to Theorem 1.1, we can write $\mu_0(\lambda) = (\lambda - \lambda_0)\nu_0(\lambda)$ with $\nu_0(\lambda_0) \neq 0$ and the sign to be associated with λ_0 is positive or negative according as $\nu_0(\lambda_0) > 0$ or $\nu_0(\lambda_0) < 0$. If the negative sign applies, then $\mu_0(\lambda)$ is decreasing at λ_0 and so $\mu_0(\lambda_0+) < 0$. But $L_\ell > 0$ and by Theorem 2.1 $\mu_0(\lambda) > 0$ for $\lambda > \lambda_{\max}$. It follows then that there is $\widetilde{\lambda}_0 > \lambda_0$ such that $\mu_0(\widetilde{\lambda}_0) = 0$ and $\mu_0'(\widetilde{\lambda}_0) > 0$. Thus, the eigenvalue $\widetilde{\lambda}_0$ has an associated positive sign.

This means first that $\lambda_0 < \lambda_{\max}$ (i.e., $n(\lambda_{\max}) = 0$ and so $\lambda_{\max}$ is of positive type) and, also, that for each eigenvalue of $L(\lambda)$ in the interval $[\alpha, \lambda_{\max})$ of negative type there is a larger eigenvalue in $(\alpha, \lambda_{\max}]$ of positive type. Noting that $\lambda_{\max}$ is necessarily of positive type, and $L(\lambda)$ has no eigenvalues in $(\lambda_{\max}, +\infty)$, we conclude that $n(\alpha) \leq p(\alpha)$ for all $\alpha \in [\lambda_{\min}, \lambda_{\max}]$, as desired. $\qquad\square$

When $L(\lambda)$ is semisimple and of even degree, the number of eigenvalues of positive type equals the number of negative type ([1, Prop. 4.2]). In this case $n(\lambda_{\min}) = p(\lambda_{\min})$ implying that $\lambda_{\min}$ is of negative type. On the other hand, we have seen in the proof of the theorem that $\lambda_{\max}$ is of positive type.

Theorem 3.1 should be compared with (the more general) Theorem 1.11 and Example 1.5 of [2].

Theorem 3.2. *Let $L(\lambda)$ be an $n \times n$ semisimple selfadjoint matrix polynomial with $L_0 > 0$ and maximal and minimal real eigenvalues $\lambda_{\max}$ and $\lambda_{\min}$, respectively.*

- *For $\alpha < 0$ let $p_-(\alpha)$ denote the number of real eigenvalues (counting multiplicities) of $L(\lambda)$ of positive type in $(\alpha, 0]$ and $n_-(\alpha)$ the number of real eigenvalues (counting multiplicites) of $L(\lambda)$ of negative type in $[\alpha, 0)$.*
- *For $\alpha > 0$ let $p_+(\alpha)$ denote the number of real eigenvalues (counting multiplicities) of $L(\lambda)$ of positive type in $(0, \alpha]$ and $n_+(\alpha)$ the number of real eigenvalues (counting multiplicites) of $L(\lambda)$ of negative type in $[0, \alpha)$.*

Then $n_-(\alpha) \leq p_-(\alpha)$ for all $\alpha \in [\lambda_{\min}, 0)$ and $n_+(\alpha) \geq p_+(\alpha)$ for all $\alpha \in (0, \lambda_{\max})$.

Proof. The line of proof is similar to that of the previous theorem but using Theorem 2.2. Let λ_z be the eigenvalue of $L(\lambda)$ closest to zero. If $\alpha \in [\lambda_{\min}, 0)$ and $\lambda_0 \in [\alpha, 0)$ is an eigenvalue of $L(\lambda)$ of negative type, the corresponding eigenfunction is negative to the right of λ_0 (but close enough to λ_0). By Theorem 2.2, that eigenfunction is positive in $(0, \lambda_z)$ so that there must be an eigenvalue $0 > \widetilde{\lambda}_0 < \lambda_0$ of $L(\lambda)$ of positive type. Hence $n_-(\alpha) \leq p_-(\alpha)$ for $\alpha \in [\lambda_{\min}, 0)$.

Similarly, if $\alpha \in (0, \lambda_{\max}]$ and $\lambda_0 \in (0, \alpha]$ is an eigenvalue of $L(\lambda)$ of positive type, the corresponding eigenfunction is negative to the left (but near) λ_0. By Theorem 2.2, that eigenfunction is positive in $(0, \lambda_z)$ so that $\widetilde{\lambda}_0 > \lambda_z$, λ_z is of negative type and there must be an eigenvalue $0 < \lambda_z \leq \widetilde{\lambda}_0 < \lambda_0$ of $L(\lambda)$ of negative type. $\qquad\square$

We remark again that, for the matrix polynomials of Theorem 3.2, λ_z is necessarily of negative type.

Putting together the previous results we can provide an additional necessary condition that the sign characteristics of all semisimple selfadjoint matrix polynomials with positive definite leading and trailing coefficients must satisfy.

Theorem 3.3. *Let $L(\lambda)$ be an $n \times n$ semisimple selfadjoint matrix polynomial with $L_\ell > 0$ and $L_0 > 0$. With the notation of the previous theorem, the following condition holds:*

$$n_+(\lambda_{\max}) = p_+(\lambda_{\max}). \tag{6}$$

And if ℓ is even then

$$n_-(\lambda_{\min}) = p_-(\lambda_{\min}) \tag{7}$$

Proof. If $L(\lambda)$ has positive definite leading coefficient and λ_z is the real positive eigenvalue of $L(\lambda)$ closest to zero, then, by (5), $p(\lambda_z) \geq n(\lambda_z)$. That is to say, the number of eigenvalues of $L(\lambda)$ of positive type in $(\lambda_z, +\infty)$ is not smaller than the number of eigenvalues of negative type in $[\lambda_z, +\infty)$.

Now, λ_z is of negative type because the trailing coefficient is positive definite, and $\lambda_{\max}$ is of positive type because the leading coefficient is positive definite. Thus, $p(\lambda_z)$ is also the number of eigenvalues of positive type in $(0, \lambda_{\max}]$ and $n(\lambda_z)$ is the number of eigenvalues of negative type in $[0, \lambda_{\max})$. Hence, $p_+(\lambda_{\max}) = p(\lambda_z)$ and $n_+(\lambda_{\max}) = n(\lambda_z)$. Since $p(\lambda_z) \geq n(\lambda_z)$ and by Theorem 3.2, $n_+(\lambda_{\max}) \geq p_+(\lambda_{\max})$, we conclude that this is indeed an equality.

As mentioned above, if $L(\lambda)$ is of even degree and semisimple, then the number of eigenvalues of positive type and negative type is the same. It follows from $n_+(\lambda_{\max}) = p_+(\lambda_{\max})$ that the number of eigenvalues of $L(\lambda)$ of positive type in $[\lambda_{\min}, 0]$ equals the number of eigenvalues of negative type in that interval. Taking into account that 0 is not an eigenvalue of $L(\lambda)$ and that $\lambda_{\min}$ is of negative type, the number of eigenvalues of $L(\lambda)$ of negative type in $[\lambda_{\min}, 0)$ is $n_-(\lambda_{\min})$ and that the number of eigenvalues of positive type in $[\lambda_{\min}, 0]$ is $p_-(\lambda_{\min})$. In conclusion $n_-(\lambda_{\min}) = p_-(\lambda_{\min})$ as claimed. □

4. Conclusions

Using the notion of the "sign characteristic" of real eigenvalues of selfadjoint matrix polynomials, Theorems 2 and 3 establish new results on the ordering of real eigenvalues with respect to these signs. The results take into account the inertia of the (invertible) leading coefficient, and the trailing coefficient, respectively.

In Section 3, these results have been applied to semisimple matrix polynomials with positive definite leading coefficient (Theorem 4), and with positive definite trailing coefficient (Theorem 5).

These results will be used in [6] to provide solutions for the inverse real symmetric quadratic eigenvalue problem, in the semisimple case, when the leading and trailing coefficients are prescribed to be hold some definiteness constraints.

References

[1] Gohberg I., Lancaster P., and Rodman L., Spectral analysis of self-adjoint matrix polynomials. *Research paper* 419 (1979) Dept. Mathematics and Statistics, University of Calgary, Canada.

[2] Gohberg I., Lancaster P., and Rodman L., Spectral analysis of self-adjoint matrix polynomials. *Annals of Math.*, 112, (1980), 33–71.

[3] Gohberg I., Lancaster P., and Rodman L., *Matrix Polynomials*, Academic Press, New York, 1982 and SIAM, Philadelphia, 2009.

[4] Horn R.A., Johnson Ch.R., *Matrix Analysis*, Cambridge University Press, New York, 1985.

[5] Lancaster P., Prells U., Zaballa I., An Orthogonality property for real symmetric matrix polynomials, *Operators and Matrices*, to appear (avaliable on line http://files.ele-math.com/preprints/oam-07-21-pre.pdf).

[6] Lancaster P., Zaballa I., On the Inverse Symmetric Quadratic Eigenvalue Problem. *In preparation.*

Peter Lancaster
Dept. of Mathematics and Statistics
University of Calgary
Calgary
Alberta, T2N 1N4, Canada
e-mail: lancaste@ucalgary.ca

Ion Zaballa
Departamento de Matemática Aplicada y EIO
Euskal Herriko Unibertsitatea (UPV/EHU)
Apdo 644
E-48080 Bilbao, Spain
e-mail: ion.zaballa@ehu.es

Operator Theory:
Advances and Applications, Vol. 237, 197–219
© 2013 Springer Basel

Quadratic Operators in Banach Spaces and Nonassociative Banach Algebras

Yu.I. Lyubich

Dedicated to Leonia Lerer on the occasion of his 70th birthday

Abstract. A survey of a general theory of quadratic operators in Banach spaces with close relations to the nonassociative Banach algebras is presented. Some applications to matrix and integral quadratic operators in classical Banach spaces are given.

Mathematics Subject Classification (2010). 46H70, 45G10.

Keywords. Cubic matrices, integral operators, Bernstein algebras.

1. Introduction

Let Φ be a field, and let X be a linear space over Φ, finite- or infinite-dimensional. If, in addition, a bilinear mapping W from the Cartesian square $X \times X$ into X is given then X is called an *algebra* over Φ. In this setting the vector $W(x, y)$ is called the *product* of x and y and usually denoted by xy. Accordingly, the mapping W is called a *multiplication*. Its bilinearity means the *distributive laws*

$$(x + y)z = xz + yz, \quad z(x + y) = zx + zy \quad (x, y, z \in X)$$

and the *homogeneity of degree 1*, i.e.,

$$(\alpha x)y = x(\alpha y) = \alpha(xy) \equiv \alpha xy, \quad (x, y \in X, \quad \alpha \in \Phi).$$

Neither associativity nor unitality are assumed in this definition. For the general theory of nonassociative algebras the basic reference is the book [20].

Given an algebra X, the diagonal restriction of the multiplication, i.e., the mapping $V : X \to X$ defined as $Vx = xx \equiv x^2$ is called a *quadratic mapping* or *quadratic operator*. Its simplest properties are

$$V(\alpha x) = \alpha^2 V x, \quad V(-x) = V x, \quad V(0) = 0. \tag{1.1}$$

If the algebra is commutative then we have the elementary identities

$$x^2 - y^2 = (x - y)(x + y) \tag{1.2}$$

and

$$(x \pm y)^2 = x^2 + y^2 \pm 2xy. \tag{1.3}$$

Later on char$(\Phi) \neq 2$. From (1.3) it follows that

$$xy = \frac{1}{2} \left(V(x + y) - Vx - Vy \right), \tag{1.4}$$

or in a more elegant form

$$xy = \frac{1}{4} \left(V(x + y) - V(x - y) \right). \tag{1.5}$$

As a result, *with a fixed underlying linear space X the correspondence between commutative algebras and quadratic operators is one-to-one.* Furthermore, *a subspace $Y \subset X$ is invariant for V if and only if this is a subalgebra of the corresponding commutative algebra.* The quadratic operator corresponding to this subalgebra is the restriction $V|Y$. In particular, a one-dimensional subspace is a subalgebra if and only if its basis vector is an eigenvector of V. Note that if $Vx = \lambda x$, $\lambda \in \Phi$, and $\lambda \neq 0$ then $z = x/\lambda$ is a fixed point of V that is the same as an *idempotent* of the corresponding algebra: $z^2 = z$.

Even without commutativity the mapping $x \mapsto x^2$ is quadratic by definition. However, $x^2 = x \circ x$ where the new multiplication

$$x \circ y = \frac{xy + yx}{2}. \tag{1.6}$$

is commutative. But for a given quadratic mapping a noncommutative multiplication is not unique. The general form of such multiplications is

$$xy = x \circ y + [x, y] \tag{1.7}$$

where the second summand is anticommutative, i.e., $[x, y] = -[y, x]$ for all x, y. Indeed, the anticommutativity is equivalent to the identity $[z, z] = 0$. Note that (1.7) is a unique decomposition of a given algebra into the sum of a commutative algebra and an anticommutative one. If the algebra is associative then the commutative summand is an Jordan algebra, while the anticommutative one is a Lie algebra, and both are not associative, in general.

The quadratic operators come from the classical analysis, calculus of variations and differential geometry, etc., as the second differentials of mappings of class C^2. On the other hand, they appear as the generators of nonlinear dynamical systems in form of cubic matrices (finite or infinite), the integral operators, etc. In this setting the corresponding algebras are very useful for studying of the dynamics. A bright example is the Mendelian algebra in genetics connected with the Hardy–Weinberg quadratic mapping, see, e.g., [10, Sections 3 and 8]. This algebra is three-dimensional but, in general, the genetic situations are multidimensional. In this way many interesting classes of nonassociative algebras appear after the pioneering works of Serebrovskii [19], Glivenko [7] and Etherington [6]. Much earlier Bernstein [1] suggested the quadratic operators in $\mathbb{R}^n$ as an adequate language for a fundamental problem in the population dynamics. A synthesis of operator and algebra approaches turned out to be extraordinary fruitful, see book [14] and the references therein, especially, [18]. For purely algebraic aspects we refer to the book [21].

In [10] the author investigated the *Bernstein quadratic operators*, and after that the corresponding algebras were introduced in [11] as a powerful tool for the Bernstein problem. Eventually, in this way the problem has been completely solved [10, 11, 12, 8]. Note that the name *Bernstein algebras* appeared in [13] for the first time. The main structure theorem for these ones was first obtained in the operator form [10, Theorem 4.1]. In [9] this result was reproduced in an explicit algebraic form as a base for a further investigation. Nowadays the Bernstein algebra theory is a well-developed area of the modern algebra.

In the present paper we consider the quadratic operators in infinite-dimensional spaces with a special attention to the operators of *finite rank*, i.e., with finite-dimensional images (Section 2). If a quadratic operator in a Banach space is continuous then the corresponding algebra is a Banach one, and vice versa. We study this relation in Sections 3 and then proceed to a compact situation (Section 6) focusing on integral quadratic operators. These considerations culminate in Section 7 devoted mainly to the Bernstein quadratic operators and algebras. In the compact case they are of finite rank [15]. Moreover, the Bernstein integral operators in $C[0,1]$ with positive kernels are of rank 1. This remarkable result announced by Bernstein [1] was never proved until [16] where a proof was given with help of the corresponding Bernstein algebra introduced in [15].

2. Algebraic preliminaries

Obviously, the set Q_X of all quadratic operators in a linear space X is a linear space isomorphic to the space of all commutative multiplications in X. Some quadratic operators come naturally from the linear ones. For example, if T is a linear operator and φ is a linear functional then

$$Vx = \varphi[x]Tx \tag{2.1}$$

is a quadratic operator corresponding to the multiplication

$$xy = \frac{1}{2}(\varphi[x]Ty + \varphi[y]Tx). \tag{2.2}$$

Also note that if T is a linear operator then the superposition TV and VT are quadratic operators for every quadratic V. Indeed, $T(xy)$ and $(Tx)(Ty)$ are bilinear for any algebra. A quadratic operator V is called *elementary* if

$$Vx = v[x]e, \quad x \in X, \tag{2.3}$$

where $e \in X$ and $v[x]$ is a *quadratic functional*, i.e., $v[x] = w[x,x]$ where $w[x,y]$ is a (unique) symmetric bilinear functional, the *polar* of v.

Accordingly,

$$xy = w[x,y]e, \quad x \in X, \tag{2.4}$$

The simplest example is

$$xy = \varphi[x]\varphi[y]e \tag{2.5}$$

where φ is a linear functional. This algebra is associative.

For any quadratic operator V we consider the invariant subspace $S(V) =$ Span$(\operatorname{Im} V)$ of the space X. In more detail,

$$S(V) = \left\{\sum\nolimits_{j=1}^{m} \alpha_j V x_j : x_j \in X, \quad \alpha_j \in \Phi, \quad m \geq 1\right\}.$$

The corresponding subalgebra (even an ideal) of the algebra X is

$$X' = \left\{\sum\nolimits_{j=1}^{m} \alpha_j x_j y_j : x_j, y_j \in X, \quad \alpha_j \in \Phi, \quad m \geq 1\right\}.$$

Indeed, $S(V) \subset X'$, obviously. On the other hand $X' \subset S(V)$ by *polarization* (1.4) or (1.5). Hence, $S(V) = X'$. Up to the end of this section we keep all in terms of V but, of course, everything can be immediately polarized.

The dimension $r(V)$, $0 \leq r(V) \leq \infty$, of the space $S(V)$ is called the *rank* of the quadratic operator V. Obviously, $r(V) = 0$ if and only if $V = 0$. Denote by FQ_X the set of all quadratic operators of finite rank.

Proposition 2.1. *The set FQ_X is a subspace of Q_X.*

Proof. For every $\alpha \in \Phi/\{0\}$ we have $S(\alpha V) = S(V)$. Furthermore, $S(V_1 + V_2) \subset S(V_1) + S(V_2)$. Thus,

$$r(\alpha V) = r(V) \quad (\alpha \neq 0), \quad r(V_1 + V_2) \leq r(V_1) + r(V_2). \tag{2.6}$$

$\square$

In addition, $r(TV) \leq R(V)$ and $r(VT) \leq r(V)$ for any linear operator T. Hence, the subspace FQ_X is invariant for the mappings $V \mapsto TV$ and $V \mapsto VT$.

Proposition 2.2. *Each $V \in FQ_X$ is the sum of $r(V)$ elementary operators.*

Proof. Let $(e_j)_{j=1}^{r}$ be a basis of $S(V)$, so $r = r(V)$, and let $(e_j^*)_{j=1}^{r}$ be the dual basis in $S(V)^*$. Since $Vx \in S(V)$ for every $x \in X$, we have

$$Vx = \sum\nolimits_{j=1}^{r} v_j[x] e_j, \quad x \in X, \tag{2.7}$$

where $v_j[x] = e_j^*[Vx]$ are quadratic functionals. $\square$

Corollary 2.3. *An operator $V \in Q_X \backslash \{0\}$ is of rank 1 if and only if it is elementary.*

Corollary 2.4. *For $V \in FQ_X$ the rank $r(V)$ is the minimal number of elementary quadratic operators sum of which is equal to V.*

Proof. This follows from Proposition 2.2 because of the inequality (2.6). $\square$

We call *minimal* any decomposition of $V \in Q_X \setminus \{0\}$ into a sum of $r(V)$ elementary quadratic operators. To investigate this case we use the following general

Lemma 2.5. *If $(\omega_i(t))_{i=1}^n$ is a linearly independent system of scalar functions on a set S then there are n points $t_1, \ldots, t_n$ in S such that*

$$\det(\omega_i(t_k))_{i,k=1}^n \neq 0. \tag{2.8}$$

Proof. Let us consider the mapping $\Omega : S \to \Phi^n$ defined as $\Omega(t) = (\omega_i(t))_{i=1}^n$, $t \in S$. The subspace $L = \mathrm{Span}(\mathrm{Im}\,\Omega)$ coincides with Φ^n. Indeed, otherwise, L would lie in a hyperplane, so there is $(\alpha_i)_{i=1}^n \in \Phi^n \setminus \{0\}$ such that

$$\sum_{i=1}^n \alpha_i \omega_i(t) = 0, \quad t \in S,$$

that contradicts the linear independence of the system $(\omega_i(t))_{i=1}^n$. As a result, there are n linearly independent vectors in $\mathrm{Im}\,\Omega$. They are $\Omega(t_k)$ with some $t_k \in S$. Thus, the columns of the matrix $(\omega_i(t_k))_{i,k=1}^n$ are linearly independent. □

Now let

$$Vx = \sum_{j=1}^s v_j[x]e_j, \quad x \in X, \tag{2.9}$$

where e_j are vectors and $v_j[x]$ are quadratic functionals.

Lemma 2.6. *If in (2.9) the v_j's are linearly independent then all $e_j \in S(V)$.*

Proof. By Lemma 2.5 there are s vectors $x_1, \ldots, x_s$ such that

$$\det(v_j[x_k])_{j,k=1}^s \neq 0.$$

Therefore, the system of linear equations

$$\sum_{j=1}^s v_j[x_k]e_j = Vx_k, \quad 1 \leq k \leq s,$$

with unknown e_j's is solvable. By Cramer's rule all $e_j \in S(V)$. □

Now we are able to characterize the minimal decompositions.

Theorem 2.7. *The following statements are equivalent:*
1) *The decomposition (2.9) is minimal, i.e., $s = r(V)$.*
2) *The systems $(e_j)_{j=1}^s$ and $(v_j)_{j=1}^s$ are both linearly independent.*
3) *The system $(e_j)_{j=1}^s$ is a basis in $S(V)$.*

Proof. 1)⇒2). Suppose to the contrary. For definiteness, let the system $(e_j)_{j=1}^s$ be linearly dependent. Then one of e_j's is a linear combination of others. The substitution of this expression into (2.9) reduces the number of summands to $s-1$ that contradicts the minimality.

2)⇒3) since all $e_j \in S(V)$ by Lemma 2.6 and $\mathrm{Span}(e_j)_{j=1}^s = S(V)$ by (2.9).

3)⇒1). $s = r(V)$ since $r(V) = \dim S(V)$ and $(e_j)_{j=1}^s$ is a basis in $S(V)$. □

Corollary 2.8. *If the decomposition (2.9) is minimal then the system $(e_j)_{j=1}^s$ is a basis in $S(V)$ and $v_j[x] = e_j^*[Vx]$, $1 \leq j \leq s$, where $(e_j^*)_{j=1}^s$ is the dual basis in $S(V)^*$.*

3. Continuous quadratic operators and Banach algebras

From now on the ground field Φ is $\mathbb{R}$ or $\mathbb{C}$, until otherwise stated. Recall that a real or complex Banach space X is said to be a *Banach algebra* if X is an algebra with continuous multiplication, i.e., the product xy is a continuous function of $(x, y) \in X \times X$.

Proposition 3.1. *Let X be an algebra and a Banach space with a norm $\|.\|$. Then the following statements are equivalent:*

(1) *X is a Banach algebra.*
(2) *The product xy is continuous at the point $(0,0)$.*
(3) *The inequality*

$$\|xy\| \leq C \|x\| \|y\| \tag{3.1}$$

holds with a constant $C > 0$.

Proof. (1)$\Rightarrow$(2) trivially.

(2)$\Rightarrow$(3). Suppose to the contrary. Then there is a sequence $(x_n, y_n)_{n=1}^{\infty} \subset X \times X$ such that $\|x_n y_n\| > n \|x_n\| \|y_n\|$. Obviously, $x_n \neq 0$ and $y_n \neq 0$. By letting

$$u_n = x_n/\sqrt{n} \|x_n\|, \quad v_n = y_n/\sqrt{n} \|y_n\|,$$

we obtain $\|u_n v_n\| > 1$, while $\|u_n\| = \|v_n\| = 1/\sqrt{n} \to 0$, a contradiction.

(3)$\Rightarrow$(1). From (3.1) it follows that

$$\|(x + u)(y + v) - xy\| = \|xv + uy + uv\| \leq C(\|x\| \|v\| + \|u\| \|y\| + \|u\| \|v\|).$$

Hence, $(x + u)(y + v) \to xy$ when $(u, v) \to 0$ in $X \times X$. $\square$

Sometimes, it can be reasonable to change the norm in a Banach algebra X to an equivalent one. By definition, the ratio of two equivalent norms lies in a segment $[a, b]$ of the semiaxis $(0, \infty)$. When a norm runs over an equivalence class, the topology remains the same, while the constant C in (3.1) takes all positive values. This is true even if we only consider the norms proportional to a fixed one: $\{\|.\|' = q \|.\| : q > 0\}$. Indeed, from (3.1) it follows that

$$\|xy\|' \leq Cq^{-1} \|x\|' \|y\|'. \tag{3.2}$$

By the way, with $q = C$ we have

$$\|xy\|' \leq \|x\|' \|y\|'. \tag{3.3}$$

It is useful to add a geometrical criterion to the Proposition 3.1. Let us consider the balls $B_r = \{z \in X : \|z\| \leq r\}$, $r > 0$, in a Banach space X. Let X be an algebra, as before.

Proposition 3.2. *For X in order to be a Banach algebra it is necessary and sufficient that the product xy is bounded on every $B_r \times B_r$ and sufficient that this product is bounded on a $B_r \times B_r$.*

Proof. Let X be a Banach algebra. Then for $(x, y) \in B_r \times B_r$ the inequality (3.1) yields $\|xy\| \le Cr^2$. Conversely, let $\|xy\| \le M$ for a $M > 0$ and $(x, y) \in B_r \times B_r$ with an r. For any $(x, y) \in X \times X$ the vectors $rx/\|x\|$ and $ry/\|y\|$ belong to B_r. Hence, $\|xy\| \le Mr^{-2}\|x\|\|y\|$. $\qquad\qquad\qquad\qquad\qquad\qquad\qquad\qquad\quad\square$

Now we proceed to the quadratic operators in a Banach space X and establish some criteria for their continuity.

Proposition 3.3. *Let* $V : X \to X$ *be a quadratic operator. The following statements are equivalent:*

(1) *V is continuous.*
(2) *The corresponding commutative algebra is a Banach algebra.*
(3) *V is bounded in the sense that the inequality*

$$\|Vx\| \le C\|x\|^2 \qquad\qquad (3.4)$$

holds with a constant $C > 0$.
(4) *The image of every ball B_r is contained in a ball $B_{R(r)}$.*
(5) *The image of a ball B_r is contained in a ball B_R.*
(6) *V is continuous at the point $x = 0$.*

Proof. (1)$\Rightarrow$(2) by polarization.

(2)$\Rightarrow$(3) by taking $y = x$ in the inequality (3.1).

(3)$\Rightarrow$(4) since if $\|x\| \le r$ then $\|Vx\| \le Cr^2$ by (3.4).

(4)$\Rightarrow$(5) trivially.

(5)$\Rightarrow$(6) since if $VB_r \subset B_R$ with some r and R then by homogeneity we have $VB_\delta \subset B_\varepsilon$ where $\delta = r\sqrt{\varepsilon/R}$.

(6)$\Rightarrow$(2) since xy is continuous at $(0, 0)$ by polarization, and then one can refer to (2)$\Rightarrow$(1) from Proposition 3.1.

(2)$\Rightarrow$(1) trivially. $\qquad\qquad\qquad\qquad\qquad\qquad\qquad\qquad\qquad\qquad\qquad\quad\square$

We denote the linear space of continuous quadratic operators (a subspace of Q_X) by BQ_X in accordance with the equivalence (1) $\Leftrightarrow$ (3). The latter allows us to introduce the norm

$$\|V\| = \sup_{x \ne 0} \frac{\|Vx\|}{\|x\|^2} = \sup_{\|x\|=1} \|Vx\|, \quad V \in BQ_X. \qquad (3.5)$$

This definition is a counterpart of that which is the standard for a linear continuous ($\Leftrightarrow$ bounded) operator T. Obviously,

$$\|TV\| \le \|T\|\|V\|, \quad \|VT\| \le \|T\|^2\|V\|.$$

At the end of Section 4 we show that the normed space BQ_X is Banach. Its (nonclosed, in general) subspace $BQ_X \cap FQ_X$ we denote by BFQ_X.

Theorem 3.4. *An operator $V \in FQ_X$ belongs to BFQ_X if in a decomposition (2.9) the quadratic functionals $v_j[x]$ are continuous. This condition is necessary if the decomposition is minimal.*

Proof. The sufficiency is obvious. The necessity follows from Corollary 2.8 since all linear functionals on the finite-dimensional space $S(V)$ are continuous.   □

In conclusion of this section we prove the following

Proposition 3.5. *A quadratic operator $V \neq 0$ of form* (2.1) *is continuous if and only if the linear functional φ and the linear operator T are both continuous.*

Proof. The "if" part is obvious. By the implication 1) ⇒ 2) in Proposition 3.3 it suffices to prove the "only if" part in terms of the multiplication (2.2). Let the latter be continuous, and let $T|\ker\varphi \neq 0$. Then there is y such that $\varphi[y] = 0$, $Ty \neq 0$, and then there is a continuous linear functional ψ such that $\psi[Ty] = 2$. From (2.2) it follows that $\varphi[x] = \psi[xy]$, hence φ is continuous.

Now let $T|\ker\varphi = 0$. Since $V \neq 0$, we have $T \neq 0$ and $\varphi \neq 0$. Then there is a vector e such that $\varphi[e] = 1$. This yields $x - \varphi[x]e \in \ker\varphi$ for all $x \in X$. Hence, $Tx = \varphi[x]Te$, so $Te \neq 0$. Taking a continuous linear functional θ such that $\theta[Te] = 1$ we get $\varphi[x] = \theta[xe]$, hence φ is continuous again.

Now to complete the proof we return to (2.2) taking any y such that $\varphi[y] = 1$. Then we obtain $Tx = 2xy - \varphi[x]Ty$. Thus, the operator T is continuous.   □

4. Intrinsic characterization of quadratic operators

Here we prove the following

Theorem 4.1. *Let X be a Banach space. A continuous mapping $V : X \to X$ is a quadratic operator if and only if for every two vectors $x, y \in X$ and every continuous linear functional ω on X the function $\omega[V(\alpha x + \beta y)]$ is a quadratic form of the scalar variables α, β.*

Proof. "Only if." In the corresponding algebra we have

$$\omega[V(\alpha x + \beta y)] = \omega[(\alpha x + \beta y)^2] = \alpha^2\omega[x^2] + \beta^2\omega[y^2] + 2\alpha\beta\omega[xy].$$

"If." Now we have

$$\omega[V(\alpha x + \beta y)] = \alpha^2 a(x, y; \omega) + \beta^2 b(x, y; \omega) + 2\alpha\beta c(x, y; \omega) \qquad (4.1)$$

where a, b, c are some scalar functions of the triple $(x, y; \omega)$. By setting $\alpha = 1, \beta = 0$ and $\alpha = 0, \beta = 1$ we obtain

$$a(x, y; \omega) = \omega[Vx], \quad b(x, y; \omega) = \omega[Vy], \qquad (4.2)$$

and then with $\alpha = 1, \beta = \pm 1$ we get

$$\omega[V(x \pm y)] = \omega[Vx] + \omega[Vy] \pm 2c(x, y; \omega).$$

Hence,

$$c(x, y; \omega) = \omega\left[W(x, y)\right] \qquad (4.3)$$

where

$$W(x, y) = \frac{V(x + y) - V(x - y)}{4}. \qquad (4.4)$$

By substitution from (4.3) and (4.2) into (4.1) and linearity of the functional ω we obtain

$$\omega[V(\alpha x + \beta y)] = \omega[\alpha^2 V x + \beta^2 V y + 2\alpha\beta W(x, y)]. \tag{4.5}$$

By the Hahn–Banach theorem the continuous linear functionals on the Banach space X separate the vectors. Therefore, from (4.5) it follows that

$$V(\alpha x + \beta y) = \alpha^2 V x + \beta^2 V y + 2\alpha\beta W(x, y). \tag{4.6}$$

In particular,

$$V(\alpha x) = \alpha^2 V x, \quad V(0) = 0.$$

Now from (4.4) it follows that $W(x, y)$ is continuous and $W(x, x) = V x$. It remains to prove that $W(x, y)$ is a multiplication. This multiplication will turn out to be commutative automatically since $W(x, y) = W(y, x)$ by (4.4). Also due to this symmetry the bilinearity of $W(x, y)$ reduces to the linearity of $W(x, .)$.

According to (4.4) we have

$$4W(\alpha x_1 + \beta x_2, y) = V(\alpha x_1 + \beta x_2 + y) - V(\alpha x_1 + \beta x_2 - y). \tag{4.7}$$

In particular,

$$4W(\alpha x, y) = V(\alpha x + y) - V(\alpha x - y) = 4\alpha W(x, y)$$

by (4.6) with $\beta = \pm 1$. Thus $W(x, .)$ is homogeneous of degree 1.

To prove the additivity of $W(x, .)$ we note that for every triple $x, y, z \in E$ the formula (4.6) implies

$$V(x + z \pm y) = V(x + z) + V(y) \pm 2W(x + z, y)$$

whence

$$V(x + z + y) + V(x + z - y) = 2\{V(x + z) + V(y)\}$$

and then

$$V(x - z + y) + V(x - z - y) = 2\{V(x - z) + V(y)\}.$$

By subtraction we get

$$\{V(x+z+y) - V(x-z+y)\} + \{V(x+z-y) - V(x-z-y)\} = 2\{V(x+z) - V(x-z)\}$$

that can be rewritten as

$$W(x + y, z) + W(x - y, z) = 2W(x, z) = W(2x, z).$$

By substitution $x = (u + v)/2, y = (u - v)/2$ we finally obtain

$$W(u, z) + W(v, z) = W(u + v, z). \qquad \square$$

In essence, the proof of the additivity above is a version of an argument which shows that in a normed space the parallelogram identity implies the Euclidean structure, see [4, Ch. 7, Section 3]. On the other hand, the topological aspect of the proof can be ignored that yields the following general result, cf. [2], n°3, Ex. 8.

Theorem 4.2. *Let X and Y be some linear spaces over a field Φ, char$(\Phi) \neq 2$. A mapping $V : X \to Y$ is quadratic (i.e., generated by a bilinear mapping $X \times X \to Y$) if and only if such are the restrictions of V to all two-dimensional subspaces of X.*

Taking $Y = \Phi$ we obtain

Corollary 4.3. *A scalar function on a linear space X over Φ is a quadratic functional if and only if such are its restrictions to all two-dimensional subspaces.*

As an application of Theorem 4.1 (with some elements of its proof) we prove

Proposition 4.4. *With the norm (3.5) the space BQ_X of continuous quadratic operators in a Banach space X is a Banach space.*

Proof. Let $(V_n)_{n=1}^\infty$ be a Cauchy sequence in BQ_X. Then $(V_n x)_{n=1}^\infty$ is a Cauchy sequence in X for every $x \in X$. Since X is a Banach space, the self-mapping $Vx = \lim_{n\to\infty} V_n x$ is well defined on X. This is continuous since the convergence is uniform on every ball. This is quadratic since one can pass to the limit in

$$V_n(\alpha x + \beta y) = \alpha^2 V_n x + \beta^2 V_n y + 2\alpha\beta W_n(x,y)$$

where

$$W_n(x,y) = \frac{1}{4}\left(V_n(x+y) - V_n(x-y)\right). \qquad \square$$

5. Cubic matrices and quadratic integral operators

Assume that a Banach space X has a Schauder basis $(e_j)_{j=1}^\nu$, $1 \leq \nu \leq \infty$. By definition, every vector $x \in X$ can be uniquely represented as the sum of a convergent series (or as a finite sum if $\nu < \infty$) of the form

$$x = \sum_{j=1}^\nu \xi_j e_j \qquad (5.1)$$

with scalar coefficients ξ_j. The latter are the *coordinates* of x at the given basis. The functionals $\xi_j = e_j^*[x]$ are linear and continuous.

If X is an algebra then we have the *table of multiplication*

$$e_i e_k = \sum_{j=1}^\nu c_{ik}^j e_j. \qquad (5.2)$$

The scalar coefficients c_{ik}^j are called the *structural constants* of the algebra regarding the basis $(e_j)_{j=1}^\nu$. The structural constants constitute a $(\nu \times \nu \times \nu)$- *cubic matrix* c which also can be treated as a sequence of length ν of square $(\nu \times \nu)$-matrices $c^j = [c_{ik}^j]_{i,k=1}^\nu$. The cubic matrix c is called the *structural matrix* of the algebra at the given basis.

Example 5.1. The structural coefficients of the algebra (2.2) are

$$c_{ik}^j = \frac{1}{2}(\varphi[e_i]\tau_{jk} + \varphi[e_k]\tau_{ji}) \tag{5.3}$$

where $[\tau_{jk}]_{j,k=1}^{\nu}$ is the matrix of the operator T at the same basis.

Theorem 5.2. *Let X be a Banach algebra with a Schauder basis $(e_j)_{j=1}^{\nu}$, and let ξ_i and η_k be the coordinates of vectors x and y, respectively. Then the coordinates of the product xy are the bilinear forms*

$$\zeta_j = \sum_{i,k=1}^{\nu} c_{ik}^j \xi_i \eta_k \tag{5.4}$$

if $\nu < \infty$ and

$$\zeta_j = \lim_{n\to\infty} \sum_{i,k=1}^{n} c_{ik}^j \xi_i \eta_k \tag{5.5}$$

if $\nu = \infty$.

Proof. For $\nu < \infty$ the formula (5.4) follows immediately by bilinearity of the multiplication and definition of the structural constants. Now let $\nu = \infty$, and let

$$x_n = \sum_{i=1}^{n} \xi_i e_i, \quad y_n = \sum_{k=1}^{n} \eta_k e_k.$$

Then $x_n \to x$ and $y_n \to y$ as $n \to \infty$. Since the multiplication is continuous, we have $x_n y_n \to xy$. Since the coordinate functionals e_j^* are continuous, we get

$$\zeta_j = e_j^*[xy] = \lim_{n\to\infty} e_j^*[x_n y_n],$$

and the case reduces to the previous one. □

As an immediate consequence we get

Corollary 5.3. *A Banach algebra with a Schauder basis is commutative if and only if the corresponding structural constants c_{ik}^j are symmetric with respect i,k, i.e., all matrices $c^j = [c_{ik}^j]_{i,k=1}^{\nu}$ are symmetric.*

Note that the transformation (1.6) of an algebra to a commutative one corresponds to the standard symmetrization $[c_{ik}^j] \mapsto \frac{1}{2}([c_{ik}^j] + [c_{ki}^j])$.

Corollary 5.4. *Let V be a continuous quadratic operator in a Banach space X with a Schauder basis $(e_j)_{j=1}^{\nu}$, $1 \leq \nu \leq \infty$, and let c_{ik}^j be the structural constants of the corresponding commutative Banach algebra. Then if ξ_i are the coordinates of a vector $x \in X$ then the jth coordinate of the vector Vx is the quadratic form*

$$\zeta_j = \sum_{i,k=1}^{\nu} c_{ik}^j \xi_i \xi_k \tag{5.6}$$

if X is finite dimensional or

$$\zeta_j = \lim_{n \to \infty} \sum_{i,k=1}^{n} c_{ik}^j \xi_i \xi_k \tag{5.7}$$

if X is infinite dimensional.

In the finite-dimensional space X the formulas (5.4) are obviously valid for any multiplication, irrespective to a topology. On the other hand, a linear topology in X is unique and can be defined by any norm. Using the l_1-norm

$$\|x\| = \sum_{i=1}^{\nu} |\xi_i| \tag{5.8}$$

we see that (5.4) implies the inequality (3.1) with

$$C = \sum_{j=1}^{\nu} \max_{i,k} \left| c_{ik}^j \right|.$$

This results in the following

Proposition 5.5. *With respect to the linear topology all finite-dimensional algebras are Banach, and all quadratic operators in finite-dimensional spaces are continuous.*

This fails in any infinite-dimensional Banach space X.

Example 5.6. The multiplication (2.5) is not continuous if such is φ and $e \neq 0$. Such a φ can be obtained as the linear extension of any unbounded scalar function on an algebraic basis (a Hamel basis) Γ such that $\|g\| = 1$ for $g \in \Gamma$.

With finite ν any cubic matrix $[c_{ik}^j]_{i,k,j=1}^{\nu}$ is the structural matrix of the multiplication defined by (5.4) at an arbitrary basis $(e_j)_{j=1}^{\nu}$. Indeed, in this setting

$$e_l e_m = \sum_{j=1}^{\nu} \left(\sum_{i,k=1}^{\nu} c_{ik}^j \delta_{li} \delta_{mk} \right) e_j = \sum_{j=1}^{\nu} c_{lm}^j e_j \tag{5.9}$$

where δ is the Kronecker delta.

In the infinite-dimensional case a cubic matrix must satisfy some conditions in order to be the structural matrix of a continuous multiplication. Of course, these conditions depend on the underlying Banach space. Let us consider the multiplications in the classical spaces l_p, $1 \leq p \leq \infty$. As usual, we set $q = p/(p-1)$ for $1 < p < \infty$, $q = \infty$ for $p = 1$ and $q = 1$ for $p = \infty$. We denote the norm of $x \in l_p$ by $\|x\|_p$ and say that a square matrix $a = [a_{ik}]_{i,k=1}^{\infty}$ is of *class l_q* if

$$\|a\|_q \equiv \left(\sum_{i,k=1}^{\infty} |a_{ik}|^q \right)^{1/q} < \infty$$

for $q < \infty$ and

$$\|a\|_\infty \equiv \sup_{i,k \geq 1} |a_{ik}| < \infty.$$

Now, given a cubic matrix $c = [c_{ik}^j]_{i,k,j=1}^\infty$, we say that it is of *class* $l_{p,q}$ if all square matrices $c^j = [c_{ik}^j]_{i,k=1}^\infty$ are of class l_q and the sequence $(\|c^j\|_q)_{j=1}^\infty$ belongs to l_p. In this case we set

$$\|c\|_{p,q} = \left\| \left(\|c^j\|_q \right)_{j=1}^\infty \right\|_p. \tag{5.10}$$

Theorem 5.7. *Let* $1 \le p < \infty$. *In the space* l_p *every cubic matrix* $c = [c_{ik}^j]_{i,k,j=1}^\infty$ *of class* $l_{p,q}$ *being assigned to the Schauder basis* $\delta_j = (\delta_{ji})_{i=1}^\infty$, $1 \le j < \infty$, *is the structural matrix of a continuous multiplication.*

Proof. For every pair $x = (\xi_i)_{i=1}^\infty$, $y = (\eta_k)_{k=1}^\infty \in l_p$ the values

$$\zeta_j = \sum_{i,k=1}^\infty c_{ik}^j \xi_i \eta_k, \quad j \ge 1, \tag{5.11}$$

are determined since these series converge, even absolutely. Indeed,

$$\sum_{i,k=1}^\infty \left| c_{ik}^j \xi_i \eta_k \right| \le \|c^j\|_q \|x\|_p \|y\|_p$$

by the Hölder inequality for $p > 1$ and trivially for $p = 1$. Moreover, the sequence $z = (\zeta_j)_{j=1}^\infty$ belongs to l_p since

$$\sum_{j=1}^\infty |\zeta_j|^p \le \sum_{j=1}^\infty \sum_{i,k=1}^\infty \left| c_{ik}^j \xi_i \eta_k \right|^p \le (\|c\|_{p,q} \|x\|_p \|y\|_p)^p$$

by (5.10). The relation $z = xy$ defines a multiplication in l_p, and the last inequality can be rewritten as

$$\|xy\|_p \le \|c\|_{p,q} \|x\|_p \|y\|_p. \tag{5.12}$$

By Propositions 3.1 the multiplication is continuous. It remains to note that, according to (5.11), we have

$$\delta_l \delta_m = \sum_{j=1}^\infty c_{lm}^j \delta_j \tag{5.13}$$

similarly to (5.9). □

Remark 5.8. In fact, the proof of Theorem 5.7 does not refer to the Schauder basis in l_p. The only important is that this a sequence space, so the multiplication can be just introduced by formula (5.11), so that

$$\delta_l \delta_m = (c_{lm}^j)_{j=1}^\infty,$$

instead of (5.13). For this reason *Theorem 5.7 extends to* $p = \infty$ *in the same way as before.* In this case we have the inequality (5.12) with

$$\|c\|_{\infty,1} = \sup_{j \ge 1} \sum_{i,k=1}^\infty \left| c_{ik}^j \right|. \tag{5.14}$$

By the way, there is no Schauder basis in l_∞ since this Banach space is not separable.

Corollary 5.9. *Let H be a separable Hilbert space. Under condition*

$$\sum_{i,k,j=1}^{\infty} \left|c_{ik}^j\right|^2 < \infty \tag{5.15}$$

a cubic matrix $c = [c_{ik}^j]_{i,k,j=1}^{\infty}$ assigned to any orthonormal basis $(e_j)_{j=1}^{\infty}$ in H is the structural matrix of a continuous multiplication.

Proof. One can assume that $H = l_2$ and $e_j = \delta_j$. Then the sum in (5.15) turns into the square of $\|c\|_{2,2}$. $\square$

Remark 5.10. By the Parseval equality we have

$$\|e_i e_k\|^2 = \sum_{j=1}^{\infty} \left|c_{ik}^j\right|^2 .$$

Hence, (5.15) can be rewritten as

$$\sum_{i,k=1}^{\infty} \|e_i e_k\|^2 < \infty. \tag{5.16}$$

Let us omit the obvious reformulations of the last series of statements in setting of quadratic operators, cf. Corollary 5.4.

The condition $c \in l_{p,q}$ is not necessary for the structural matrices c of continuous multiplications in l_p. We show this for $p = q = 2$, i.e., in the case of Corollary 5.9. To this end it suffices to consider the algebras of form (2.2) with the structural coefficients (5.3). Recall that a linear operator T in a Hilbert space is said to be a *Hilbert–Schmidt operator* if

$$\|T\|_{HS} \equiv \left(\sum_{j=1}^{\infty} \|Te_i\|^2\right)^{1/2} < \infty$$

for an (and then for every) orthonormal basis $(e_i)_{i=1}^{\infty}$. All Hilbert–Schmidt operators are continuous since $\|T\| \leq \|T\|_{HS}$. However, for example, the unit operator is not Hilbert–Schmidt. In addition to Proposition 3.5 we have

Proposition 5.11. *Let T be a linear operator in a Hilbert space H with an orthonormal basis $(e_i)_{i=1}^{\infty}$. For the algebra (2.2) the condition (5.15) is fulfilled for all linear continuous functionals φ if and only if T is a Hilbert–Schmidt operator.*

Proof. Let us deal with the equivalent condition (5.16). The functional $\varphi[x]$ can be represented as the inner product $\langle x, h \rangle$ with an $h \in H$. Accordingly,

$$4\|e_i e_k\|^2 = \|\langle e_i, h\rangle Te_k + \langle e_k, h\rangle Te_i\|^2$$
$$\leq |\langle e_i, h\rangle|^2 \|Te_k\|^2 + |\langle e_k, h\rangle|^2 \|Te_i\|^2 + 2|\langle e_i, h\rangle\langle h, e_k\rangle| \|Te_i\| \|Te_k\|,$$

whence

$$2 \sum_{i,k=1}^{n} \|e_i e_k\|^2 \leq \sum_{i=1}^{n} |\langle e_i, h \rangle|^2 \sum_{k=1}^{n} \|Te_k\|^2 + \left(\sum_{i=1}^{n} |\langle e_i, h \rangle| \|Te_i\| \right)^2. \qquad (5.17)$$

Applying the Cauchy–Bunyakovski inequality to the last term in (5.17) we obtain

$$\sum_{i,k=1}^{n} \|e_i e_k\|^2 \leq \sum_{i=1}^{n} |\langle e_i, h \rangle|^2 \sum_{k=1}^{n} \|Te_k\|^2 \leq \|h\|^2 \sum_{k=1}^{n} \|Te_k\|^2.$$

Therefore,

$$\sum_{i,k=1}^{\infty} \|e_i e_k\|^2 \leq \|T\|_{HS}^2 \|h\|^2 < \infty$$

if T is a Hilbert–Schmidt operator. Conversely, if T is not a Hilbert–Schmidt operator then starting with $h = e_1$ we get

$$2 \sum_{i,k=1}^{n} \|e_i e_k\|^2 = \sum_{k=1}^{n} \|Te_k\|^2 + \|Te_1\|^2 \to \infty$$

as $n \to \infty$. $\qquad \square$

Now we consider an integral counterpart of the structural matrix. This is the *kernel* $K(s,t;u)$ of the multiplication

$$(f \circ g)(u) = \int_{\mathbb{R}} \int_{\mathbb{R}} K(s,t;u) f(s) g(t) \, ds \, dt, \quad u \in \mathbb{R}. \qquad (5.18)$$

In order to realize this formal construction we assume that, at least, the kernel is measurable and the measurable functions f, g run over a linear space E such that in (5.18) the integrand belongs to $L_1(\mathbb{R}^2)$, and the integral belongs to E. If $K(s,t;u)$ is symmetric in s,t ("symmetric" for brevity) then the multiplication (5.18) is commutative. For $f = g$ we get the *quadratic integral operator* $V_K : E \to E$:

$$(V_K f)(u) = \int_{\mathbb{R}} \int_{\mathbb{R}} K(s,t;u) f(s) f(t) \, ds \, dt, \quad u \in \mathbb{R}, \qquad (5.19)$$

and here the kernel K can be changed for a symmetric one: $V_K = V_{\hat{K}}$ where

$$\hat{K}(s,t;u) = \frac{1}{2} \left(K(s,t;u) + K(t,s;u) \right).$$

In role of E one can consider $L_p(\mathbb{R})$, $1 \leq p \leq \infty$.

Theorem 5.12. *If the function $K^u(s,t) = K(s,t;u)$ belongs to $L_q(\mathbb{R}^2)$ for almost every u and the function $\kappa^u = \|K^u\|_q$ of u belongs to $L_p(\mathbb{R})$ then the multiplication (5.18) is defined and continuous in $L_p(\mathbb{R})$. Moreover,*

$$\|f \circ g\|_p \leq \kappa_{p,q} \|f\|_p \|g\|_p \qquad (5.20)$$

where $\kappa_{p,q}$ is the L_p-norm of κ^u.

Proof. The same as for Theorem 5.7 but with integrals instead of sums. $\qquad \square$

Note that the conditions of Theorem 5.12 are symmetric since such is $K^u(s,t)$.

Corollary 5.13. *Under conditions of Theorem 5.12 the quadratic operator* (5.19) *is defined and continuous in $L_p(\mathbb{R})$. Moreover,*

$$\|V_K f\|_p \le \kappa_{p,q} \|f\|_p^2. \tag{5.21}$$

Corollary 5.14. *If $K(s,t;u)$ is a* Hilbert–Schmidt *kernel, i.e.,*

$$\kappa_{2,2}^2 \equiv \int_{\mathbb{R}}\int_{\mathbb{R}}\int_{\mathbb{R}} |K(s,t;u)|^2 \, ds\, dt\, du < \infty,$$

then in $L_2(\mathbb{R})$ the quadratic operator (5.19) *is defined and continuous. Moreover,*

$$\|V_K f\|_2 \le \kappa_{2,2} \|f\|_2^2. \tag{5.22}$$

It is interesting to compare this result to Corollary 5.9. Let $(e_j(u))_{j=1}^\infty$ be an orthonormal basis in $L_2(\mathbb{R})$. Then $(e_i(s)e_k(t)e_j(u))_{i,k,j=1}^\infty$ is an orthonormal basis in $L_2(\mathbb{R}^3)$. The Hilbert–Schmidt kernel $K(s,t;u)$ belongs to $L_2(\mathbb{R}^3)$ by assumption. If its coordinates are c_{ik}^j then for all $f, g \in L_2(\mathbb{R})$ the coordinates of $f \circ g$ are

$$\zeta_j = \sum_{i,k=1}^\infty c_{ik}^j \varphi_i \gamma_k$$

where φ_i and γ_k are the coordinates of f and g. Hence, $[c_{ik}^j]_{i,k,j=1}^\infty$ is the structural matrix of this algebra. Moreover, $c_{2,2} = \kappa_{2,2}$ by Parseval's equality. Thus, (5.15) is equivalent to that $K(s,t;u)$ is the Hilbert Schmidt kernel.

An important case is a kernel $K(s,t;u)$ with finite support, say, $K(s,t;u) = 0$ outside the cube $0 \le s,t,u \le 1$, so, accordingly,

$$(f \circ g)(u) = \int_0^1 \int_0^1 K(s,t;u) f(s) g(t) \, ds\, dt, \quad 0 \le u \le 1. \tag{5.23}$$

This yields the following important Banach algebra [15].

Proposition 5.15. *With a continuous kernel $K(s,t;u)$, $0 \le s,t,u \le 1$, the multiplication* (5.23) *is defined and continuous in $C[0,1]$. Moreover,*

$$\|f \circ g\| \le N(K)\|f\|\|g\|, \quad N(K) = \max_{u \in [0,1]} \int_0^1 \int_0^1 |K(s,t;u)| \, ds\, dt. \tag{5.24}$$

Proof. For $f, g \in C[0,1]$ we have $f \circ g \in C[0,1]$ since $K(s,t;u)$ is uniformly continuous. The rest is obvious. □

The symmetric continuous kernels $K(s,t;u)$ form a subspace $L \subset C([0,1]^3)$. We endow it with the norm $N(K)$ weaker than the standard one.

Corollary 5.16. *The linear mapping $J: K \mapsto V_K$ is a continuous embedding of the N-normed space L into the space of continuous quadratic operators in $C[0,1]$.*

Proof. J is injective since if $V_K = 0$ then $f \circ g = 0$ for all $f, g \in C[0,1]$, and then $K = 0$ because of the completeness of the set of products fg in $C([0,1]^2)$. Furthermore, $\|V_K\| \le N(K)$ because of (3.5) and (5.24), so J is continuous. □

6. Compact quadratic operators and algebras

The compactness of some linear and nonlinear operators is a powerful tool in the functional analysis, especially in the theory of integral and differential equations, see, e.g., [17] and the references therein. According to [15] a commutative Banach algebra is called *compact* if such is the corresponding quadratic operator. In turn, a quadratic operator V in a Banach space X is called *compact* if the image of a ball B_r is relatively compact. In this case the image of every bounded set is relatively compact. From Proposition 3.3 it follows that every compact quadratic operator is continuous. Every continuous quadratic operator V of finite rank is compact. Therefore, *every commutative Banach algebra X with finite-dimensional X' is compact.*

We denote by CQ_X the linear space of all compact quadratic operators, so $BFQ_X \subset CQ_X \subset BQ_X$. The subspace CQ_X is closed. Furthermore, we have

Theorem 6.1. *Let a Banach space X has a Schauder basis $(e_j)_{j=1}^{\nu}$. Then the subspace BFQ_X is dense in CQ_X.*

Proof. If $\nu < \infty$ then $BFQ_X = CQ_X$. Let $\nu = \infty$, let $V \in CQ_X$, and let

$$P_n x = \sum_{j=1}^{n} e_j^*[x]e_j, \quad x \in X, \quad n \geq 1.$$

Then

$$(P_n V)x = \sum_{j=1}^{n} e_j^*[Vx]e_j,$$

so, $P_n V$ is a continuous quadratic operator at most of rank n. For $n \to \infty$ we have $P_n x \to x$ for all $x \in X$. Hence, $(P_n V)x \to Vx$ for all x. This convergence is uniform on the ball B_1 since $V B_1$ is relatively compact. Thus, $\|P_n V - V\| \to 0$. □

Remark 6.2. A similar theorem for the compact linear operators is well known, and the proof is the same. However, there exists a Banach space in which not every compact linear operator can be approximated by continuous linear operators of finite rank [5]. The question arises: *are the approximation properties for the linear and for the quadratic operators equivalent?*

Now let us turn to the quadratic integral operators. For such an operator of rank r its symmetric kernel is called *of rank r* as well.

Theorem 6.3. *In the space $C[0,1]$ general form of a symmetric continuous kernel of a finite rank r is*

$$K(s,t;u) = \sum_{j=1}^{r} K_j(s,t)e_j(u) \tag{6.1}$$

where all functions $e_j(u)$ are continuous, all partial kernels $K_j(s,t)$ are symmetric and continuous, and the systems $(e_j)_{j=1}^{r}$ and $(K_j)_{j=1}^{r}$ are both linearly independent.

Proof. Every function $K(s, t; u)$ of form (6.1) with continuous $e_j(u)$ and symmetric continuous $K_j(s, t)$ is the symmetric continuous kernel of the operator

$$(V_K f)(u) = \sum_{j=1}^{r} e_j(u) \int_0^1 K_j(s, t) f(s) f(t) \, ds \, dt \tag{6.2}$$

in the space $X = C[0, 1]$. By Theorem 2.7 this is a minimal decomposition of V_K into a sum of elementary quadratic operators. Thus, $r(V_K) = r$.

Conversely, let a symmetric continuous kernel $K(s, t; u)$ be such that

$$(V_K f)(u) = \sum_{j=1}^{r} v_j[f] e_j(u)$$

where $r = r(V_K)$, $e_j \in C[0, 1]$ and v_j are quadratic functionals on $C[0, 1]$. By Corollary 2.8 and the general form of continuous linear functionals on $C[0, 1]$,

$$v_j[f] = e_j^*[V_K f] = \int_0^1 d\sigma_j(u) \int_0^1 \int_0^1 K(s, t; u) f(s) f(t) \, ds \, dt$$

where $\sigma_j(u)$ are some functions of bounded variation. This yields (6.1) with

$$K_j(s, t) = \int_0^1 K(s, t; u) \, d\sigma_j(u), \quad 1 \le j \le r.$$

The partial kernels $K_j(s, t)$ are symmetric and continuous. They are linearly independent, as well as $e_j(u)$, by Theorem 2.7.   □

Theorem 6.4. *In the space $C[0, 1]$ any quadratic integral operator*

$$(V_K f)(u) = \int_0^1 \int_0^1 K(s, t; u) f(s) f(t) \, ds \, dt, \quad 0 \le u \le 1, \tag{6.3}$$

with continuous kernel $K(s, t; u)$ is compact.

Proof. Let $f \in C[0, 1]$, $\|f\| \le 1$, and let $g = V f$. Then

$$|g(u + \tau) - g(u)| \le \int_0^1 \int_0^1 |K(s, t; u + \tau) - K(s, t; u)| \, ds \, dt, \quad \tau \in \mathbb{R},$$

hence, the set $V B_1$ is uniformly continuous. Furthermore, $V B_1$ is bounded since

$$|g(u)| \le \int_0^1 \int_0^1 |K(s, t; u)| \, ds \, dt, \quad \tau \in \mathbb{R}.$$

Thus, the set $V B_1$ is relatively compact by the Arzelà–Ascoli theorem.   □

Further all kernels of integral operators in $C[0, 1]$ are assumed continuous.

Theorem 6.5. *If $K(s, t; u)$ is a Hilbert–Schmidt kernel then in $L_2(\mathbb{R})$ the quadratic operator (5.19) is compact.*

Proof. According to Theorem 20 from [3, Chapter 4], we have to check the following three properties of the set $G = \{Vf : \|f\| \leq 1\}$.

(1) G is bounded. This is obvious since V is bounded.

(2) If $a > 0$, $h_a(u) = 0$ for $|u| < a$ and $h_a(u) = 1$ for $|u| \geq a$ then $\sup_{g \in G} \|gh_a\|$ tends to zero as $a \to \infty$. This follows from the inequality

$$\|gh_a\|^2 \leq \int_{\mathbb{R}} \int_{\mathbb{R}} \int_{\mathbb{R}} |K(s,t,u)|^2 \, h_a(u) \, ds \, dt \, du$$

since the integrand pointwise tends to zero under the L_1-majorant $|K(s,t,u)|^2$.

(3) If $\tau \in \mathbb{R}$ and $g_\tau(u) = g(u + \tau)$ then $\sup_{g \in G} \|g_\tau - g\|$ tends to zero as $\tau \to 0$. This follows from the inequality

$$\|g_\tau - g\|^2 \leq \int_{\mathbb{R}} \int_{\mathbb{R}} \int_{\mathbb{R}} |K(s,t;u+\tau) - K(s,t;u)|^2 \, ds \, dt \, du$$

the right-hand side of which can be rewritten as

$$\int_{\mathbb{R}} \int_{\mathbb{R}} \int_{\mathbb{R}} \left| \hat{K}(\xi, \eta, \zeta)(e^{i<\zeta,\tau>} - 1) \right|^2 \, d\xi \, d\eta \, d\zeta$$

by Parseval's equality for the Fourier–Plansherel transform $K \mapsto \hat{K}$.   □

Remark 6.6. Theorems 6.4 and 6.5 can be extended to the operators of form

$$(V_K f)(u) = \int_S \int_S K(s,t,u) f(s) f(t) \, ds \, dt, \quad u \in S, \tag{6.4}$$

in the space $C(S)$ on a compact topological space S with a Radon measure ds or, respectively, in $L_2(S)$ on a locally compact Abelian group S with a Haar measure.

7. Baric and Bernstein algebras and quadratic operators

We start with a purely algebraic theory originated from the population genetics. A scalar function $\omega \neq 0$ on an algebra X is called a *weight* if it is a linear multiplicative functional or, the same, a homomorphism of X into the ground field. If a weight exists and fixed then X is called a *baric algebra* [6]. More rigorously, a baric algebra is a pair (X, ω) where X is an algebra and ω is a weight on X. The hyperplane $H_\omega^0 = \ker \omega = \{x \in X : \omega[x] = 0\}$ is an ideal (*the barideal*) of X.

With $\varphi[e] = 1$ the algebra (2.5) is baric with the weight φ. With $\varphi \neq 0$ the algebra (2.2) is baric with the weight φ if and only if $\varphi[Tx] = \varphi[x]$ for all $x \in X$.

A linear functional $\omega \neq 0$ is called a *weight* of a quadratic operator V if

$$\omega[Vx] = \omega^2[x], \quad x \in X. \tag{7.1}$$

Lemma 7.1. *For a quadratic operator V and for the corresponding commutative algebra X the sets Ω_V and Ω_X of weights coincide.*

Proof. Obviously, $\Omega_X \subset \Omega_V$. The converse inclusion follows by polarization.   □

The following is a refinement of Proposition 2.2.

Proposition 7.2. *If $r = r(V) < \infty$ and $\omega \in \Omega_V$ then*

$$Vx = \omega^2[x]e_1 + \sum_{j=2}^{r} v_j[x]e_j, \quad x \in X, \tag{7.2}$$

where $e_1 \in \operatorname{Im} V$, $\omega[e_1] = 1$, $(e_j)_{j=2}^{r}$ is a basis in $S(V) \cap H_\omega^0$, $v_j[x]$ are quadratic functionals linearly independent along with $\omega^2[x]$.

Proof. In the proof of Proposition 2.2 one can take any $e_1 = Ve_0$ with $\omega[e_0] = 1$. Then $\omega[e_1] = \omega[Ve_0] = \omega^2[e_0] = 1$. Let $(e_j)_{j=2}^{r}$ be a basis in $S(V) \cap H_\omega^0$, and let $(e_j^*)_{j=2}^{r}$ be its dual basis extended to a basis in $S(V)^*$ by setting $e_j^*[e_1] = 0$, $2 \leq j \leq r$, and by joining $e_1^* = \omega$. In the corresponding decomposition of V we have $v_1[x] = e_1^*[Vx] = \omega[Vx] = \omega^2[x]$. $\qquad\square$

An important example is the operator (6.3) in the space $C[0,1]$ under condition

$$\int_0^1 K(s,t;u)\,du = 1, \quad 0 \leq s, t \leq 1. \tag{7.3}$$

In this case the functional

$$\omega[f] = \int_0^1 f(u)\,du \tag{7.4}$$

is a weight. Then

$$K(s,t;u) = e_1(u) + \sum_{j=2}^{r} K_j(s,t)e_j(u) \tag{7.5}$$

where

$$e_1(u) = (V_K \mathbf{1})(u) = \int_0^1 \int_0^1 K(s,t;u)\,ds\,dt, \quad \int_0^1 e_j(u)\,du = 0, \quad 2 \leq j \leq r, \tag{7.6}$$

the sets $\{e_j(u)\}_{j=2}^{r}$ and $\{K_j(s,t)\}_{j=2}^{r} \cup \{\mathbf{1}(s,t)\}$ are both linearly independent.

A commutative baric algebra (X, ω) is called a *Bernstein algebra* if

$$(x^2)^2 = \omega^2[x]x^2, \quad x \in X. \tag{7.7}$$

For example, the baric algebra (2.5) is Bernstein. The baric algebra (2.2) is Bernstein if and only if $T^2 = T$, i.e., T is a projection. (In both cases $\omega = \varphi$.)

In terms of the *Bernstein quadratic operators* the identity (7.7) is

$$V^2x = \omega^2[x]Vx, \quad x \in X, \tag{7.8}$$

or, equivalently, if $V_1 = V|H_\omega^1$ where $H_\omega^1 = \{x \in X : \omega[x] = 1\}$ then

$$V_1^2 = V_1. \tag{7.9}$$

Lemma 7.3. *Every Bernstein quadratic operator (or Bernstein algebra) with a weight ω has a fixed point (an idempotent, respectively) e such that $\omega[e] = 1$.*

Proof. Take $e = Vx$ with $\omega[x] = 1$. Then $\omega[e] = 1$ and $Ve = V^2x = Vx = e$. $\qquad\square$

Below e is an idempotent, $\omega[e] = 1$. Every vector $x \in X$ can be uniquely represented as

$$x = \omega[x]e + z, \quad z \in H_\omega^0. \tag{7.10}$$

In the barideal H_ω^0 we have the linear operator $P_e x = 2ex$.

Lemma 7.4. *The operator P_e is a projection.*

Proof. For $\omega[x] = 0$ the line $e + tx$, $t \in \mathbb{R}$, belongs to H_ω^1. Hence, $((e+tx)^2)^2 = (e + tx)^2$, i.e., $e + 4te(ex) + \cdots = e + 2tex + \cdots$. Thus, $4e(ex) = 2ex$, i.e., $P_e^2 = P_e$. $\quad\square$

As a consequence, we have the direct decomposition

$$H_\omega^0 = U \oplus W \quad (U = \operatorname{Im} P_e, \quad W = \ker P_e). \tag{7.11}$$

As a result,

$$X = \operatorname{Span}\{e\} \oplus U \oplus W. \tag{7.12}$$

The following lemma is an immediate corollary of Theorem 3.4.8 from [14].

Lemma 7.5. *For every $x \in X$ the W-component of x^2 is u^2 with an $u \in U$.*

Now let X be a Banach space, and let $\omega \neq 0$ be a continuous linear functional. With the weight ω we consider a continuous Bernstein quadratic operator and the corresponding *Banach–Bernstein algebra* [15].

Theorem 7.6. *A Banach–Bernstein algebra X is compact if and only if the subalgebra X' is finite dimensional.*

Proof. "If" is obvious. For the "only if" note that the linear operator P_e is compact since such is V and

$$P_e x = \frac{1}{2}\left(V(x+e) - V(x-e)\right)$$

Since P_e is a projection and $P_e|U = \mathbf{1}$, the subspace U is finite dimensional. Let $(u_i)_{i=1}^l$ be its basis. By decomposition (7.12) and Lemma 7.5 we have

$$X' \subset \operatorname{Span}\{e, u_1, \ldots, u_l, u_1^2, \ldots, u_l^2, u_1 u_2, \ldots, u_{l-1} u_l\}. \quad\square$$

The operator form of Theorem 7.6 is

Theorem 7.7. *A continuous Bernstein quadratic operator is compact if and only if it is of finite rank.*

Combining this result with Lemma 7.3 and Proposition 7.2 we obtain

Corollary 7.8. *Every compact Bernstein quadratic operator is of form*

$$V x = \omega^2[x]e_1 + \sum_{j=2}^r v_j[x]e_j, \quad x \in X, \tag{7.13}$$

where $r = r(V) < \infty$, $V e_1 = e_1$, $\omega[e_1] = 1$, $(e_j)_{j=2}^r$ is a basis in $S(V) \cap H_\omega^0$, $v_j[x]$ are continuous quadratic functionals linearly independent along with $\omega^2[x]$.

In addition,
$$v_j[Vx] = \omega^2[x]v_j[x], \quad 2 \le j \le r. \tag{7.14}$$
This follows by $x \mapsto Vx$ in (7.13) and by applying (7.8) to the result.

Theorem 7.7 is applicable to the Bernstein quadratic integral operators in $C[0,1]$ since they are compact by Theorem 6.4. The standard weight in this case is (7.4). Thus, all *Bernstein kernels* are of form (7.5) where the functions e_j and $K_j(s,t)$ are such as aforesaid.

The following theorem announced in [1] was proved for the first time in [16] applying Theorem 7.7. The latter (as well as Theorem 7.6) was proved in [15].

Theorem 7.9. *If $K(s,t;u) > 0$ is a Bernstein kernel then K only depends on u.*

In other words, K is of rank 1. Actually, in [16] the assumption is $K(s,t;u) \ge 0$ with $K(s,s;u) > 0$. Also note that all our results on continuous kernels remain true for $C(S,\mu)$ where S is a compact and μ is a Radon measure, $\text{supp}\,\mu = S$.

References

[1] S.N. Bernstein. Solution of a mathematical problem related to the theory of inheritance, Uchenye Zapiski n.-i. kafedr Ukrainy 1 (1924), 83–115 (in Russian).

[2] N. Bourbaki. Éléments de mathématique, Algèbre, Ch. 9: Formes sesquilinéaires et formes quadratiques, Hermann, Paris, 1959 (in French).

[3] N. Danford and J.T. Schwartz. Linear operators, Part 1: General theory, Intersci. Publ., 1958.

[4] M. Day. Normed linear spaces, Springer-Verlag, 1958.

[5] P. Enflo. A counterexample to the approximation problem in Banach spaces, Acta Math. 130 (1973), 309–317.

[6] I.M.H. Etherington. Genetic algebras, Proc. Roy. Soc. Edinburgh, A 59 (1939), 242–258.

[7] V.I. Glivenko. Mendelian algebra, Doklady Akad. Nauk SSSR 8, No.4 (1936), 371–372 (in Russian).

[8] J.C. Gutiérrez. Solution of the Bernstein problem in the non-regular case, J. Algebra 223, No.1 (2000), 226–132.

[9] P. Holgate. Genetic algebras satisfying Bernstein's stationarity principle. J. London Math. Soc. (2) 9 (1975), 612–624.

[10] Yu.I. Lyubich. Basis concepts and theorems of evolutionary genetics for free populations, Russian Math. Surveys 26, No. 5 (1971), 51–123.

[11] Yu.I. Lyubich, Two-level Bernstein populations, Math. USSR Sb. 24, No. 1 (1974), 593–615.

[12] Yu.I. Lyubich. Proper Bernstein populations, Probl. Inform. Transmiss. (1978), 228–235.

[13] Yu.I. Lyubich. Bernstein algebras, Uspekhi Mat. Nauk 32, No. 6 (1977), 261–263 (in Russian).

[14] Yu.I. Lyubich. Mathematical structures in population genetics, Springer-Verlag, 1992. (Translated from the Russian original, 1983, Naukova Dumka.)

[15] Yu.I. Lyubich. Banach–Bernstein algebras and their applications. Nonassociative algebra and its applications, Lecture Notes in Pure and Appl. Math., 211, Dekker, 2000, pp. 205–210.

[16] Yu.I. Lyubich. A theorem on Bernstein quadratic integral operators, Nonassociative algebra and its applications, Lect. Notes Pure and Appl. Math., 246, Chapman–Hall, 2006, pp. 245–251.

[17] L. Nirenberg. Topics in nonlinear functional analysis, Courant Inst, 1974.

[18] O. Reiersöl. Genetic algebras studied recursively and by means of differential operators, Math. Scand., 10 (1962), 25–44.

[19] A.S. Serebrovskii. Properties of Mendelian equalities, Doklady Akad. Nauk SSSR 2, No. 1 (1934), 33–36.

[20] R.D. Schafer. An introduction to nonassociative algebras, Acad. Press, 1966.

[21] A. Wörz-Busekros. Algebras in Genetics, Springer-Verlag, 1980.

Yu.I. Lyubich
Department of Mathematics
Technion, 32000
Haifa, Israel
e-mail: lyubich@tx.technion.ac.il

Operator Theory:
Advances and Applications, Vol. 237, 221–239
© 2013 Springer Basel

Strong Stability of Invariant Subspaces of Quaternion Matrices

Leiba Rodman

Dedicated to Leonid Lerer on occasion of his 70th birthday

Abstract. Classes of matrices and their invariant subspaces with various robustness properties are described, in the context of matrices over real quaternions and quaternionic subspaces. Robustness is understood in the sense of being close to the original invariant subspace under small perturbation of the matrix.

Mathematics Subject Classification (2010). 15B33.

Keywords. Real quaternions, matrices over quaternions, invariant subspaces, robustness, stability.

1. Introduction

In the present paper we study certain classes of invariant subspaces of matrices that change little (in various senses) after small perturbations of the matrix. Such classes have been much studied in recent 35 years, for complex and real matrices, starting with the grounbreaking papers [3, 5]. The literature on the subject is extensive, and we mention here only chapters and parts in books [2, Chapter 8], [6, Chapter I.5], [7, Chapter 15], [8, Chapter II.4], [4, Part 4].

In [13], the results on stable and α-stable invariant subspaces for complex matrices have been extended to matrices over the real quaternions and quaternionic invariant subspaces. An A-invariant subspace is said to be stable if, loosely speaking, every matrix B close to A has an invariant subspace $\mathcal{N}$ close to $\mathcal{M}$; α-stability is characterized by the stronger property that the gap between $\mathcal{N}$ and $\mathcal{M}$ is a Hölder function of $\|B - A\|$ of exponent α. (See the precise definitions in Section 2.)

Here, we continue this line of investigation. We extend to quaternionic matrices the characterizations of strongly stable and strongly α-stable invariant subspaces of complex matrices obtained in [12]. Strong stability means that stability holds for every B-invariant subspace $\mathcal{N}$ (in the above notation), provided the obvious dimensionality conditions are satisfied, and similarly for strong α-stability. From the standpoint of basic backward error analysis, the notion of strong stability is perhaps more appropriate than the notion of stability as a measure of robustness of invariant subspaces of matrices.

Besides the introduction, the paper consists of five sections. In Section 2 we present some known results on matrix analysis for quaternionic matrices, for the readers' benefit. Our main Theorems 3.3 and 6.1 are stated in Sections 3 and 6 respectively. We prove Theorem 3.3 in Sections 4 and 5.

2. Quaternionic linear algebra

In this section we review briefly some known facts in linear algebra over the skew field of real quaternions H, and introduce notation to be used throughout the paper.

Denote by i, j, k the standard quaternion imaginary units. For $x \in \mathsf{H}$, $x = a_0 + a_1 i + a_2 j + a_3 k$ with $a_j \in \mathsf{R}$ (the real field), let $\mathfrak{R}(x) = a_0$ and $\mathfrak{V}(a) = a_1 i + a_2 j + a_3 k$ be the real and the vector part of x, respectively, $x^* = a_0 - a_1 i - a_2 j - a_3 k$ be the conjugate quaternion, and $|x| = \sqrt{a_0^2 + a_1^2 + a_2^2 + a_3^2}$ be the length of x.

Let $\mathsf{H}^{n \times m}$ (abbreviated to H^n if $m = 1$) be the set of $n \times m$ matrices with entries in H. The quaternionic conjugation naturally leads to the concept of conjugate transposed matrices; we denote by $A^* \in \mathsf{H}^{m \times n}$ the conjugate transpose of a matrix $A \in \mathsf{H}^{n \times m}$. H^n is considered as a right quaternionic vector space, endowed with the quaternion-valued inner product $\langle u, v \rangle = v^* u$, $u, v \in \mathsf{H}^n$ and norm $\|v\| = \sqrt{\langle v, v \rangle}$, $v \in \mathsf{H}^n$. Matrices $A \in \mathsf{H}^{n \times m}$ will be also considered as linear transformations $\mathsf{H}^m \to \mathsf{H}^n$ by way of matrix-vector multiplication $x \mapsto Ax$. We use the operator norm for matrices:

$$\|A\| = \max\{\|Au\| \ : \ u \in \mathsf{H}^n, \ \|u\| = 1\},$$

where $A \in \mathsf{H}^{m \times n}$. Note that $\|A\|$ coincides with the largest singular value of A.

Consider now the complex matrix representation of quaternions. For $x = a_0 + i a_1 + j a_2 + k a_3 \in \mathsf{H}$, $a_0, a_1, a_2, a_3 \in \mathsf{R}$, define

$$\omega(x) = \begin{bmatrix} a_0 + i a_1 & a_2 + i a_3 \\ -a_2 + i a_3 & a_0 - i a_1 \end{bmatrix} \in \mathsf{C}^{2 \times 2},$$

($\mathsf{C}^{m \times n}$ stands for the set of $m \times n$ complex matrices) and extend ω entrywise to a map

$$\omega_{m,n} : \mathsf{H}^{m \times n} \to \mathsf{C}^{2m \times 2n}, \quad \omega_{m,n}\left([x_{i,j}]_{i,j=1}^{m,n}\right) = [\omega(x_{i,j})]_{i,j=1}^{m,n}, \qquad x_{i,j} \in \mathsf{H}.$$

We have:

(i) $\omega_{n,n}$ is a unital homomorphism of real algebras;
(ii) if $X \in \mathsf{H}^{m \times n}$, $Y \in \mathsf{H}^{n \times p}$, then $\omega_{m,p}(XY) = \omega_{m,n}(X)\omega_{n,p}(Y)$;
(iii) $\omega_{n,m}(X^*) = (\omega_{m,n}(X))^*$, $\quad \forall \ X \in \mathsf{H}^{m \times n}$;

(iv) there exist positive constants $c_{m,n}$, $C_{m,n}$ such that

$$c_{m,n}\|\omega_{m,n}(X)\| \leq \|X\| \leq C_{m,n}\|\omega_{m,n}(X)\| \qquad (2.1)$$

for every $X \in \mathsf{H}^{m \times n}$.

Often, we will abbreviate $\omega_{m,n}$ to ω (with m, n understood from context).

Let $A \in \mathsf{H}^{n \times n}$. A vector $u \in \mathsf{H}^n \setminus \{0\}$ is said to be an *eigenvector* of A corresponding to the *eigenvalue* $\alpha \in \mathsf{H}$ if

$$Av = v\alpha. \qquad (2.2)$$

The set of all eigenvalues of A is denoted $\sigma(A)$, the *spectrum* of A. Note that $\sigma(A)$ is closed under similarity of quaternions: If (2.2) holds, then $A(v\lambda) = (v\lambda)(\lambda^{-1}\alpha\lambda)$, $\lambda \in \mathsf{H} \setminus \{0\}$, so $v\lambda$ is an eigenvector of A corresponding to the eigenvalue $\lambda^{-1}\alpha\lambda$. Note also that the similarity orbit $\{\lambda^{-1}\alpha\lambda : \lambda \in \mathsf{H} \setminus \{0\}\}$ of α consists exactly of those quaternions μ for which $\Re(\mu) = \Re(\alpha)$ and $|\mathfrak{V}(\mu)| = |\mathfrak{V}(\alpha)|$. In particular, if $\alpha \in \sigma(A)$, then also $\alpha^* \in \sigma(A)$.

The Jordan form is valid for quaternionic matrices:

Theorem 2.1. *Let $A \in \mathsf{H}^{n \times n}$. Then there exists an invertible $S \in \mathsf{H}^{n \times n}$ such that $S^{-1}AS$ has the form*

$$S^{-1}AS = J_{m_1}(\lambda_1) \oplus \cdots \oplus J_{m_p}(\lambda_p), \qquad \lambda_1, \ldots, \lambda_p \in \mathsf{H}, \qquad (2.3)$$

where $J_m(\lambda)$ is the $m \times m$ (upper triangular) Jordan block having eigenvalue λ. The form (2.3) is uniquely determined by A up to an arbitrary permutation of blocks and up to a replacement of $\lambda_1, \ldots, \lambda_p$ with $\alpha_1^{-1}\lambda_1\alpha_1, \ldots, \alpha_p^{-1}\lambda_p\alpha_p$, respectively, where $\alpha_j \in \mathsf{H} \setminus \{0\}$, $j = 1, 2, \ldots, p$.

A proof is given in [16]; the result goes back to [15].

We will need the following transformation properties of bases and Jordan forms under ω:

Proposition 2.2. *Let $u_1, \ldots, u_k$ be a basis (resp., orthogonal basis, orthonormal basis, spanning set) of a subspace $\mathcal{U} \subseteq \mathsf{H}^n$. Then:*

(1) *The columns of $\omega(u_1), \ldots, \omega(u_k)$ form a basis (resp., orthogonal basis, orthonormal basis) of the subspace $\omega(\mathcal{U}) \subseteq \mathsf{C}^{2n}$.*

(2) *The subspace $\omega(\mathcal{U})$ is independent of the choice of the basis (resp., orthogonal basis, orthonormal basis) of $\mathcal{U}$.*

Proof. Part (1) is [13, Proposition 2.1]. If $u_1, \ldots, u_k$ and $u'_1, \ldots, u'_k$ are two bases for $\mathcal{U}$, then $[\, u_1 \ \ldots \ u_k \,]S = [\, u'_1 \ \ldots \ u'_k \,]$ for some invertible matrix S. Applying ω to this equality, (2) follows. $\qquad \square$

Proposition 2.3. *Let $A \in \mathsf{H}^{n \times n}$, and let $\mathcal{N}$ of H^n be an A-invariant subspace. Denote by J the Jordan form of $A|_{\mathcal{N}}$ specialized so that $J \in \mathsf{C}^{n \times n}$. Then the Jordan form (over the complexes) of $\omega(A|_{\mathcal{N}})$ is $J \oplus \overline{J}$, where the overline stands for the complex conjugation.*

Proof. Let $u_1, \ldots, u_k$ be a Jordan basis for $A|_{\mathcal{N}}$ in $\mathcal{N}$, such that

$$A \begin{bmatrix} u_1 & u_2 & \ldots & u_k \end{bmatrix} = \begin{bmatrix} u_1 & u_2 & \ldots & u_k \end{bmatrix} J. \tag{2.4}$$

By Proposition 2.2, the columns of $[\omega(u_1),\ \omega(u_2),\ \ldots, \omega(u_k)]$ form a basis for $\omega(\mathcal{N})$ (cf. [13, Proposition 2.4]). It remains to apply the map ω to (2.4). □

The distance between two subspaces in H^n will be measured by the gap. Define the *gap* between two subspaces $\mathcal{M}$ and $\mathcal{N}$ of H^n by

$$\theta(\mathcal{M}, \mathcal{N}) = \|P_{\mathcal{M}} - P_{\mathcal{N}}\|,$$

where $P_{\mathcal{X}}$ is the orthogonal projection on a subspace $\mathcal{X}$. If $\{u_1, \ldots, u_p\}$ is an orthonormal basis for $\mathcal{X}$, then

$$P_{\mathcal{X}} = \begin{bmatrix} u_1 & u_2 & \ldots & u_p \end{bmatrix} \begin{bmatrix} u_1 & u_2 & \ldots & u_p \end{bmatrix}^*. \tag{2.5}$$

The gap is a metric on the set of all subspaces of H^n that turns the set into a compact complete metric space; this is well known in the context of complex subspaces, and can be proved for quaternion subspaces in the same way (see, e.g., the proof of [7, Theorem 13.4.1]). Many basic properties of subspaces in H^n as they relate to the gap metric and are familiar in the setting of subspaces of the real vector space R^n or of C^n, remain valid in the setting of quaternion subspaces, with essentially the same proofs, for example, [7, Theorems 13.1.1, 13.1.2, 13.1.3, 13.4.1, 13.4.2, 13.4.3]. Some are proved with complete details in [13, Theorem 2.11]. We will use these properties in the sequel as necessary, and present here only a few of them (Theorem 2.4 below; parts (1) and (3) are standard, (2) is proved in [13], and a short proof of (4) is supplied). We denote by $d(x, Z) = \inf_{t \in Z} \|x - t\|$ the *distance* from $x \in \mathsf{H}^n$ to a set $Z \subseteq \mathsf{H}^n$.

Theorem 2.4.
(1) *If $\theta(\mathcal{M}, \mathcal{N}) < 1$, then $\dim \mathcal{M} = \dim \mathcal{N}$. (The dimensions of subspaces in H^n are understood in the quaternionic sense.)*
(2) *Assume $\mathcal{M}', \mathcal{M}$ are subspaces of H^n such that the sum $\mathcal{M}' \dotplus \mathcal{M}$ is direct. Then there exists $\delta > 0$ (which depends on $\mathcal{M}$ and $\mathcal{M}'$ only) such that, if $\mathcal{N}, \mathcal{N}'$ are subspaces of H^n and*

$$\max\{\theta(\mathcal{M}, \mathcal{N}), \theta(\mathcal{M}', \mathcal{N}')\} \le \delta, \tag{2.6}$$

then the sum $\mathcal{N}' \dotplus \mathcal{N}$ is also direct.
*   Assume in addition that $\mathcal{M}' \dotplus \mathcal{M} = \mathsf{H}^n$. Then there exists $\delta_1 > 0$ such that if (2.6) holds (with δ replaced by δ_1) for subspaces $\mathcal{N}, \mathcal{N}'$ of H^n, then $\mathcal{N}' \dotplus \mathcal{N} = \mathsf{H}^n$ and*

$$\max\{\theta(\mathcal{M}, \mathcal{N}), \theta(\mathcal{M}', \mathcal{N}')\} \le \|\widetilde{P}_{\mathcal{M}, \mathcal{M}'} - \widetilde{P}_{\mathcal{N}, \mathcal{N}'}\| \tag{2.7}$$

$$\le \left(4(1 + 2\|\widetilde{P}_{\mathcal{M}, \mathcal{M}'}\|) \max_{x \in \mathcal{M}', \, \|x\|=1} \{d(x, \mathcal{M})^{-1}\} \right)$$

$$\times \left(\theta(\mathcal{M}, \mathcal{N}) + \theta(\mathcal{M}', \mathcal{N}') \right), \tag{2.8}$$

where the matrix $\widetilde{P}_{\mathcal{M}, \mathcal{M}'}$ projects H^n onto $\mathcal{M}$ along $\mathcal{M}'$, whereas $\widetilde{P}_{\mathcal{N}, \mathcal{N}'}$ projects H^n onto $\mathcal{N}$ along $\mathcal{N}'$.

(3) *For all subspaces $\mathcal{N}_1, \mathcal{N}_2 \subseteq \mathsf{H}^n$ and all invertible matrices $S \in \mathsf{H}^{n \times n}$, the inqualities*

$$\left(\|S\| \, \|S^{-1}\| \right)^{-1} \theta \left(\mathcal{N}_1, \mathcal{N}_2 \right) \leq \theta \left(S\mathcal{N}_1, S\mathcal{N}_2 \right) \leq \|S\| \, \|S^{-1}\| \theta \left(\mathcal{N}_1, \mathcal{N}_2 \right)$$

hold.

(4) *Let $\mathcal{Q}_1 \dotplus \mathcal{Q}_2 = \mathsf{H}^n$, a direct sum of subspaces. Then there exists a constant $C > 0$ that depends on $\mathcal{Q}_1, \mathcal{Q}_2$ only such that*

$$\theta(\mathcal{M}_1, \mathcal{M}_2) \leq C\theta(\mathcal{M}_1 \dotplus \mathcal{Q}_2, \mathcal{M}_2 \dotplus \mathcal{Q}_2)$$

for every pair of subspaces $\mathcal{M}_1, \mathcal{M}_2 \subseteq \mathcal{Q}_1$.

Proof of (4). Let $S \in \mathsf{H}^{n \times n}$ be an invertible matrix such that $S\mathcal{Q}_1$ and $S\mathcal{Q}_2$ are orthogonal. Then for subspaces $\mathcal{M}_1, \mathcal{M}_2 \subseteq \mathcal{Q}_1$ the equality

$$\theta(S\mathcal{M}_1, S\mathcal{M}_2) = \theta(S\mathcal{M}_1 \dotplus S\mathcal{Q}_2, S\mathcal{M}_2 \dotplus S\mathcal{Q}_2) \tag{2.9}$$

is obvious. Now (3) gives

$$\theta(\mathcal{M}_1, \mathcal{M}_2) \leq \|S\| \, \|S^{-1}\| \theta \left(S\mathcal{M}_1, S\mathcal{M}_2 \right) \leq \text{ by (2.9)}$$
$$\leq \left(\|S\| \, \|S^{-1}\| \right)^2 \theta \left(\mathcal{M}_1 \dotplus \mathcal{Q}_2, \mathcal{M}_2 \dotplus \mathcal{Q}_2 \right). \qquad \square$$

The complex representation ω keeps the gap between subspaces within universal bounds (for fixed n):

$$c_{n,n}\theta \left(\omega \left(\mathcal{U} \right), \omega \left(\mathcal{V} \right) \right) \leq \theta \left(\mathcal{U}, \mathcal{V} \right) \leq C_{n,n}\theta \left(\omega \left(\mathcal{U} \right), \omega \left(\mathcal{V} \right) \right) \tag{2.10}$$

for all subspaces $\mathcal{U}, \mathcal{V} \subseteq \mathsf{H}^n$, where the positive constants $c_{n,n}, C_{n,n}$ are taken from (2.1). Indeed, letting $u_1, \ldots, u_k$ and $v_1, \ldots, v_\ell$ be orthonormal bases for $\mathcal{U}$ and $\mathcal{V}$, respectively, we have (see Proposition 2.2)

$$\theta(\mathcal{U}, \mathcal{V}) = \| [u_1 \ \ldots \ u_k]^* [u_1 \ \ldots \ u_k] - [v_1 \ \ldots \ v_\ell]^* [v_1 \ \ldots \ v_\ell] \|,$$
$$\theta(\omega \left(\mathcal{V} \right), \omega \left(\mathcal{V} \right)) = \| [\omega \left(u_1 \right) \ \ldots \ \omega \left(u_k \right)]^* [\omega \left(u_1 \right) \ \ldots \ \omega \left(u_k \right)]$$
$$- [\omega \left(v_1 \right) \ \ldots \ \omega \left(v_\ell \right)]^* [\omega \left(v_1 \right) \ \ldots \ \omega \left(v_\ell \right)] \|,$$

and so (2.10) follows from (2.1).

The *rank* of $A \in \mathsf{H}^{m \times n}$ is, by definition, the dimension of the range of A, or equivalently, the dimension of the column space (understood as a right quaternion vector space) of A.

For $A \in \mathsf{H}^{m \times n}$, the *pseudoinverse*, or *Moore–Penrose inverse*, is defined as a matrix $A^+ \in \mathsf{H}^{n \times m}$ that is the unique solution of the following system of equations:

$$AA^+A = A, \quad A^+AA^+ = A^+, \quad (AA^+)^* = AA^+, \quad (A^+A)^* = A^+A.$$

Let

$$\mathcal{T}_{m,n,k} := \{ A \in \mathsf{H}^{m \times n} : \text{rank } A = k \}, \qquad k = 1, 2, \ldots, \min\{m, n\}$$

be sets of quaternion matrices of fixed rank.

Theorem 2.5. *The pseudoinverse is a (local) Lipschitz function on each of the sets* $\mathcal{T}_{m,n,k}$, *namely: Given* $A \in \mathsf{H}^{m \times n}$, *there exist positive constants* δ, K *depending on* A *only such that*

$$\|B^+ - A^+\| \leq K\|B - A\|$$

for all $B \in \mathcal{T}_{m,n,k}$ *with* $\|B - A\| \leq \delta$.

For a proof see [16, 14] for the complex matrices; it can be extended easily to quaternion matrices using the complex representation ω.

Formulas for orthogonal projections on intersections and sums of subspaces can be given in terms of the relevant pseudoinverses:

Proposition 2.6. *Let* $\mathcal{M}, \mathcal{N}$ *be subspaces in* H^n. *Then*

$$P_{\mathcal{M} \cap \mathcal{N}} = 2P_{\mathcal{N}}(P_{\mathcal{N}} + P_{\mathcal{M}})^+ P_{\mathcal{M}} \quad \text{and} \quad P_{\mathcal{M} + \mathcal{N}} = (P_{\mathcal{M}} + P_{\mathcal{N}})(P_{\mathcal{M}} + P_{\mathcal{N}})^+. \quad (2.11)$$

This result is proved in [9] for complex matrices; extension to quaternion matrices is immediate (same proofs apply). For the second formula in (2.11) see also [1].

Next, we discuss briefly root subspaces of quaternionic matrices. Let $A \in \mathsf{H}^{n \times n}$, and let $p^{(A)}(x)$ be the minimal polynomial with real coefficients and leading coefficient 1 for A; in other words, $p^{(A)}(x)$ is the monic real polynomial of minimal degree such that $p^{(A)}(A) = 0$. One easily verifies that the (real and complex) roots of $p^{(A)}(x)$ are exactly the eigenvalues of A that belong to C. Write

$$p^{(A)}(x) = p_1(x)^{m_1} \cdots p_k(x)^{m_k},$$

where the $p_j(x)$'s are distinct monic irreducible real polynomials (i.e., of the form $x - a$, a real, or of the form $x^2 + px + q$, $p, q \in \mathsf{R}$, with no real roots), and the m_j's are positive integers. The subspace

$$\mathcal{M}_j := \{u : u \in \mathsf{H}^n, \quad p_j(A)^{m_j}u = 0\}, \quad j = 1, 2, \ldots, k,$$

is called the *root subspace* of A corresponding to the roots of $p_j(x)$. Obviously, the root subspaces of A are A-invariant. We refer the reader to [13] for some elementary properties of the minimal polynomials and subspaces. In particular, $\mathcal{M} = \sum_{j=1}^{k}(\mathcal{M} \cap \mathcal{M}_j)$ for every A-invariant subspace $\mathcal{M}$. Also, root subspaces, and more generally, their sums, are Lipschitz functions of a matrix:

Proposition 2.7.

(a) *The roots of* $p^{(A)}(x)$ *depend continuously on* A: *Fix* $A \in \mathsf{H}^{n \times n}$, *and let* $\lambda_1, \ldots, \lambda_s$ *be all the distinct roots of* $p^{(A)}(x)$ *in the closed upper complex half-plane* C_+. *Then for every* $\epsilon > 0$ *there exists* $\delta > 0$ *such that if* $B \in \mathsf{H}^{n \times n}$ *satisfies* $\|B - A\| < \delta$, *then the roots of* $p^{(B)}(x)$ *in* C_+ *are contained in the union*

$$\cup_{j=1}^{s}\{z \in \mathsf{C}_+ : |z - \lambda_j| < \epsilon\}.$$

(b) *Sums of root subspaces of* A *are Lipschitz functions of* A: *Given* A *and* $\lambda_1, \ldots, \lambda_s$ *as in part (a), there exist* $\delta_0, K_0 > 0$ *such that for every* $B \in \mathsf{H}^{n \times n}$ *satisfying* $\|B - A\| < \delta_0$, *it holds that if* T *is any nonempty subset of* $\lambda_1, \ldots, \lambda_s$

and T' is the set of all eigenvalues of B contained in $\cup_{j\in T}\{z \in \mathbb{C}_+ : |z-\lambda_j| < \delta_0\}$, then the sum of root subspaces $\mathcal{M}'$ of B corresponding to $T'\cup\overline{T'}$ and the sum of root subspaces $\mathcal{M}$ of A corresponding to $T\cup\overline{T}$ satisfy the inequality

$$\theta(\mathcal{R},\mathcal{R}') \le K_0\,\|B-A\|. \tag{2.12}$$

For the reader's convenience, we quote the main result (Theorem 2.8 below) on stability of invariant subspaces from [13], where a complete and detailed proof is given.

Let $A \in \mathsf{H}^{n\times n}$. An A-invariant subspace $\mathcal{M} \subseteq \mathsf{H}^n$ is called *stable* if for every $\epsilon > 0$ there exists $\delta > 0$ such that every $B \in \mathsf{H}^{n\times n}$ satisfying $\|B-A\| < \delta$ has a B-invariant subspace $\mathcal{N} \subseteq \mathsf{H}^n$ with the property that $\theta(\mathcal{M},\mathcal{N}) < \epsilon$. For a fixed $\alpha \ge 1$, an A-invariant subspace $\mathcal{M}$ is called α-*stable*, if there exist $\delta_0, K_0 > 0$ such that for every $B \in \mathsf{H}^{n\times n}$ satisfying $\|B-A\| \le \delta_0$ there exists a B-invariant subspace $\mathcal{N}$ with the property that

$$\theta(\mathcal{M},\mathcal{N}) \le K_0\|B-A\|^{1/\alpha}. \tag{2.13}$$

Noting that $\theta(\mathcal{M},\mathcal{N}) \le 1$ for all subspaces $\mathcal{M},\mathcal{N} \in \mathsf{H}^n$, an equivalent definition of α-stable A-invariant subspace $\mathcal{M}$ is obtained by requiring that there exists $K_0' > 0$ such that for every $B \in \mathsf{H}^{n\times n}$ there is a B-invariant subspace $\mathcal{N}$ satisfying inequality

$$\theta(\mathcal{M},\mathcal{N}) \le K_0'\|B-A\|^{1/\alpha}. \tag{2.14}$$

Indeed, if (2.13) holds for all $B \in \mathsf{H}^{n\times n}$ with $\|B-A\| \le \delta_0$, then (2.14) holds with $K_0' = \max\{K_0, \delta_0^{-1/\alpha}\}$ for all $B \in \mathsf{H}^{n\times n}$ where in the case $\|B-A\| > \delta_0$ we take any B-invariant subspace for $\mathcal{N}$.

1-stable A-invariant subspaces are called *Lipschitz stable*.

For two positive integers $k < p$, define

$$s(k,p) = \begin{cases} p & \text{if no } k \text{ distinct } p\text{th roots of 1 sum up to zero;} \\ p-1 & \text{if there are } k \text{ distinct } p\text{th roots of 1 that sum up to zero.} \end{cases} \tag{2.15}$$

In the following theorem as well as in later statements, the following property of a matrix $A \in \mathsf{H}^{n\times n}$ and its invariant subspace $\mathcal{M}$ will be used:

($\aleph(A, \mathcal{M})$) the intersection of $\mathcal{M}$ with any root subspace $\mathcal{M}_j$ of A corresponding to a real eigenvalue satisfies

$$\dim(\mathcal{M}\cap\mathcal{M}_j) \le 1 \quad \text{or} \quad \dim(\mathcal{M}\cap\mathcal{M}_j) \ge \dim\mathcal{M}_j - 1. \tag{2.16}$$

A root subspace $\mathcal{M}_j$ of $A \in \mathsf{H}^{n\times n}$ is said to be of *geometric multiplicity one* if there is only one eigenvector (up to scaling) of A in $\mathcal{M}_j$.

Theorem 2.8. *Let $A \in \mathsf{H}^{n\times n}$, and let $\mathcal{M} \ne \{0\}$ be an A-invariant subspace. Then:*

(a) *$\mathcal{M}$ is Lipschitz stable if and only if $\mathcal{M}$ is a sum of root subspaces.*

(b) *$\mathcal{M}$ is stable if and only if for every root subspace $\mathcal{M}_j$ of A that contains at least two linearly independent eigenvectors of A, we have $\mathcal{M}\cap\mathcal{M}_j = \{0\}$ or $\mathcal{M}_j \subseteq \mathcal{M}$.*

(c) *if $\mathcal{M}$ is α-stable, then $\mathcal{M}$ is stable and for every root subspace $\mathcal{M}_j$ of A of geometric multiplicity one and such that $\{0\} \neq \mathcal{M} \cap \mathcal{M}_j \neq \mathcal{M}_j$, we have $\alpha \geq s(\dim (\mathcal{M} \cap \mathcal{M}_j), \dim \mathcal{M}_j)$.*

Conversely, assume $(\aleph(A, \mathcal{M}))$ holds. Then, if $\mathcal{M}$ is stable, and if for every root subspace $\mathcal{M}_j$ of A of geometric multiplicity one and such that $\{0\} \neq \mathcal{M} \cap \mathcal{M}_j \neq \mathcal{M}_j$, we have $\alpha \geq s(\dim (\mathcal{M} \cap \mathcal{M}_j), \dim \mathcal{M}_j)$, then $\mathcal{M}$ is α-stable.

Remark 2.9. The result of Theorem 2.8 holds for complex matrices and complex invariant subspaces without the hypothesis $(\aleph(A, \mathcal{M}))$; see [3, 5, 12].

Open Problem 2.10. Is the converse statement in Theorem 2.8(c) valid under the following hypothesis $(\aleph_0(A, \mathcal{M}))$ which is weaker than $(\aleph(A, \mathcal{M}))$?

$(\aleph_0(A, \mathcal{M}))$ the intersection of $\mathcal{M}$ with any root subspace $\mathcal{M}_j$ of geometric multiplicity one of A corresponding to a real eigenvalue satisfies

$$\dim (\mathcal{M} \cap \mathcal{M}_j) \leq 1 \quad \text{or} \quad \dim (\mathcal{M} \cap \mathcal{M}_j) \geq \dim \mathcal{M}_j - 1.$$

Analogous question with respect to Theorem 3.3(a) and Theorem 6.1.

3. Main results

Let $A \in \mathsf{H}^{n \times n}$, and let let $\lambda_1, \ldots, \lambda_s$ be all the distinct roots of $p^{(A)}(x)$ in the closed upper complex half-plane C_+. Select $\epsilon_0 > 0$ so that the disks

$$D_{\epsilon_0}(\lambda_j) := \{z \in \mathsf{C} : |z - \lambda_j| < \epsilon_0\}, \qquad j = 1, 2, \ldots, s,$$

do not intersect. By Proposition 2.7, there exists $\delta_0 > 0$ such that if $B \in \mathsf{H}^{n \times n}$, $\|B - A\| < \delta_0$, then all roots of $p^{(B)}(x)$ in the closed upper half-plane are contained in $\cup_{j=1}^s D_{\epsilon_0}(\lambda_j)$.

Fix $\alpha \geq 1$. An A-invariant subspace $\mathcal{M} \subseteq \mathsf{H}^n$ is said to be *strongly α-stable* if there exist $\epsilon \leq \delta_0$, $K > 0$ (that depend on A and $\mathcal{M}$ only) with the following property: For every $B \in \mathsf{H}^{n \times n}$ such that $\|B - A\| < \epsilon$, and every B-invariant subspace $\mathcal{N}$ such that

$$\dim(\mathcal{N} \cap \mathcal{R}_j(B)) = \dim(\mathcal{M} \cap \mathcal{R}_j(A)), \quad j = 1, 2, \ldots, s, \qquad (3.1)$$

the inequality

$$\theta(\mathcal{N}, \mathcal{M}) \leq K\|B - A\|^{1/\alpha}$$

holds; here we denote by $\mathcal{R}_j(X)$ the sum of the root subspaces of the matrix X corresponding to the roots of its minimal polynomial in the set $D_\epsilon(\lambda_j) \cup \overline{D_\epsilon(\lambda_j)}$, $j = 1, 2, \ldots, s$. Note that B-invariant subspaces $\mathcal{N}$ such that (3.1) holds do exist; indeed, it follows from the condition $\|B - A\| < \epsilon \leq \delta_0$ and the properties of δ_0 (see the first paragraph of this section) that

$$\dim \mathcal{R}_j(B) = \dim \mathcal{R}_j(A), \quad j = 1, 2, \ldots, s;$$

then existence of such $\mathcal{N}$ becomes obvious.

When $\alpha = 1$, we say that strongly 1-stable A-invariant subspace is *strongly Lipschitz stable*. Finally, an A-invariant subspace $\mathcal{M} \subseteq \mathsf{H}^n$ is said to be *strongly stable* if for every $\epsilon > 0$ there exists $\delta > 0$ (which can be taken $\leq \delta_0$) such that for every $B \in \mathsf{H}^{n \times n}$ satisfying $\|B - A\| < \delta$ and every B-invariant subspace $\mathcal{N}$ satisfying (3.1) the inequality $\theta(\mathcal{M}, \mathcal{N}) < \epsilon$ holds. Again, B-invariant subspaces $\mathcal{N}$ satisfying (3.1) do exist (if δ is taken sufficiently small).

Clearly, strong α-stability (for any $\alpha \geq 1$) implies strong stability; the converse is generally false (see Theorem 3.3 below). The concepts of strong α-stability and strong stability for invariant subspaces of complex matrices were introduced and studied in [12] and for a particular situation in [10].

Proposition 3.1. *If an A-invariant subspace is strongly α-stable, then it is also α-stable. If an A-invariant subspace is strongly stable, then it is stable.*

Proof. The result follows from the already mentioned fact that the set of B-invariant subspaces $\mathcal{N}$ for which (3.1) holds is nonempty provided B is sufficiently close to A (cf. Proposition 2.7(b) and Theorem 2.4(1)). $\qquad\square$

The following characterization of strong α-stability is proved in [12] for complex matrices. The definition of strongly α-stable invariant subspaces in the context of complex matrices is analogous to the above definition, with all considerations restricted to complex matrices and invariant subspaces in C^n.

Theorem 3.2. *Let $A \in \mathsf{C}^{n \times n}$. An A-invariant subspace $\mathcal{M} \subseteq \mathsf{C}^n$ is strongly α-stable in the context of complex matrices if and only if for every (complex) eigenvalue λ of A that has at least two linearly independent associated (complex) eigenvectors, we have*

$$\mathcal{M} \cap \{v \in \mathsf{C}^n : (A - \lambda I)^n v = 0\} = \{0\}$$

or

$$\mathcal{M} \supseteq \{v \in \mathsf{C}^n : (A - \lambda I)^n v = 0\},$$

and for every (complex) eigenvalue λ of A that has a unique up to scaling (complex) eigenvector and such that

$$\{0\} \neq \mathcal{M} \cap \{v \in \mathsf{C}^n : (A - \lambda I)^n v = 0\} \neq \{v \in \mathsf{C}^n : (A - \lambda I)^n v = 0\}$$

we have

$$\alpha \geq \dim \{v \in \mathsf{C}^n : (A - \lambda I)^n v = 0\}.$$

We prove analogous result in the context of quaternion matrices, imposing the additional condition $(\aleph(A, \mathcal{M}))$, as necessary:

Theorem 3.3. *Let $A \in \mathsf{H}^{n \times n}$, and let $\mathcal{M}$ be an A-invariant subspace. Then:*

(a) *if $\mathcal{M}$ is strongly α-stable, then $\mathcal{M}$ is strongly stable and for every root subspace $\mathcal{M}_j$ of A of geometric multiplicity one and such that $\{0\} \neq \mathcal{M} \cap \mathcal{M}_j \neq \mathcal{M}_j$, we have $\alpha \geq \dim \mathcal{M}_j$.*

Conversely, assume property $(\aleph(A, \mathcal{M}))$ *holds. Then, if* $\mathcal{M}$ *is strongly stable, and for every root subspace* $\mathcal{M}_j$ *of* A *of geometric multiplicity one and such that* $\{0\} \neq \mathcal{M} \cap \mathcal{M}_j \neq \mathcal{M}_j$, *we have* $\alpha \geq \dim \mathcal{M}_j$, *then* $\mathcal{M}$ *is strongly* α-*stable.*

(b) $\mathcal{M}$ *is strongly Lipschitz stable if and only if* $\mathcal{M}$ *is a sum of root subspaces of* A.
(c) $\mathcal{M}$ *is strongly stable if and only if* $\mathcal{M}$ *is stable.*

Remark 3.4. We do not know whether or not the property $(\aleph(A, \mathcal{M}))$ is essential in the converse part of Theorem 3.3(a).

The proof of Theorem 3.3 will be given in the next two sections.

4. Preliminaries for the proof of Theorem 3.3

The following fact is the key; it allows us to reduce the proof to the case of just one root subspace.

Fact 4.1. *Let* $A \in \mathsf{H}^{n \times n}$, *and let* $\mathcal{M}$ *be an* A-*invariant subspace. Fix* $\alpha \geq 1$. *Then* $\mathcal{M}$ *is strongly stable or strongly* α-*stable if and only if for every sum of root subspaces* $\mathcal{R} \subseteq \mathsf{H}^n$ *for* A *the intersection* $\mathcal{M} \cap \mathcal{R}$ *is strongly stable or strongly* α-*stable, respectively, as an* $A|_{\mathcal{R}}$-*invariant subspace.*

We provide details of the proof for the case of strong α-stability only (the case of strong stability can be proved analogously). The proof will be accomplished by proving Steps 1, 2, 3, 4 below (often parallel to the proof of [13, Fact 4.2]).

Let $\lambda_1, \ldots, \lambda_s$ be all the distinct roots of $p^{(A)}(x)$ in C_+, the closed upper complex half-plane. By Proposition 2.7, there exists $\delta_0 > 0$ such that if $B \in \mathsf{H}^{n \times n}$, $\|B - A\| < \delta_0$, then all roots of $p^{(B)}(x)$ in the closed upper half-plane are contained in $\cup_{j=1}^s D_{\epsilon_0}(\lambda_j)$, where $\epsilon_0 > 0$ is selected so that the disks $D_{\epsilon_0}(\lambda_j)$, $j = 1, 2, \ldots, s$, do not intersect.

Assuming $X \in \mathsf{H}^{n \times n}$ satisfies $\|X - A\| < \delta_0$, we let $\mathcal{R}_j(X)$ be the sum of root subspaces of X corresponding to the roots of the minimal polynomial of X in $D_{\epsilon_0}(\lambda_j) \cup \overline{D_{\epsilon_0}(\lambda_j)}$, $j = 1, 2, \ldots, s$.

Fix $\mathcal{R}$, a sum of root subspaces for A, and let $\mathcal{R}_1, \ldots, \mathcal{R}_r$ be all the root subspaces of A contained in $\mathcal{R}$. We denote by $K_0, K_1, \ldots$ positive constants that depend on A and $\mathcal{N}$ only.

Step 1. *If an* A-*invariant subspace* $\mathcal{N} \subseteq \mathcal{R}$ *is strongly* α-*stable as a* $A|_{\mathcal{R}}$-*invariant subspace, then* $\mathcal{N}$ *is also strongly* α-*stable as an* A-*invariant subspace.*

Proof of Step 1. Suppose not. Then there exists a sequence $\{S_m\}_{m=1}^\infty$, $S_m \in \mathsf{H}^{n \times n}$, such that $\|S_m - A\| < m^{-1}$, $m = 1, 2, \ldots$, and for some S_m-invariant subspace $\mathcal{N}_m$ such that

$$\dim(\mathcal{N}_m \cap \mathcal{R}_j(S_m)) = \dim(\mathcal{N} \cap \mathcal{R}_j(A)), \quad j = 1, 2, \ldots, s, \qquad (4.1)$$

we have

$$\theta(\mathcal{N}_m, \mathcal{N}) \geq m \|S_m - A\|^{1/\alpha}, \qquad m = 1, 2, \ldots. \qquad (4.2)$$

Denote

$$\widetilde{\mathcal{R}}_m = \sum_{j=1}^{r} \mathcal{R}_j(S_m), \quad m = 1, 2, \dots .$$

For sufficiently large m, the subspace $\widetilde{\mathcal{R}}_m$ is a direct complement of $\mathcal{R}^\perp$, the orthogonal complement of $\mathcal{R}$ in H^n ((2.12) and Theorem 2.4(2)). For such m, we define the linear transformation $U_m : \mathsf{H}^n \to \mathsf{H}^n$ by

$$U_m = \begin{bmatrix} I & (-P_{\widetilde{\mathcal{R}}_m, \mathcal{R}^\perp} + P_\mathcal{R})|_\mathcal{R} \\ 0 & I \end{bmatrix},$$

with respect to the orthogonal decomposition $\mathsf{H}^n = \mathcal{R}^\perp \oplus \mathcal{R}$, where $P_{\widetilde{\mathcal{R}}_m, \mathcal{R}^\perp}$ is the projection on $\widetilde{\mathcal{R}}_m$ along $\mathcal{R}^\perp$. $((-P_{\widetilde{\mathcal{R}}_m, \mathcal{R}^\perp} + P_\mathcal{R})|_\mathcal{R}$ is known as the *angular operator*, cf. [2].) Clearly, $U_m \mathcal{R}^\perp = \mathcal{R}^\perp$. Also, $U_m \widetilde{\mathcal{R}}_m = \mathcal{R}$. Now

$$\|U_m - I\| = \|(-P_{\widetilde{\mathcal{R}}_m, \mathcal{R}^\perp} + P_\mathcal{R})|_\mathcal{R}\| \le \|(-P_{\widetilde{\mathcal{R}}_m, \mathcal{R}^\perp} + P_\mathcal{R})\|$$

$$\text{(by Theorem} 2.4(2)) \le K_0 \theta\,(\widetilde{\mathcal{R}}_m, \mathcal{R}) \tag{4.3}$$

$$\text{(by Proposition 2.7(b))} \le K_1 \|S_m - A\|.$$

Next, let $T_m = U_m S_m U_m^{-1}$; then $U_m \mathcal{N}_m$ and $\mathcal{R}$ are T_m-invariant. It is easy to see (in view of (4.3)) that

$$\|T_m - A\| \le K_2 \|S_m - A\|. \tag{4.4}$$

Now

$$\theta(U_m \mathcal{N}_m, \mathcal{N}_m) = \max\left\{ \sup_{x \in \mathcal{N}_m, \|x\|=1} d(x, U_m \mathcal{N}_m), \sup_{y \in U_m \mathcal{N}_m, \|y\|=1} d(y, \mathcal{N}_m) \right\}$$

$$\le \max\{\|U_m - I\|, \|U_m^{-1} - I\|\} \le K_3 \|S_m - A\|,$$

and

$$\theta(U_m \mathcal{N}_m, \mathcal{N}) \ge \theta(\mathcal{N}_m, \mathcal{N}) - \theta(U_m \mathcal{N}_m, \mathcal{N}_m)$$

$$\ge m\|S_m - A\|^{1/\alpha} - K_3\|S_m - A\|$$

$$\ge (m - K_3)\|S_m - A\|^{1/\alpha} \tag{4.5}$$

$$\ge (m - K_3)K_2^{-1/\alpha}\|T_m - A\|^{1/\alpha}.$$

Restricting j so that $\mathcal{R}_j \subseteq \mathcal{R}$, in view of (4.1) and (4.5), a contradiction with strong α-stability of $\mathcal{N}$ as an $A|_\mathcal{R}$-invariant subspace is obtained. □

Step 2. *If an A-invariant subspace $\mathcal{N} \subseteq \mathcal{R}$ is strongly α-stable as an A-invariant subspace, then $\mathcal{N}$ is also strongly α-stable as an $A|_\mathcal{R}$-invariant subspace.*

Proof of Step 2. Suppose not. Then there exists a sequence $\{S_{m,\mathcal{R}}\}_{m=1}^{\infty}$, $S_{m,\mathcal{R}}$ a linear transformation on $\mathcal{R}$, and there exists an $S_{m,\mathcal{R}}$-invariant subspace $\mathcal{N}_m$, with the following properties:

(1) $\|S_{m,\mathcal{R}} - A|_\mathcal{R}\| < m^{-1}$, $m = 1, 2, \dots$;
(2) $\dim\,(\mathcal{N}_m \cap \mathcal{R}_j(S_{m,\mathcal{R}})) = \dim\,(\mathcal{N} \cap \mathcal{R}_j(A))$ for $j = 1, 2, \dots, r$;
(3) $\theta(\mathcal{N}_m, \mathcal{N}) \ge m\|S_{m,\mathcal{R}} - A|_\mathcal{R}\|^{1/\alpha}$ for $m = 1, 2, \dots .$

As in the proof of Step 2 of [13, Fact 4.2], let

$$S_m = \begin{bmatrix} A|_{\mathcal{R}^c} & 0 \\ 0 & S_{m,\mathcal{R}} \end{bmatrix}, \qquad m = 1, 2, \dots,$$

with respect to the decomposition $\mathsf{H}^n = \mathcal{R}^c \dotplus \mathcal{R}$, where $\mathcal{R}^c$ is the sum of root subspaces of A not contained in $\mathcal{R}$. Then $\|S_m - A\| = K_4 \|S_{m,\mathcal{R}} - A|_{\mathcal{R}}\|$, and $S_m \to A$ as $m \to \infty$. By the strong α-stability of $\mathcal{N}$, we have, for sufficiently large m, and every S_m-invariant subspace $\mathcal{N}'_m$:

$$\dim\left(\mathcal{N}'_m \cap \mathcal{R}_j(S_m)\right) = \dim\left(\mathcal{N} \cap \mathcal{R}_j(A)\right), \quad j = 1, 2, \dots, s$$

$$\implies \theta(\mathcal{N}'_m, \mathcal{N}) \le K_5 \|S_m - A\|^{1/\alpha} \qquad (4.6)$$

$$\le K_5 K_4^{1/\alpha} \|S_{m,\mathcal{R}} - A|_{\mathcal{R}}\|^{1/\alpha}.$$

Applying (4.6) with $\mathcal{N}'_m = \mathcal{N}_m$, we obtain a contradiction with item (3). □

Step 3. *If an A-invariant subspace $\mathcal{N}$ is strongly α-stable, then $\mathcal{N} \cap \mathcal{R}$ is strongly α-stable as an $A|_{\mathcal{R}}$-invariant subspace.*

Proof of Step 3. Let $S : \mathcal{R} \to \mathcal{R}$ be a linear transformation sufficiently close to $A|_{\mathcal{R}}$, and let $\mathcal{N}'$ be an S-invariant subspace such that

$$\dim\left(\mathcal{N}' \cap \mathcal{R}_j(S)\right) = \dim\left((\mathcal{N} \cap \mathcal{R}) \cap \mathcal{R}_j(A)\right), \quad j = 1, 2, \dots, r. \qquad (4.7)$$

Define $\widetilde{S}x = Sx$ if $x \in \mathcal{R}$, and $\widetilde{S}x = Ax$ if $x \in \mathcal{R}^c$. By the strong α-stability of $\mathcal{N}$ we have

$$\theta(\mathcal{N}, \mathcal{N}' \dotplus (\mathcal{N} \cap \mathcal{R}^c)) \le K_6 \|\widetilde{S} - A\|^{1/\alpha}.$$

It is easy to see that

$$\|\widetilde{S} - A\| \le K_7 \|S - A|_{\mathcal{R}}\|.$$

Now

$$\theta(\mathcal{N} \cap \mathcal{R}, \mathcal{N}') \le \text{ by Theorem 2.4(4), with } \mathcal{Q}_1 = \mathcal{R}, \mathcal{Q}_2 = \mathcal{N} \cap \mathcal{R}^c,$$

$$\le K_8 \, \theta(\mathcal{N}, \mathcal{N}' \dotplus (\mathcal{N} \cap \mathcal{R}^c)) \le K_8 K_6 K_7^{1/\alpha} \|S - A_{\mathcal{R}}\|^{1/\alpha},$$

and strong α-stability of $\mathcal{N} \cap \mathcal{R}$ follows. □

Step 4. *Let $\mathcal{R}_1, \dots, \mathcal{R}_s$ be all the distinct root subspaces of A, and assume that an A-invariant subspace $\mathcal{N}$ is such that every intersection $\mathcal{N} \cap \mathcal{R}_j$ is strongly α-stable as an $A|_{\mathcal{R}_j}$-invariant subspace, $j = 1, 2, \dots, s$. Then $\mathcal{N}$ is strongly α-stable as an A-invariant subspace.*

Proof of Step 4. By Step 1, $\mathcal{N} \cap \mathcal{R}_j$ is strongly α-stable as an A-invariant subspace, for $j = 1, 2, \dots, s$. Now, repeat the arguments in the proof of [13, Step 4 of Fact 4.2], using [13, Theorem 2.12]. □

5. Proof of Theorem 3.3

Part (b) of Theorem 3.3 follows easily from (a). Indeed, if $\mathcal{M}$ is strongly Lipschitz stable, then by part (a) we have that $\mathcal{M}$ is a sum of root subspaces. Conversely, assume $\mathcal{M}$ is a sum of root subspaces. By Fact 4.1 it follows that $\mathcal{M}$ is strongly Lipschitz stable.

Consider now part (a). In view of Fact 4.1 we may (and do) assume that H^n is a root subspace for A. If H^n contains two linearly independent eigenvectors of A, then by Theorem 2.8 a nontrivial (i.e., not equal to $\{0\}$ or to H^n) A-invariant subspace cannot be stable, hence it cannot be strongly stable by Proposition 3.1, and we are done in this case: There are no nontrivial strongly α-stable A-invariant subspaces.

Thus, suppose that A has only one eigenvector (up to scaling). Using Theorem 2.1, we may assume without loss of generality that $A = J_n(\lambda)$, where $\lambda \in \mathsf{C}$ has nonnegative imaginary part. Then part (a) amounts to the following two statements; here and in the sequel we denote by e_p the vector having 1 in the pth position and zeros elsewhere:

Statement 5.1. *Assume $(\aleph(A, \mathcal{M}))$ holds. Then there exist $\delta, K > 0$ such that for every $B \in \mathsf{H}^{n\times n}$ with $\|B - A\| < \delta$ and every k-dimensional B-invariant subspace $\mathcal{N} \subseteq \mathsf{H}^n$ we have*

$$\theta(\mathcal{N}, \operatorname{Span}\{e_1, \ldots, e_k\}) \leq K\|B - A\|^{1/n}. \tag{5.1}$$

Statement 5.2. *For every $k = 1, 2, \ldots, n-1$, and for every $\alpha < n$, there exists a sequence $\{B_m\}_{m=1}^\infty$, $B_m \in \mathsf{H}^{n\times n}$, and there exists a k-dimensional B_m-invariant subspace $\mathcal{M}_m \subseteq \mathsf{H}^n$, $m = 1, 2, \ldots$, such that $\lim_{m\to\infty} B_m = A$ and*

$$\theta(\mathcal{M}_m, \operatorname{Span}\{e_1, \ldots, e_k\}) \geq m\|B - A\|^{1/\alpha}.$$

Proof of Statement 5.1. Consider first the case when λ is nonreal. We use the complex representation ω of quaternion matrices; note that

$$P\omega(A)P^{-1} = \begin{bmatrix} J_n(\lambda) & 0 \\ 0 & J_n(\overline{\lambda}) \end{bmatrix} \in \mathsf{C}^{2n\times 2n},$$

for a suitable permutation matrix P.

By Theorem 3.2 the following claim holds true: *there exist $\delta', K' > 0$ such that for every $B' \in \mathsf{C}^{2n\times 2n}$ with $\|B' - P\omega(A)P^{-1}\| < \delta'$ and every $2k$-dimensional B'-invariant subspace $\mathcal{M}' \subseteq \mathsf{C}^{2n}$ we have*

$$\theta(\mathcal{M}', \operatorname{Span}\{e_1, \ldots, e_\ell, e_{n+1}, \ldots e_{n+2k-\ell}\}) \leq K'\|B' - P\omega(A)P^{-1}\|^{1/n} \tag{5.2}$$

for some ℓ, $\max\{2k-n, 0\} \leq \ell \leq \min\{n, 2k\}$, which may depend on $\mathcal{M}'$. Note the fact that, for $B \in \mathsf{H}^{n\times n}$ and its invariant subspace $\mathcal{N}$, the subspace $\omega(\mathcal{N}) \subseteq \mathsf{C}^{2n}$ (see Proposition 2.2 for the definition) is $\omega(B)$-invariant. We apply the claim to matrices B' of the form $B' = P\omega(B)P^{-1}$ for some $B \in \mathsf{H}^{n\times n}$ and their invariant subspaces of the form

$$\mathcal{M}' = P\left(\operatorname{Col}\left(\omega(v_1), \ldots, \omega(v_k)\right)\right),$$

where $\mathrm{Col}\,(\omega(v_1),\ldots,\omega(v_k))$ is subspace spanned by the columns of $\{\omega(v_1),\ldots,\omega(v_k)\}$, and $\{v_1,\ldots,v_k\}$ is a basis of a B-invariant subspace $f\mathcal{N}$. In view of Proposition 2.3, the Jordan form of such matrices B' restricted to $\mathcal{M}'$ is symmetric with respect to the real axis, so in fact it must be $\ell = k$ in (5.2). We now have, for a B-invariant k-dimensional subspace $\mathcal{N} \subseteq \mathsf{H}^n$:

$$
\begin{aligned}
\theta(\mathcal{N}, &\mathrm{Span}\,\{e_1,\ldots,e_k\}) \quad \text{(by Proposition 2.2 and 2.1)} \\
&\leq C_{n,n}\theta(\omega\,(\mathcal{N}),\omega\,(\mathrm{Span}\,\{e_1,\ldots,e_k\})) \\
&= C_{n,n}\theta(P\,(\omega\,(\mathcal{N})),P\,(\omega\,(\mathrm{Span}\,\{e_1,\ldots,e_k\}))) \\
&\leq \text{(by (5.2))}\ \ C_{n,n}K'\|\omega\,(B) - \omega\,(A)\|^{1/n} \\
&\leq \text{(by (2.1))}\ \ c_{n,n}^{-1}C_{n,n}K'\|B - A\|^{1/n},
\end{aligned}
$$

and Statement 5.1 follows for the case of a nonreal λ. Consider the case when λ is real; then the result follows from Theorem 2.8. $\qquad\square$

Proof of Statement 5.2. In view of Theorem 3.2, there are sequences of matrices $\{B_m\}_{m=1}^{\infty}$, $B_m \in \mathsf{C}^{n\times n}$ and of complex subspaces $\{\mathcal{M}_m\}_{m=1}^{\infty}$, $\mathcal{M}_m \subseteq \mathsf{C}^n$ with the required properties. $\qquad\square$

Finally, consider (c). Assume that $\mathcal{M}$ is stable. We need to prove that $\mathcal{M}$ is strongly stable. By Fact 4.1, we may assume that H^n is the (sole) root subspace of A. Ignoring the trivial cases $\mathcal{M} = \{0\}$ and $\mathcal{M} = \mathsf{H}^n$, in view of description of nontrivial stable A-invariant subspaces (Theorem 2.8(b)) we may further assume that $A = J(\lambda)$ for some $\lambda \in \mathsf{C}$. Arguing by contradiction, assume $\mathcal{M}$ is not strongly stable. Letting $k = \dim \mathcal{M}$, there exists $\epsilon_0 > 0$ such that for some sequence $\{B_m\}_{m=1}^{\infty}$, $B_m \in \mathsf{H}^{n\times n}$, we have $\|B_m - A\| < m^{-1}$ and $\theta(\mathcal{M}_m, \mathcal{M}) \geq \epsilon_0$ for some B_m-invariant subspace $\mathcal{M}_m$ of dimension k. Passing to a subsequence, we may assume that $\{\mathcal{M}_m\}$ converges in the gap norm: $\lim_{m\to\infty}\mathcal{M}_m = \mathcal{N}$ for some subspace $\mathcal{N}$. We then have $\theta(\mathcal{M},\mathcal{N}) \geq \epsilon_0$ and $\dim \mathcal{N} = k$.

Moreover, $\mathcal{N}$ is A-invariant. Indeed, let $u_1,\ldots,u_k$ be a basis for $\mathcal{N}$. Then there exist $u_{m,1},\ldots,u_{m,k}$ such that $u_{m,\ell} \in \mathcal{M}_m$, $m = 1,2,\ldots$, and

$$
\lim_{m\to\infty} u_{m,\ell} = u_{\ell}, \qquad \ell = 1,2,\ldots,k.
$$

Clearly, for sufficiently large m (which will be assumed), the vectors $u_{m,1},\ldots,u_{m,k}$ form a basis for $\mathcal{M}_m$. Since $\mathcal{M}_m$ is B_m-invariant, we have

$$
B_m\,[u_{m,1} \ \cdots \ u_{m,k}] = [u_{m,1} \ \cdots \ u_{m,k}]\,V_m \tag{5.3}
$$

for some matrix V_m. In fact,

$$
V_m = (Q_m^*Q_m)^{-1}B_mQ_m,
$$

where $Q_m = [u_{m,1} \ \cdots \ u_{m,k}]$, and the invertibility of $Q_m^*Q_m$ follows from the linear independence of the columns of Q_m. Passing to the limit when $m \to \infty$ in (5.3), the A-invariance of $\mathcal{N}$ follows.

But now we obtain a contradiction, because A has only one invariant subspace of the fixed dimension k (cf. [13, Proposition 2.10]).

6. Strong α-stability: alternative formulation

In this section we re-cast strong α-stability property of invariant subspaces in a different form that does not involve equalities (3.1), and is more in the spirit of the definition of α-stability. The next theorem provides the alternative formulation for the strong stability property.

Theorem 6.1. *Fix $\alpha \geq 1$. In the following statements, (1) implies (2) for $A \in \mathsf{H}^{n \times n}$ and an A-invariant subspace $\mathcal{M} \subseteq \mathsf{H}^n$:*

(1) *$\mathcal{M}$ is strongly α-stable;*
(2) *there are positive constants δ_1, δ_1', K_1 that depend on A and $\mathcal{M}$ only such that the set of all B-invariant subspaces $\mathcal{N}$ for which the inequality $\theta(\mathcal{N}, \mathcal{M}) \leq \delta_1'$ holds is non-empty for every $B \in \mathsf{H}^{n \times n}$ satisfying $\|B - A\| \leq \delta_1$, and*

$$\theta(\mathcal{N}, \mathcal{M}) \leq K_1 \|B - A\|^{1/\alpha} \tag{6.1}$$

holds for every such $\mathcal{N}$.

Assume in addition that the property $(\aleph(A, \mathcal{M}))$ is satisfied. Then the conditions (1) and (2) are equivalent.

We do not state a parallel version for strong stability, because by Theorem 3.3 strong stability is equivalent to stability, and therefore the definition of stability can be thought of as an alternative version of strong stability.

The rest of this section is devoted to the proof of Theorem 6.1. We need two lemmas.

Lemma 6.2. *Let $\mathcal{Q}_1, \mathcal{Q}_2, \mathcal{Z}_1, \mathcal{Z}_2$ be subspaces in H^n such that*

$$\mathcal{Q}_1 \dot{+} \mathcal{Q}_2 = \mathsf{H}^n, \quad \mathcal{Z}_1 \subseteq \mathcal{Q}_1, \quad \mathcal{Z}_2 \subseteq \mathcal{Q}_2.$$

Then there exists $\delta_3 > 0$ depending on $\mathcal{Q}_1, \mathcal{Q}_2, \mathcal{Z}_1, \mathcal{Z}_2$ only with the following property: If $\mathcal{Y}, \mathcal{Q}_1', \mathcal{Q}_2' \subseteq \mathsf{H}^n$ are subspaces such that

$$\theta(\mathcal{Y}, \mathcal{Z}_1 \dot{+} \mathcal{Z}_2) < \delta_3, \quad \theta(\mathcal{Q}_1', \mathcal{Q}_1) < \delta_3, \quad \theta(\mathcal{Q}_2', \mathcal{Q}_2) < \delta_3,$$

and

$$\mathcal{Y} = (\mathcal{Y} \cap \mathcal{Q}_1') + (\mathcal{Y} \cap \mathcal{Q}_2'), \tag{6.2}$$

then

$$\dim (\mathcal{Y} \cap \mathcal{Q}_j') = \dim \mathcal{Z}_j, \quad j = 1, 2.$$

Proof. By Theorem 2.4 (2) we have $\mathcal{Q}_1' \dot{+} \mathcal{Q}_2' = \mathsf{H}^n$ (if δ_3 is taken sufficiently small), hence $(\mathcal{Y} \cap \mathcal{Q}_1') \dot{+} (\mathcal{Y} \cap \mathcal{Q}_2')$ is a direct sum, and

$$\dim \mathcal{Z}_1 + \dim \mathcal{Z}_2 = \dim \mathcal{Y} = \dim (\mathcal{Y} \cap \mathcal{Q}_1') + \dim (\mathcal{Y} \cap \mathcal{Q}_2').$$

Thus, it suffices to prove that

$$\dim (\mathcal{Y} \cap \mathcal{Q}_j') \leq \dim \mathcal{Z}_j, \quad j = 1, 2.$$

We prove this for $j = 1$; the case $j = 2$ is analogous. Assume not. Then there exist sequences of subspaces $\{\mathcal{Y}_m, \mathcal{Q}'_{1,m}, \mathcal{Q}'_{2,m}\}_{m=1}^{\infty}$ such that

$$\mathcal{Y}_m \to \mathcal{Z}_1 \dotplus \mathcal{Z}_2, \qquad \mathcal{Q}'_{j,m} \to \mathcal{Q}_j, \quad j = 1, 2, \tag{6.3}$$

as $m \to \infty$,

$$\mathcal{Y}_m = (\mathcal{Y}_m \cap \mathcal{Q}'_{1,m}) + (\mathcal{Y}_m \cap \mathcal{Q}'_{2,m}), \qquad m = 1, 2, \ldots,$$

but

$$\dim(\mathcal{Y}_m \cap \mathcal{Q}'_{1,m}) > \dim \mathcal{Z}_1, \qquad m = 1, 2, \ldots. \tag{6.4}$$

Using compactness of the metric space of subspaces in H^n, we may assume that the sequence $\{\mathcal{Y}_m \cap \mathcal{Q}'_{1,m}\}_{m=1}^{\infty}$ converges to a subspace $\mathcal{W}$. Take $x \in \mathcal{W}$. Then there is a sequence $\{x_m\}_{m=1}^{\infty}$, $x_m \in \mathcal{Y}_m \cap \mathcal{Q}'_{1,m}$, such that $\lim_{m \to \infty} x_m = x$. By (6.3) we also have $x \in \mathcal{Q}_1$, $x \in \mathcal{Z}_1 \dotplus \mathcal{Z}_2$, hence (because $\mathcal{Z}_1 \subseteq \mathcal{Q}_1$) $x \in \mathcal{Z}_1$. Thus $\mathcal{W} \subseteq \mathcal{Z}_1$, a contradiction with our assumption (6.4) (note that $\dim(\mathcal{Y}_m \cap \mathcal{Q}'_{1,m}) = \dim \mathcal{W}$ for large m by Theorem 2.4(1)). $\qquad\square$

Lemma 6.3. *Given A and $\mathcal{M}$ as in Theorem 6.1, for every $\epsilon_4 > 0$ there exists $\delta_4 > 0$ with the property that for each $B \in \mathsf{H}^{n \times n}$ and for each B-invariant subspace $\mathcal{N}$ such that*

$$\|B - A\| < \delta_4, \qquad \theta(\mathcal{N}, \mathcal{M}) < \delta_4, \tag{6.5}$$

the inequality

$$\max_{j=1,2,\ldots,s} \{\theta(\mathcal{N} \cap \mathcal{R}_j(B), \mathcal{M} \cap \mathcal{R}_j(A))\} \leq \epsilon_4$$

holds, where the maximum is taken over all sums of root subspaces $\mathcal{R}_1(A)$, ..., $\mathcal{R}_s(A)$ for A, and where the sums of root subspaces $\mathcal{R}_1(B), \ldots, \mathcal{R}_s(B)$ for B are such that the eigenvalues of B to which $\mathcal{R}_j(B)$ corresponds are in a neighborhood of the eigenvalues of A to which $\mathcal{R}_j(A)$ corresponds, for $j = 1, 2, \ldots, s$.

Proof. Denote by $\mathcal{R}_j(A)^c$, resp. $\mathcal{R}_j(B)^c$, the sum of root subspaces for A, resp. B, which is a direct complement to $\mathcal{R}_j(A)$, resp. $\mathcal{R}_j(B)$, in H^n, and let $\widetilde{P}_{\mathcal{R}_j(A), \mathcal{R}_j(A)^c}$, resp. $\widetilde{P}_{\mathcal{R}_j(B), \mathcal{R}_j(B)^c}$, be the projection on $\mathcal{R}_j(A)$ along $\mathcal{R}_j(A)^c$, resp. on $\mathcal{R}_j(B)$ along $\mathcal{R}_j(B)^c$. In what follows, we denote by $K_0, K_1, \ldots$ positive constants that depend only on A and $\mathcal{M}$. By Theorem 2.4(2)

$$\begin{aligned} \left\| \widetilde{P}_{\mathcal{R}_j(B), \mathcal{R}_j(B)^c} - \widetilde{P}_{\mathcal{R}_j(A), \mathcal{R}_j(A)^c} \right\| & \\ \leq K_0(\theta(\mathcal{R}_j(B), \mathcal{R}_j(A)) + \theta(\mathcal{R}_j(B)^c, \mathcal{R}_j(A)^c)) & \\ \leq K_0 K_1 \|B - A\|, & \end{aligned} \tag{6.6}$$

where the second inequality follows from Proposition 2.7(b). Proposition 2.6 gives:

$$\begin{aligned} \theta(\mathcal{N} \cap \mathcal{R}_j(B), \mathcal{M} \cap \mathcal{R}_j(A)) & \\ = 2 \left\| P_{\mathcal{N}}(P_{\mathcal{N}} + P_{\mathcal{R}_j(B)})^+ P_{\mathcal{R}_j(B)} - P_{\mathcal{M}}(P_{\mathcal{M}} + P_{\mathcal{R}_j(A)})^+ P_{\mathcal{R}_j(A)} \right\|; \end{aligned} \tag{6.7}$$

here $\mathcal{N}$ is a B-invariant subspace. On the other hand, taking δ_4 sufficiently small and assuming (6.5) holds, Lemma 6.2 (applied with $\mathcal{Q}_1 = \mathcal{R}_j(A)$, $\mathcal{Q}_2 = \mathcal{R}_j(A)^c$,

$\mathcal{Z}_1 = \mathcal{M} \cap \mathcal{R}_j(A)$, $\mathcal{Z}_2 = \mathcal{M} \cap \mathcal{R}_j(A)^c$, $\mathcal{Q}'_1 = \mathcal{R}_j(B)$, $\mathcal{Q}'_2 = \mathcal{R}_j(B)^c$, $\mathcal{Y} = \mathcal{N}$), together with (6.6) and Theorem 2.4(1), yields that

$$\dim\left(\mathcal{N} \cap \mathcal{R}_j(B)\right) = \dim\left(\mathcal{M} \cap \mathcal{R}_j(A)\right), \qquad j = 1, 2, \ldots, s,$$

and hence

$$\dim\left(\mathcal{N} + \mathcal{R}_j(B)\right) = \dim\left(\mathcal{M} + \mathcal{R}_j(A)\right), \qquad j = 1, 2, \ldots, s.$$

Since the range of $P_{\mathcal{N}} + P_{\mathcal{R}_j(B)}$ is equal to $\mathcal{N} + \mathcal{R}_j(B)$ (see [9, Corollary 2], for example), and similarly for A, we now have

$$\operatorname{rank}\left(P_{\mathcal{N}} + P_{\mathcal{R}_j(B)}\right) = \operatorname{rank}\left(P_{\mathcal{M}} + P_{\mathcal{R}_j(A)}\right), \qquad j = 1, 2, \ldots, s.$$

Using Theorem 2.5, formula (6.7) gives the result of Lemma 6.3. □

Proof of Theorem 6.1, (1) $\Rightarrow$ (2). Assume (1) holds. Then by Lemma 6.3 (taking $\epsilon_4 < 1$ and using Theorem 2.4(1)) equalities (3.1) are guaranteed, and we have

$$\theta(\mathcal{N}, \mathcal{M}) \leq K_1 \|B - A\|^{1/\alpha}$$

for every B-invariant subspace $\mathcal{N}$ provided the inequality

$$\max\{\theta(\mathcal{N}, \mathcal{M}), \|B - A\|\} \leq \delta_4/2$$

holds. It remains to prove that for some δ_1, $0 < \delta_1 \leq \delta_4/2$, the set of all B-invariant subspaces $\mathcal{N}$ for which the inequality $\theta(\mathcal{N}, \mathcal{M}) \leq \delta_4/2$ holds is non-empty for every $B \in \mathsf{H}^{n \times n}$ satisfying $\|B - A\| \leq \delta_1$. To this end we take advantage of the fact that under (1) $\mathcal{M}$ is α-stable. Thus, there exist $\delta', K' > 0$ such that

$$\|B - A\| \leq \delta' \quad \Rightarrow \quad \exists \ B\text{-invariant } \mathcal{N} \text{ such that } \quad \theta(\mathcal{N}, \mathcal{M}) \leq K' \|B - A\|^{1/\alpha}.$$

Now take

$$\delta_1 = \min\{\delta', \delta_4/2, (\delta_4/(2K'))^\alpha\}. \qquad \square$$

Proof of Theorem 6.1 in the complex case. The implication (1) $\Rightarrow$ (2) is verified as in the quaternionic case. Assume now that (2) holds. Then $\mathcal{M}$ is in particular α-stable, and using the description of α-stability in the complex case (Theorem 2.8, Remark 2.9), we easily reduce the proof to the case when $A = J_n(0)$. Theorem 3.2 shows that $\mathcal{M}$ is strongly n-stable. Thus, if $\alpha \geq n$, we are done. However, if $\alpha < n$, then (6.1) cannot hold (unless $\mathcal{M}$ is trivial: $\mathcal{M} = \{0\}$ or $\mathcal{M} = \mathsf{H}^n$).

Indeed, let k, $1 \leq k \leq n - 1$, be the (complex) dimension of

$$\mathcal{M} := \operatorname{Range} \begin{bmatrix} I_k \\ 0_{n-k, k} \end{bmatrix},$$

and let $\tau_1, \ldots, \tau_k$ be a set of distinct nth roots of unity that do not sum up to zero. For $\eta > 0$, let $B_\eta \in \mathsf{C}^{n \times n}$ be the matrix obtained from $A = J_n(0)$ by adding η in the lower left corner. Clearly, $\|B_\eta - A\| = \eta$. Letting $\mathcal{N}_\eta$ to be the B_η-invariant subspace generated by the eigenvectors B_η corresponding to the eigenvalues $\tau_1 \eta^{1/n}, \ldots, \tau_k \eta^{1/n}$, one verifies that

$$\mathcal{N}_\eta = \operatorname{Range} \begin{bmatrix} I_k \\ X_\eta \end{bmatrix},$$

where $X_\eta \in \mathbb{C}^{(n-k) \times k}$ has $\tau_1 \eta^{1/n} + \cdots + \tau_k \eta^{1/n}$ in its top left corner (see the proof of [7, Lemma 16.5.2], also [13, Lemma 4.16]). Now [13, Lemma 5.2(a3)] guarantees existence of a constant $K_{0,1} > 0$ which depends on n only such that

$$\theta (\mathcal{M}, \mathcal{N}_\eta) \geq K_{0,1} |\tau_1 + \cdots + \tau_k| \eta^{1/n} \tag{6.8}$$

as long as $\eta \leq 1$. Letting $\eta \to 0$, a contradiction with (6.1) is obtained. $\square$

Proof of Theorem 6.1, (2) $\Rightarrow$ (1). The proof of (2) $\Rightarrow$ (1), under the additional hypothesis that $(\aleph(A, \mathcal{M}))$ holds, is essentially the same as in the complex case (use Theorem 3.3(a) instead of Theorem 3.2). $\square$

References

[1] W.N. Anderson, Jr., and R.J. Duffin. Series and parallel addition of matrices, *J. Math. Anal. Appl.*, **26** (1969), 576–594.

[2] H. Bart, I. Gohberg, and M.A. Kaashoek. Minimal factorization of matrix and operator functions. *Operator Theory: Advances and Applications*, Vol **1**, Birkhäuser, 1979.

[3] H. Bart, I. Gohberg, and M.A. Kaashoek. Stable factorizations of monic matrix polynomials and stable invariant subspaces, *Integral Equations Operator Theory*, **1** (1978), no. 4, 496–517.

[4] H. Bart, I. Gohberg, M.A. Kaashoek, and A.C.M. Ran. Factorization of matrix and operator functions: the state space method, *Operator Theory: Advances and Applications*, Vol **178**, Birkhäuser, 2008.

[5] S. Campbell and J. Daughtry. The stable solutions of quadratic matrix equations, *Proc. Amer. Math. Soc.*, **74** (1979), no. 1, 19–23.

[6] I. Gohberg, P. Lancaster, and L. Rodman. *Matrix Polynomials*, Academic Press, 1982; republication, SIAM, 2009.

[7] I. Gohberg, P. Lancaster, and L. Rodman. *Invariant Subspaces of Matrices with Applications*, John Wiley, 1986; republication, SIAM, 2006.

[8] I. Gohberg, P. Lancaster, and L. Rodman. Matrices and Indefinite Scalar Products, *Operator Theory: Advances and Applications*, Vol. **8**, Birkhäuser, Basel and Boston, 1983.

[9] R. Piziak, P.L. Odell, and R. Hahn. Constructing projections for sums and intersections, *Computers and Mathematics with Applications*, **17** (1999), 67–74.

[10] A.C. M. Ran and L. Rodman. Stability of invariant Lagrangian subspaces I, *Operator Theory: Advances and Applications*, **32** (1988), 181–218.

[11] A.C. M. Ran and L. Rodman. The rate of convergence of real invariant subspaces, *Linear Algebra Appl.*, **207** (1994), 197–224.

[12] A.C.M. Ran and L. Roozemond. On strong α-stability of invariant subspaces of matrices, *Operator Theory: Advances and Applications*, **40** (1989), 427–435.

[13] L. Rodman. Stability of invariant subspaces of quaternion matrices, *Complex Analysis and Operator Theory*, **6** (2012), 1069–1119.

[14] G.W. Stewart. On the continuity of the generalized inverse, *SIAM J. Appl. Math.*, **17** (1969), 33–45.

[15] N.A. Wiegmann. Some theorems on matrices with real quaternion entries, *Canadian J. of Math.*, **7** (1955), 191–201.

[16] F. Zhang. Quaternions and matrices of quaternions, *Linear Algebra Appl.*, **251** (1997), 21–57.

Leiba Rodman
Department of Mathematics
College of William and Mary
Williamsburg, VA 23187-8795, USA
e-mail: lxrodm@gmail.com

Operator Theory:
Advances and Applications, Vol. 237, 241–246
© 2013 Springer Basel

Determinantal Representations of Stable Polynomials

Hugo J. Woerdeman

Dedicated to Leonia Lerer on the occasion of his seventieth birthday

Abstract. For every stable multivariable polynomial p, with $p(0) = 1$, we construct a determinantal representation

$$p(z) = \det(I - M(z)),$$

where $M(z)$ is a matrix-valued rational function with $\|M(z)\| \le 1$ and $\|M(z)^n\| < 1$ for $z \in \mathbb{T}^d$ and $M(az) = aM(z)$ for all $a \in \mathbb{C} \setminus \{0\}$.

Mathematics Subject Classification (2010). 15A15; 11C20, 47A13, 47A48.

Keywords. Determinantal representation; multivariable polynomial; stable polynomial.

1. Introduction

A polynomial $p(z) = p(z_1, \ldots, z_d)$ is called *stable* if $p(z) \ne 0$ for $z \in \overline{\mathbb{D}}^d$, where $\mathbb{D}$ is the open unit disk in $\mathbb{C}$ and $\overline{\mathbb{D}}$ is its closure. We shall also use the notation $\mathbb{T}$ for the unit circle. The polynomial $p(z)$ is called *semi-stable* when $p(z) \ne 0$ for $z \in \mathbb{D}^d$. It is an open question whether every multivariable stable polynomial $p(z)$ with $p(0) = 1$ can be written as

$$p(z) = \det(I - KZ(z)),$$

where $Z(z)$ is a diagonal matrix with coordinate variables on the diagonal and K is a strict contraction. For one and two variable polynomials such a representation always exists; in one variable it is an easy consequence of the fundamental theorem of algebra, while in two variables the result follows from [8, Theorem 1] and [7]. The question of existence of representations of the form $\det(I - KZ(z))$ was the topic of the paper [6], where such representation was shown to lead to rational

The author was partially supported by NSF grant DMS-0901628.

inner functions in the Schur–Agler class. As a consequence of these results, it can be seen that for $5/6 < r < 1$, the stable polynomial

$$q(z_1, z_2, z_3) = 1 + \frac{r}{5} z_1 z_2 z_3 \left(z_1^2 z_2^2 + z_2^2 z_3^2 + z_3^2 z_1^2 - 2z_1 z_2 z_3^2 - 2z_1 z_2^2 z_3 - 2z_1^2 z_2 z_3 \right)$$

can not be represented as $\det(I - KZ(z))$ where K is a 9×9 (or smaller size) contraction. It is an open question whether such a representation exists with a larger size contraction K. Our main result shows, however, that we can find a representation $q(z) = \det(I - M(z))$, with $M(z)$ a rational matrix function satisfying $\|M(z)\| \leq 1$ and $\|M(z)^7\| < 1$, $z \in \mathbb{T}^3$, and $M(az) = aM(z)$. In fact, for this particular polynomial q, one may choose $M(z)$ to be the 7×7 rational matrix function

$$M(z) = \begin{pmatrix} 0 & z_1 & 0 & & & & \\ & 0 & z_1 & 0 & & & \\ & & 0 & z_1 & 0 & & \\ & & & 0 & z_1 & 0 & \\ & & & & 0 & z_1 & 0 \\ 0 & & & & & 0 & z_1 \\ \frac{q(z)-1}{z_1^6} & 0 & & & & & 0 \end{pmatrix}.$$

Our main result is the following determinantal characterization of stable polynomials. Recall that the *total degree* tdeg$p$ of a polynomial p is the maximum among the total degrees of all its terms, where the total degree of the monomial $z_1^{n_1} \cdots z_d^{n_d}$ is $n_1 + \cdots + n_d$.

Theorem 1.1. *Let p be a polynomial in d variables. Put $n = \text{tdeg}p$. Then p is stable with $p(0) = 1$ if and only if for some $k \in \mathbb{N}$ there exists a $k \times k$ rational matrix-valued function $M(z)$ which has no singularities on the set $\cup_{r>0} r\mathbb{T}^d$ so that*

(i) $p(z) = \det(I_k - M(z))$,
(ii) $M(z)$ *is contractive and* $M(z)^n$ *is strictly contractive for all* $z \in \mathbb{T}^d$,
(iii) $M(az) = aM(z)$ *for all* $a \in \mathbb{C} \setminus \{0\}$ *and* $z \in \mathbb{T}^d$.

In [13] real zero polynomials were studied, for which a desirable determinantal representation is $\det(I + x_1 A_1 + \cdots + x_k A_k)$ with $A_1, \ldots, A_k$ symmetric. Not all real zero polynomials have such a representation; see [12]. In [13, Theorem 3.1] it was shown, however, that every square-free real zero polynomial can be written as $\det(I - M(x))$, where $M(x)$ is a symmetric rational matrix function and $M(ax) = aM(x)$, $a \in \mathbb{R}$, $x \in \mathbb{R}^d$. Our Theorem 1.1 can be seen as an analog of [13, Theorem 3.1] to the setting of stable polynomials.

2. Proof of main result

We first need a couple of lemmas.

Lemma 2.1. *Let $Q(z)$ be a positive definite $n \times n$ matrix-valued trigonometric polynomial on $\mathbb{T}^d$, so that in the Laurent expansion the (i, j)th entry is homogeneous of degree $i - j$. Then there exists a factorization $Q(z) = P(z)P(z)^*$, with $P(z)$ a* *rational matrix function of size $n \times k$, say, so that in the Laurent expansion the ith row of the P is homogeneous of degree $i - 1$. The rational matrix function P may be chosen to be polynomial in at least $d - 1$ variables.*

Proof. By [3, Corollary 5.2] a polynomial matrix function R exists so that $Q = RR^*$. Write now $R = \sum_{j=0}^{N} R_j$, where R_j is homogeneous of degree j. Next, write $R_j = \mathrm{row}(p_{kj})_{k=1}^{n}$. Observe that

$$Q_{ij} = \left(\sum_{k=0}^{N} p_{ik}\right)\left(\sum_{r=0}^{N} p_{jr}^*\right) = \sum_{k-r=i-j} p_{ik}p_{jr}^* + \sum_{k-r\neq i-j} p_{ik}p_{jr}^*,$$

but as the last term equals 0 due to Q_{ij} being homogeneous of degree $i - j$, we actually have

$$Q_{ij} = \sum_{k-r=i-j} p_{ik}p_{jr}^*.$$

Define now

$$P = \begin{pmatrix} p_{1,N}z_1^{-N} & \cdots & & p_{1,0}z_1^0 & & 0 \\ 0 & p_{2,N}z_1^{-N+1} & & \cdots & & p_{2,0}z_1^1 \\ \ddots & & \ddots & & & \ddots \\ & 0 & & p_{n,N}z_1^{-N+n-1} & \cdots & p_{n,0}z_1^{n-1} \end{pmatrix}.$$

Then the ith row of P is homogeneous of degree $i - 1$ and $Q = PP^*$. □

Lemma 2.2. *Let $a(z) = a_0 + a_1 z + \cdots + a_n z^n$ with $a_0 = 1$ be a one variable stable polynomial, and let $Q := AA^* - B^*B$, where*

$$A = \begin{pmatrix} a_0 & & \\ \vdots & \ddots & \\ a_{n-1} & \cdots & a_0 \end{pmatrix}, \quad B = \begin{pmatrix} a_n & \cdots & a_1 \\ & \ddots & \vdots \\ & & a_n \end{pmatrix}.$$

In addition, let

$$C = \begin{pmatrix} -a_1 & 1 & & 0 \\ \vdots & & \ddots & \\ -a_{n-1} & 0 & & 1 \\ -a_n & 0 & \cdots & 0 \end{pmatrix}.$$

Then $Q > 0$,

$$\begin{pmatrix} Q^{-1} & C^*Q^{-1} \\ Q^{-1}C & Q^{-1} \end{pmatrix} \geq 0, \quad \begin{pmatrix} Q^{-1} & C^{n*}Q^{-1} \\ Q^{-1}C^n & Q^{-1} \end{pmatrix} > 0. \tag{2.1}$$

Proof. Let $f(z) = \frac{1}{|a(z)|^2}$, and write $f(z) = \sum_{j=-\infty}^{\infty} f_j z^j$, $|z| = 1$. Introduce

$$T = (f_{i-j})_{i,j=0}^{n-1}, \tilde{T} = (f_{i-j+1})_{i,j=0}^{n-1}, \hat{T} = (f_{i-j+n})_{i,j=0}^{n}.$$

By the Schur–Cohn criterion [9, Section 13.5] we have that $Q > 0$ and by the Gohberg–Semencul formula [5] we have that $Q^{-1} = T$. In addition, it is easy to check that $TC = \tilde{T}$. Next, observe that

$$\begin{pmatrix} Q^{-1} & C^*Q^{-1} \\ Q^{-1}C & Q^{-1} \end{pmatrix} = \begin{pmatrix} T & \tilde{T}^* \\ \tilde{T} & T \end{pmatrix},$$

has many identical rows and columns; indeed, columns k and $k + t - 1$ are equal for $k = 2, \ldots, n$, and by selfadjointness the same holds for the rows. Removing columns and rows $n + 1, \ldots, 2n - 1$, we remain with the positive definite matrix $(f_{i-j})_{i,j=0}^n$. But then the first inequality in (2.1) follows.

In addition, one may check that $TC^n = \hat{T}$ (this was observed in [4, Proof of Theorem 2.1]; see also [1, Equation (2.3.14)]). It remains to observe that

$$\begin{pmatrix} Q^{-1} & C^{n*}Q^{-1} \\ Q^{-1}C^n & Q^{-1} \end{pmatrix} = \begin{pmatrix} T & \hat{T}^* \\ \hat{T} & T \end{pmatrix} = (f_{i-j})_{i,j=0}^{2n-1} > 0. \qquad \square$$

Proof of Theorem 1.1. "If": Suppose that $M(z)$ as described exists, and that $p(z) = 0$ for some $z \in \mathbb{T}^d$. Then 1 is an eigenvalue of $M(z)$. But as $\|M(z)^n\| < 1$, this can not happen. Thus $p(z) \neq 0$ for $z \in \mathbb{T}^d$. Now using that $M(az) = aM(z)$, we get that $\|M(z)\| < 1$ for any $z \in r\mathbb{T}^d$ where $0 < r < 1$. Thus $p(z) \neq 0$ for any $z \in r\mathbb{T}^d$ where $0 < r \leq 1$. In addition, for $z \in \mathbb{T}^d$ one has

$$p(0) = \lim_{r \to 0+} p(rz) = \lim_{r \to 0+} \det(I - M(rz)) = \lim_{r \to 0+} \det(I - rM(z)) = 1.$$

But now the stability of p follows from Theorem 1' in [2].

"Only if": Let $p(z) = p(z_1, \ldots, z_d)$ be a stable multivariable polynomial with $p(0) = 1$. As $n = \text{tdeg}\,p$, we may write $p(z) = 1 + p_1(z) + \cdots + p_n(z)$, where p_j is a homogeneous multivariable polynomial of degree j; i.e., $p_j(az) = a^j p_j(z)$, where $a \in \mathbb{C}$. Introduce now,

$$C(z) = \begin{pmatrix} -p_1(z) & 1 & & 0 \\ \vdots & & \ddots & \\ -p_{n-1}(z) & 0 & & 1 \\ -p_n(z) & 0 & \cdots & 0 \end{pmatrix}.$$

Then $\det(I_n - C(z)) = p(z)$. In addition, if we let $D_a = \text{diag}(a^j)_{j=0}^{n-1}$, then

$$C(az) = aD_aC(z)D_a^{-1}, a \in \mathbb{C} \setminus \{0\}, z \in \mathbb{C}^d. \tag{2.2}$$

For fixed $z \in \overline{\mathbb{D}}^d$ put

$$g_z(a) = p(az) = 1 + ap_1(z) + \cdots + a^n p_n(z).$$

Then g_z is stable, so by the Schur–Cohn criterion (see, e.g., [11], [9, Section 13.5]) we have that

$$Q(z) := A(z)A(1/\bar{z})^* - B(1/\bar{z})^*B(z)$$

is positive definite for $z \in \mathbb{T}^d$. Here

$$A(z) = \begin{pmatrix} p_0(z) & & \\ \vdots & \ddots & \\ p_{n-1}(z) & \cdots & p_0(z) \end{pmatrix}, \quad B(z) = \begin{pmatrix} p_n(z) & \cdots & p_1(z) \\ & \ddots & \vdots \\ & & p_n(z) \end{pmatrix},$$

and $p_0(z) = 1$. The matrix Q is also called the *Bezoutian* corresponding to g_z and its reverse $\overleftarrow{g_z}(a) = a^n \overline{g_z(1/\overline{a})}$; see, for instance, [10] and [11]. It is easy to see that if we write $Q(z) = (Q_{ij}(z))_{i,j=1}^n$, then $Q_{ij}(az) = a^{i-j}Q_{ij}(z)$. But then $Q(az) = D_a Q(z) D_a^{-1}$ follows. Next, by Lemma 2.2, we have that

$$\begin{pmatrix} Q(z)^{-1} & C(z)^* Q(z)^{-1} \\ Q(z)^{-1}C(z) & Q(z)^{-1} \end{pmatrix} \geq 0, z \in \mathbb{T}^d. \tag{2.3}$$

Multiplying all rows and columns on both sides with $Q(z)$ we obtain

$$\begin{pmatrix} Q(z) & Q(z)C(z)^* \\ C(z)Q(z) & Q(z) \end{pmatrix} \geq 0, z \in \mathbb{T}^d. \tag{2.4}$$

As $Q(z)$ satisfies the conditions of Lemma 2.1 we may write $Q(z) = P(z)P(z)^*$, $z \in \mathbb{T}^d$, with the ith row of P being homogeneous of degree $i - 1$. Thus $P(az) = D_a P(z)$. Let now

$$M(z) = P(1/\overline{z})^* Q(z)^{-1} C(z) P(z).$$

Then $\det(I - M(z)) = \det(I - P(z)P(1/\overline{z})^* Q(z)^{-1}C(z)) = \det(I - C(z)) = p(z)$. Next,

$$M(az) = P(1/\overline{az})^* Q(az)^{-1} C(az) P(az)$$
$$= P(1/\overline{z})^* D_a^{-1} D_a Q(z)^{-1} D_a^{-1} a D_a C(z) D_a^{-1} D_a P(z) = aM(z).$$

Finally, for $z \in \mathbb{T}^d$, we have that

$$P(z)(I - M(z)^* M(z))P(z)^* = Q(z) - Q(z)C(z)^* Q(z)^{-1} C(z)Q(z) \geq 0,$$

which follows from (2.4).

Thus, as $\mathrm{Ran}M(z)^* \subseteq \mathrm{Ran}P(z)^*$, it follows that $\|M(z)\| \leq 1$. Using the second inequality in (2.1) one can show in a similar way that $\|M(z)^n\| < 1$. $\square$

Remark 2.3. If $M(z)$ in Theorem 1.1 can be chosen to be analytic in $\overline{\mathbb{D}}^d$ then one easily sees from (iii) that $M(z) = \sum_{i=1}^d z_i M_i$ for some constant matrices $M_1, \ldots, M_d$. When in addition, $M(z)$ satisfies that $\|M_1 \otimes T_1 + \cdots + M_d \otimes T_d\| \leq 1$ for all contractions $T_1, \ldots, T_d$ (i.e., $M(z)$ is in Schur–Agler class), then it follows from [6, Corollary 3.3] that $\det(I + \sum_{i=1}^d z_i M_i)$ may be written as $\det(I - KZ(z))$ with K a contraction. It is an open problem what happens when $M(z)$ is not in the Schur–Agler class.

Acknowledgments

The author wishes to thank Anatolii Grinshpan and Dmitry S. Kaliuzhnyi-Verbovetskyi for useful discussions and their input on earlier drafts of this paper.

References

[1] M. Bakonyi and H.J. Woerdeman. *Matrix completions, moments, and sums of Hermitian squares.* Princeton University Press, Princeton, NJ, 2011. xii+518 pp.

[2] Ph. Delsarte, Y.V. Genin, and Y.G. Kamp, A simple proof of Rudin's multivariable stability theorem. *IEEE Trans. Acoust. Speech Signal Process.* 28 (1980), no. 6, 701–705.

[3] M.A. Dritschel. On factorization of trigonometric polynomials. *Integral Equations Operator Theory* 49 (2004), no. 1, 11–42.

[4] I. Gohberg and L. Lerer, Matrix generalizations of M.G. Krein theorems on orthogonal polynomials. Orthogonal matrix-valued polynomials and applications, 137–202, *Oper. Theory Adv. Appl.*, 34, Birkhäuser, Basel, 1988.

[5] I.C. Gohberg and A.A. Semencul, The inversion of finite Toeplitz matrices and their continual analogues. (Russian) *Mat. Issled.* 7 (1972), no. 2(24), 201–223, 290.

[6] A. Grinshpan, D.S. Kaliuzhnyi-Verbovetskyi, and H.J. Woerdeman. Norm-constrained determinantal representations of multivariable polynomials. *Complex Anal. Oper. Theory* 7 (2013), no. 3, 635–654.

[7] A. Grinshpan, D.S. Kaliuzhnyi-Verbovetskyi, V. Vinnikov and H.J. Woerdeman. Stable polynomials and real zero polynomials in two variables. Preprint.

[8] A. Kummert. 2-D stable polynomials with parameter-dependent coefficients: generalizations and new results. *IEEE Trans. Circuits Systems I: Fund. Theory Appl.* 49 (2002), 725–731.

[9] P. Lancaster and M. Tismenetsky. *The theory of matrices.* Second edition. Computer Science and Applied Mathematics. Academic Press, Inc., Orlando, FL, 1985.

[10] P. Lancaster, L. Lerer, and M. Tismenetsky, Factored forms for solutions of $AX - XB = C$ and $X - AXB = C$ in companion matrices. *Linear Algebra Appl.* 62 (1984), 19–49.

[11] L. Lerer, L. Rodman, and M. Tismenetsky, Bezoutian and Schur–Cohn problem for operator polynomials. *J. Math. Anal. Appl.* 103 (1984), no. 1, 83–102.

[12] T. Netzer and A. Thom. Polynomials with and without determinantal representations. *Linear Algebra Appl.* 437 (2012), 1579–1595.

[13] T. Netzer, D. Plaumann and A. Thom. Determinantal representations and the Hermite matrix. *Michigan Math. J.* 62 (2013), 407–420.

Hugo J. Woerdeman
Department of Mathematics
Drexel University
3141 Chestnut St.
Philadelphia, PA, 19104, USA
e-mail: hugo@math.drexel.edu

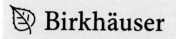 Birkhäuser | **www.birkhauser-science.com**

Operator Theory: Advances and Applications (OT)

This series is devoted to the publication of current research in operator theory, with particular emphasis on applications to classical analysis and the theory of integral equations, as well as to numerical analysis, mathematical physics and mathematical methods in electrical engineering.

Edited by
Joseph A. Ball (Blacksburg, VA, USA), Harry Dym (Rehovot, Israel),
Marinus A. Kaashoek (Amsterdam, The Netherlands), Heinz Langer (Vienna, Austria),
Christiane Tretter (Bern, Switzerland)

Printed in the United States
By Bookmasters